T0201438

Counterexamples on Uniform Convergence

Counterexamples on Uniform Convergence

Sequences, Series, Functions, and Integrals

Andrei Bourchtein
Ludmila Bourchtein

For general information on our other products and services or for technical support, please contact our Customer Care Department within the United States at (800) 762-2974, outside the United States at (317) 572-3993 or fax (317) 572-4002.

Wiley also publishes its books in a variety of electronic formats. Some content that appears in print may not be available in electronic formats. For more information about Wiley products, visit our web site at www.wiley.com.

Library of Congress Cataloging-in-Publication Data

Names: Bourchtein, Andrei. | Bourchtein, Ludmila.
Title: Counterexamples on uniform convergence : sequences, series, functions, and integrals / Andrei Bourchtein, Ludmila Bourchtein.
Description: Hoboken, New Jersey : John Wiley & Sons, Inc., c2017. | Includes bibliographical references and index.
Identifiers: LCCN 2016038687 | ISBN 9781119303381 (cloth) | ISBN 9781119303428 (epub) | ISBN 9781119303404 (epdf)
Subjects: LCSH: Mathematical analysis–Problems, exercises, etc. | Calculus–Problems, exercises, etc. | Sequences (Mathematics) | Functions. | Integrals.
Classification: LCC QA301 .B68 2017 | DDC 515–dc23
LC record available at https://lccn.loc.gov/2016038687

Cover Image credit: Abscent84/gettyimages

Set in 10/12pt Warnock by SPi Global, Pondicherry, India

Printed in the United States of America

10 9 8 7 6 5 4 3 2 1

To Haim and Maria with the warmest memories;
To Maxim with love and inspiration;
To Valentina for always yummy breakfast;
To Victoria for amazing rainbow flower

Contents

Preface

Looking for counterexamples is one of important things a mathematician does. Once a conjecture is proposed, if no proof can be found, the next step is to look for a counterexample. If no counterexample can be found, the next step is to try to find a proof again, and so on. On a more routine level, counterexamples play an important role for students as they learn new mathematical concepts. To best understand a theorem, it can be useful to see why each of the hypotheses of the theorem is necessary by finding counterexample when the hypotheses fails.

In this book, we present counterexamples related to different concepts and results on the uniform convergence usually studied in advanced calculus and real analysis courses. It includes the convergence of sequences, series and families of functions, and also proper and improper integrals depending on a parameter. The corresponding false statements are not formulated explicitly, but instead are invoked implicitly by the form of counterexamples.

The text is divided into six parts: the introductory chapter and five chapters of counterexamples. The first part contains some introductory material such as comments on notations, presentation form, and background theory. Chapter 1 considers conditions of uniform convergence. Chapter 2 deals with such properties of the limit functions as boundedness, existence of the limit and continuity. Chapter 3 analyzes the conditions of differentiability and integrability of the limit functions. Chapters 4 and 5 consider the properties of integrals (proper and improper) depending on a parameter.

The goal of the book is threefold. First, it provides a brief survey and discussion of principal results of the theory of uniform convergence in real analysis. Second, it supplies a material for a deeper study of the concepts and theorems on uniform convergence using counterexamples as a main technique. Finally, the text shows to the reader how such important mathematical tool as counterexamples can be used in different situations. We restricted our exposition to the main definitions and theorems in order to explore different versions (wrong and correct) of the fundamental concepts. Hence, many interesting (but more specific and applied) problems not related directly to the main notions and results are left out of the scope of this manuscript.

The selection and exposition of the material are directed, in the first place, to those advanced calculus and analysis students who are interested in a deeper understanding and broader knowledge of the topics of uniform convergence. We think the presented material may also be used by instructors that wish to go through the examples (or their variations) in class or assign them as homework or extracurricular projects. To this end, the main text is accompanied by the Instructor's Solutions Manual containing the detailed solutions to all the exercises proposed at the end of each chapter.

It is assumed that a reader has knowledge of a traditional university course of calculus. In order to make the majority of the examples and solutions accessible to calculus and analysis students, we tried to keep the level of reasoning as simple as possible. As in the majority of the mathematics books, the logical sequence of the material just follows the chapter sequence, that is, the content of the next chapter may depend on the previous text, but not vice-versa.

The book is not appropriate as the main textbook for a course, but rather, it can be used as a supplement that can help students to master important concepts and theorems. So we think the best way to use the book is to read its parts while taking a respective calculus/analysis course. On the other hand, the students already familiarized with the subjects of university calculus can find here deeper interpretation of the results and finer relation between concepts than in standard presentations. Also, more experienced students will better understand provided examples and ideas behind their construction.

To facilitate the reading of the main text (containing counterexamples) and make the text self-contained, and also to fix terminology, notation, and concepts, we gather the relevant definitions and results in the introductory chapter. For many examples, we make explicit references to the concepts/theorems to which they are related.

A short (but representative) list of bibliography can be found at the end of the book, including both collections of problems and textbooks in calculus/analysis. On the one hand, these references are the sources of some examples collected here, although it was out of our scope to trace all the original sources. On the other hand, they may be used for finding further information (examples and theory) on various topics. Some of these references are classic collections of the problems, such as that by Demidovich [6] and by Gelbaum and Olmsted [8]. Our preparation of the text was inspired, in the first place, by the latter book. We tried to extend its approach to the specialized topics of the uniform convergence, which frequently are sources of misunderstanding and confusion for fresh mathematics students. We hope that both students and professionals will find our book useful and (at least partly) challenging.

List of Examples

Chapter 1. Conditions of Uniform Convergence

Chapter 2. Properties of the Limit Function: Boundedness, Limits, Continuity

converges or $\lim_{x\to x_0}\sum u_n(x)$ exists, but nevertheless the remaining condition

Suppose $f_n(x)$ converges on X and $\lim_{x\to x_0}f_n(x)$ exists for each n. Even if one of the limits—$\lim_{n\to\infty}\lim_{x\to x_0}f_n(x)$ or $\lim_{x\to x_0}\lim_{n\to\infty}f_n(x)$—exists, another one may not

Assume a function $f(x,y)$ is defined on $X\times Y$, converges on X to a limit function $\varphi(x)$, as $y\to y_0$, and $\lim_{x\to x_0}f(x,y)$ exists for each $y\in Y$. Even though one of the two iterated limits—$\lim_{y\to y_0}\lim_{x\to x_0}f(x,y)$ or $\lim_{x\to x_0}\lim_{y\to y_0}f(x,y)$—exists, another one

A sequence $f_n(x)$ converges nonuniformly on X, but still $\lim_{x\to x_0}\lim_{n\to\infty}f_n(x)=$

Chapter 3. Properties of the Limit Function: Differentiability and Integrability

Chapter 4. Integrals Depending on a Parameter

Chapter 5. Improper Integrals Depending on a Parameter

Example 8. Suppose functions $f(x,y)$ and $g(x,y)$ are positive and continuous on $[a,+\infty) \times Y$, and $\lim\limits_{x \to +\infty} \frac{f(x,y)}{g(x,y)} = 1$, $\forall y \in Y$. One of the integrals $\int_a^{+\infty} f(x,y)dx$ or $\int_a^{+\infty} g(x,y)dx$ converges uniformly on Y, but another converges nonuniformly. 174

Example 9. Improper integrals $\int_a^{+\infty} f(x,y)dx$ and $\int_a^{+\infty} g(x,y)dx$ converge nonuniformly on Y, but the integral $\int_a^{+\infty} f(x,y) + g(x,y)dx$ converges uniformly on Y. 176

Improper integrals $\int_a^{+\infty} f(x,y)dx$ and $\int_a^{+\infty} g(x,y)dx$ converge nonuniformly on Y, and the integral $\int_a^{+\infty} f(x,y) + g(x,y)dx$ also converges nonuniformly on Y. 177

Example 10. Improper integrals $\int_a^{+\infty} f(x,y)dx$ and $\int_a^{+\infty} g(x,y)dx$ converge nonuniformly on Y, but the integral $\int_a^{+\infty} f(x,y)g(x,y)dx$ converges uniformly on Y. 178

An integral $\int_a^{+\infty} f(x,y)dx$ converges nonuniformly on Y, but the integral $\int_a^{+\infty} f^2(x,y)dx$ converges uniformly on Y. 179

Improper integrals $\int_a^{+\infty} f(x,y)dx$ and $\int_a^{+\infty} g(x,y)dx$ converge nonuniformly on Y, and the integral $\int_a^{+\infty} f(x,y)g(x,y)dx$ also converges nonuniformly on Y. 179

Example 11. Improper integrals $\int_a^{+\infty} f(x,y)dx$ and $\int_a^{+\infty} g(x,y)dx$ converge uniformly on Y, but the integral $\int_a^{+\infty} f(x,y)g(x,y)dx$ converges nonuniformly on Y. 179

An integral $\int_a^{+\infty} f(x,y)dx$ converges uniformly on Y, but the integral $\int_a^{+\infty} f^2(x,y)dx$ does not converge uniformly on Y. 180

Improper integrals $\int_a^{+\infty} f(x,y)dx$ and $\int_a^{+\infty} g(x,y)dx$ converge uniformly on Y, and $\int_a^{+\infty} f(x,y)g(x,y)dx$ also converges uniformly on Y. 180

Example 12. An improper integral $\int_a^{+\infty} f(x,y)dx$ converges uniformly and $\int_a^{+\infty} g(x,y)dx$ converges nonuniformly on Y, but the improper integral $\int_a^{+\infty} f(x,y)g(x,y)dx$ converges uniformly on Y. 181

An improper integral $\int_a^{+\infty} f(x,y)dx$ converges uniformly and $\int_a^{+\infty} g(x,y)dx$ converges nonuniformly on Y, but the improper integral $\int_a^{+\infty} f(x,y)g(x,y)dx$ converges nonuniformly on Y. 182

Example 13. Improper integrals $\int_a^{+\infty} f(x,y)dx$ and $\int_a^{+\infty} g(x,y)dx$ diverge on Y, but nevertheless the integral $\int_a^{+\infty} f(x,y)g(x,y)dx$ converges uniformly on Y. 182

List of Figures

About the Companion Website

This book is accompanied by a companion website:

www.wiley.com/go/bourchtein/counterexamples_on_uniform_convergence

The website includes:

- Solution Manual

Introduction

I.1 Comments

I.1.1 On the Structure of This Book

This book consists of the introductory chapter and five chapters of counterexamples. The Introduction fixes terminology and notations, and it contains some comments on presentation form and a brief overview of the background theory. The content of each of the five main chapters is as follows:

1) In Chapter 1, different conditions involving the uniform convergence of sequences, series, and functions depending on a parameter are considered—relations between different types of convergence, between convergence on a given set and its subsets, implications involving the sequences/series of the squares and products of the original terms, analysis of the conditions of Dirichlet's theorem, and Abel's theorem on uniform convergence of series.

2) In Chapter 2, the properties of the boundedness and continuity of the limit functions are investigated—relations between boundedness of the terms of a convergent sequences/series/functions depending on a parameter and that of the limit function, the conditions that do and do not guarantee the continuity of the limit function, and analysis of the conditions of Dini's theorem on uniform convergence.

3) Chapter 3 concerns the differentiability and integrability of the limit functions; first, the relations between differentiablity of the terms of a sequence/series/function depending on a parameter and that of the limit function are studied under the conditions of uniform/nonuniform convergence, including the possibility of term-by-term differentiation; then, similar results are presented for the Riemann integral.

4) In Chapter 4, the properties of the Riemann integrals depending on a parameter are considered; it includes the conditions on a function of two variables that do and do not guarantee the existence of a limit of such integral, its continuity, differentiability, and integrability with respect to

parameter, and the possibility to interchange the order of operations; some examples involve improper integrals on a finite interval, but an analysis of the uniform/nonuniform convergence of such integrals is postponed to the last part.

5) Chapter 5 deals with the properties of improper integrals depending on a parameter, many of them similar to those in preceding chapters—the nature of the convergence of improper integrals, relations between pointwise, absolute, and uniform convergence, conditions of existence/nonexistence of the limit, and the properties of continuity, differentiability, and integrability; the exposition of the material is focused on the improper integrals of the first kind (on an infinite interval) since the improper integrals of the second kind (of unbounded functions) can be reduced to the former type by a simple change of a variable.

Each chapter is divided into sections corresponding to the (conditional) division of the material in main topics. At the end of each chapter, supplementary exercises of different levels of complexity are provided, the most difficult of them with a hint to the solution. All the solutions to the proposed exercises can be found in the Instructor's Solutions Manual.

The logical sequence of the material just follows the chapter sequence, which means that the reading of a specific chapter/section does not require any knowledge of the subjects of next chapters/sections. At the same time, different concepts, results, and examples considered and analyzed in earlier chapters/sections are frequently used in the subsequent sections.

Since the use of counterexamples in the book is aimed to analyze important concepts and theorems in calculus and analysis, the selection of the material is restricted by those examples that have direct connection with main concepts. Therefore, more specific and applied problems are left out of the scope of the manuscript. The interested reader can find such exercises in different collections of problems on real analysis. Some well-known books of this type, containing problems of different levels of difficulty, are indicated in the bibliography: [2], [6], [10], [13], [14], and [15].

The following structure is chosen for the presentation of each counterexample:

1) Example statement that implicitly invokes the false general statement aimed to be disproved by this counterexample.
2) Solution (counterexample(s)), which provides a complete analytic solution to the posed problem.
3) Remark(s) (optional), which offer additional explanations, extensions of the counterexamples, and comparisons with other similar situations, and which also indicate links to a general theory and make comparisons with the correct statements.

4) Figure(s) (optional, for some examples involving more geometric and complex constructions) give geometric illustrations of the analytic solutions and clarify the analytic arguments in some complex cases.

Each example statement is formulated in the way that resembles the corresponding general false statement. Due to this choice, some formulations seem to be more complicated than they could be, but we decided to keep this form in order to make a clear association with implicitly invoked false statements. For instance, the example statement "a sequence of discontinuous functions converges uniformly on X, but the limit function is continuous on X" can be reformulated in a simpler form as "a sequence of discontinuous functions converges uniformly on X to a continuous function," but we chose the first version since it is directly associated with the corresponding false statement "if a sequence of discontinuous functions converges uniformly on X, then the limit function is discontinuous on X." For the same reasons of the logic and clarity of exposition, the well-known example of "an everywhere continuous and nowhere differentiable function" is initially formulated in the form: "a series $\sum u_n(x)$ of continuous functions converges uniformly on X, but nevertheless the sum of this series is not differentiable at any point of X." The latter form allows us to stress the connection with the analyzed concept of the uniform convergence and to recall (implicitly) the corresponding general false statement: "if a series $\sum u_n(x)$ of continuous functions converges uniformly on X, then the sum of this series is differentiable at least at one point of X." (Of course, without a reference to the series convergence, this false statement has a simpler form: "if a function is continuous everywhere, then it should be differentiable at least at one point.")

In the choice of figures for illustration of the analytic solutions, we don't try to provide as many pictures as possible or to keep an equal distribution of pictures among chapters and sections. Instead, we follow the principle of choosing the figures that can be most helpful for understanding the presented counterexamples (at least from our point of view). In the first place, it is related to the nature of each counterexample (if it is more analytic or geometric) and to the complexity and singularity of the functions employed in the solution.

I.1.2 On Mathematical Language and Notation

The language in the manuscript, just as in the majority of mathematical books, is a mixture of a natural language with mathematical terminology and symbolism used traditionally for a concise expression of numerous concepts, results, and logical reasoning. Depending on the topic, the use of mathematical symbols and terms can be more or less concentrated. In order to facilitate access to these symbols/terms inside the text, the list of notations and subject index are provided.

Throughout this work, we try to follow the standard terminology and notation used in calculus/analysis books. To avoid any misunderstanding and ambi-
guity, the main concepts and results, along with the corresponding terminology
and notation, are set out in Section I.2 containing background material.

Besides, a preliminary list of symbols (those used most frequently, somewhat
ambiguous or nontraditional) is presented below:

1) $\forall$—any, every, each, for all;
2) $\mathbb{N}$, $\mathbb{Z}$, $\mathbb{Q}$, $\mathbb{I}$, $\mathbb{R}$—set of natural, integer, rational, irrational, and real numbers,
   respectively;
3) $f_n(x)$—sequence of functions;
4) $\sum u_n(x)$—series of functions;
5) $f(x, y)$—function of two variables or function depending on a parameter y;
6) $D(x)$—Dirichlet's function, $D(x) = \begin{cases} 1, x \in \mathbb{Q} \\ 0, x \in \mathbb{I} \end{cases}$;
7) $R(x)$—Riemann function, $R(x) = \begin{cases} \frac{1}{n}, x = \frac{m}{n} \in \mathbb{Q} \\ 0, x \in \mathbb{I} \end{cases}$, where m is integer, n is
   natural, and $\frac{m}{n}$ is in lowest terms;
8) f_x—(first-order) partial or ordinary derivative in x;
9) $const$—constant value, for instance, $f(x) = const$ means a constant function.

I.2 Background (Elements of Theory)

The background material contains some basic concepts and results from the
theory of (infinite) sequences and series of functions and also functions and
integrals depending on a parameter. Some parts of this material can be found in
different analysis books, and more complete treatments can be found in monographs on sequences and series of functions. In particular, one can consult the
analysis textbooks indicated in the bibliography list, which treat the subjects
with different levels of strictness, abstraction, and generalization: [1], [3], [7],
[9], [12], [16], and [17]. The classical monographs on the subject include [4],
[5], and [11].

I.2.1 Sequences of Functions

I.2.1.1 Basic Definitions
Remark. In this text, we consider only the real-valued functions.

Definition. A sequence of functions. If each $n \in \mathbb{N}$ is associated with a function $f_n(x)$ defined on a set $X \subset \mathbb{R}$, then $f_n(x)$ is called a *sequence of functions*
defined on X. More formally, a sequence of functions is a mapping from $\mathbb{N}$ into
a set of functions Φ that assigns to each $n \in \mathbb{N}$ exactly one function in Φ.

Remark. The definition is usually extended to domains that include additional integers or exclude some naturals. The main point in these extensions is to keep the domain with the properties of ordering of $\mathbb{N}$: all the elements of a sequence should be indexed (with an integer index), the first element with an initial index must exist, and each following element has the index equal to the index of the preceding element plus one.

Definition. Convergence of a sequence of functions at a point. Let a sequence of functions $f_n(x)$ be defined on a set X. The sequence $f_n(x)$ is said to be *convergent/divergent at a point* $x_0 \in X$ if the numerical sequence $f_n(x_0)$ is convergent/divergent. Accordingly, x_0 is called a point of convergence/divergence.

Definition. Convergence of a sequence of functions on a set. Let a sequence of functions $f_n(x)$ and a function $f(x)$ be defined on a set X. The sequence $f_n(x)$ is said to be *convergent pointwise* (or simply convergent) on X to $f(x)$ if for each $x \in X$ the numerical sequence $f_n(x)$ converges to $f(x)$. Restated in more detailed $\epsilon - N$ form, the (pointwise) convergence of the sequence $f_n(x)$ on X means that for every fixed $x \in X$ and any $\epsilon > 0$, there exists a number $N \in \mathbb{N}$ (dependent on ϵ and, perhaps, on x) such that the condition $n > N$ implies $|f_n(x) - f(x)| < \epsilon$.

The function $f(x)$ is called the (pointwise) *limit function* of the sequence $f_n(x)$ on X.

Definition. Uniform convergence of a sequence of functions. A sequence $f_n(x)$ *converges uniformly* on a set X to a limit function $f(x)$ if, for any $\epsilon > 0$, there exists a number $N \in \mathbb{N}$ (dependent on ϵ, but independent of x) such that the condition $n > N$ implies $|f_n(x) - f(x)| < \epsilon$ simultaneously for all $x \in X$.

The function $f(x)$ is called the *uniform limit* of $f_n(x)$ on X.

Evidently, the uniform convergence implies the pointwise convergence.

Definition. Nonuniform convergence of a sequence of functions. A sequence $f_n(x)$ *converges pointwise but nonuniformly* on a set X to a limit function $f(x)$, if there exists $\epsilon > 0$ such that for any N there are $n_N > N$ and $x_N \in X$ for which $|f_{n_N}(x_N) - f(x_N)| \geq \epsilon$.

Remark. In some cases, the following more simple condition may be applied to show that the convergence is nonuniform: there exists $\epsilon > 0$ such that for any n there exists $x_n \in X$ for which $|f_n(x_n) - f(x_n)| \geq \epsilon$. Evidently, this condition implies that given in the definition of nonuniform convergence.

I.2.1.2 Conditions of the Uniform Convergence

Criterion of uniform convergence of a sequence. A sequence $f_n(x)$ converges uniformly on X to a function $f(x)$ if and only if $\sup_{x \in X} |f_n(x) - f(x)| \to 0$ as $n \to \infty$, that is, the numerical sequence of the maximum deviations of $f_n(x)$ from $f(x)$ on X converges to zero.

Cauchy criterion for uniform convergence of a sequence. A sequence of functions $f_n(x)$ converges uniformly on X if and only if for every $\epsilon > 0$ there exists N (independent of x) such that for all $n > N$ and all $p > 0$ the inequality $|f_{n+p}(x) - f_n(x)| < \epsilon$ holds for all $x \in X$ simultaneously.

Dini's theorem on uniform convergence. Let $f_n(x)$ be a sequence of continuous functions on a compact set X. If $f_n(x)$ converges to a continuous function $f(x)$ on X, and if this convergence is monotone with respect to n for every fixed $x \in X$, then $f_n(x)$ converges uniformly on X.

I.2.1.3 Properties of the Uniformly Convergent Sequences

Arithmetic properties of uniform convergence.

1) If $f_n(x)$ and $g_n(x)$ converge uniformly on X, then $f_n(x) + g_n(x)$ also converges uniformly on X.
2) If $f_n(x)$ converges uniformly on X and $g(x)$ is a bounded function on X, then $g(x)f_n(x)$ converges uniformly on the same set.

Theorem. Passage to limit term by term. If a sequence of functions $f_n(x)$ converges uniformly on X to $f(x)$ and for every n there exists a finite limit $\lim_{x \to x_0} f_n(x) = a_n$, then the limit function $f(x)$ also has a limit at x_0, the numerical sequence a_n is also convergent and $\lim_{x \to x_0} \lim_{n \to \infty} f_n(x) = \lim_{n \to \infty} \lim_{x \to x_0} f_n(x)$.

Theorem. Continuity and uniform convergence. If a sequence $f_n(x)$ converges uniformly on X to $f(x)$, and if each $f_n(x)$ is continuous at $x_0 \in X$, then $f(x)$ is also continuous at x_0.

If a sequence of continuous on X functions $f_n(x)$ converges uniformly on X to $f(x)$, then the limit function $f(x)$ is also continuous on X.

Theorem. Uniform continuity and uniform convergence. If a sequence of uniformly continuous on X functions $f_n(x)$ converges uniformly on X to $f(x)$, then the limit function $f(x)$ is also uniformly continuous on X.

Theorem. Interchange of the limit and integration signs. If a sequence of integrable (by Riemann) on $[a, b]$ functions $f_n(x)$ converges uniformly on $[a, b]$ to a function $f(x)$, then $f(x)$ is integrable (by Riemann) on $[a, b]$ and

$$\int_a^b \lim_{n \to \infty} f_n(x)dx = \lim_{n \to \infty} \int_a^b f_n(x)dx.$$

That is, one can *interchange the order of integration and limit calculation.* This formula is also called *term-by-term integration for sequences.*

Theorem. Interchange of the limit and differentiation signs. Let $f_n(x)$ be a sequence of differentiable on a finite interval I functions, and assume $f_n'(x)$ converges uniformly on I to a function $g(x)$. If there exists a point $x_0 \in I$ at which $f_n(x_0)$ is convergent, then $f_n(x)$ converges uniformly on I. Moreover, the limit function $f(x) = \lim\limits_{n\to\infty} f_n(x)$ is differentiable and satisfies $f'(x) = g(x)$, that is,

$$\left(\lim_{n\to\infty} f_n(x) \right)' = \lim_{n\to\infty} f_n'(x), \forall x \in I.$$

That is, one can pass to the *derivative under the limit sign.* This formula is also called *term-by-term differentiation for sequences.*

I.2.2 Series of Functions

Remark. The definitions for a series of functions and statements on its convergence follow the usual scheme, which reduces the series behavior to that of the sequences of its partial sums. In this way, almost all definitions and results stated for the convergence of sequences of functions can be reformulated for a series of functions.

I.2.2.1 Basic Definitions

Definition. A series of functions. The sum of all the elements of a sequence of functions $u_n(x)$ is called *series of functions.* The standard notation is $\sum_{n=1}^{+\infty} u_n(x)$ or $\sum_{n=i}^{+\infty} u_n(x)$, where i is the initial index, or simply $\sum u_n(x)$. The function $u_n(x)$ is called the *general term* of a series.

Definition. Partial sums. As in the case of numerical series, the sum of the first n terms of a series is called the nth *partial sum:* $f_n(x) = \sum_{k=1}^n u_k(x)$ or $f_n(x) = \sum_{k=i}^{n+i-1} u_k(x), n \geq 1.$

Remark. Even when a series starts from the index i, the nth partial sum can be defined as $f_n(x) = \sum_{k=i}^n u_k(x), n \geq i$. The choice of one of the two options for nominating the partial sums has no influence on the main properties of series, including convergence/divergence properties.

Definition. Convergence of a series of functions at a point. Let a sequence of functions $u_n(x)$ be defined on a set X. The series $\sum u_n(x)$ is *convergent/divergent at a point* $x_0 \in X$ if the sequence of its partial sums $f_n(x)$ is convergent/divergent at x_0, that is, if the numerical sequence $f_n(x_0)$ is convergent/divergent.

Definition. Convergence of a series of functions on a set. Let a sequence of functions $u_n(x)$ be defined on a set X. The series $\sum u_n(x)$ is *(pointwise) convergent* on X if the sequence of its partial sums $f_n(x)$ is (pointwise) convergent on X. In $\epsilon - N$ terms, it can be formulated as follows: for each $x \in X$ and for every $\epsilon > 0$, there exists a number N (perhaps, dependent on x) such that $|r_n(x)| = |f(x) - f_n(x)| = \left|\sum_{k=n+1}^{\infty} u_k(x)\right| < \epsilon$ whenever $n > N$.

If a series converges, the limit of the partial sums $f(x)$ is called the *sum of the series*: $\sum u_n(x) = f(x)$. The difference between $f(x)$ and $f_n(x)$ is called the nth *residual*: $r_n(x) = f(x) - f_n(x)$.

Remark. For the sake of brevity, in the text, the sum of the series is also called the limit function (as for a sequence), and the difference $f(x) - f_n(x)$ between the limit function of a convergent sequence and its general term is called the *residual* (as for a series).

Definition. Uniform convergence of a series of functions. The series $\sum u_n(x)$ is *uniformly convergent* on X if the sequence of the partial sums $f_n(x)$ is uniformly convergent on X to its sum $f(x)$. It means that for every $\epsilon > 0$, there exists a number N (independent of x) such that $|r_n(x)| = |f(x) - f_n(x)| = \left|\sum_{k=n+1}^{\infty} u_k(x)\right| < \epsilon$ for all $x \in X$ simultaneously whenever $n > N$.

Evidently, the uniform convergence implies the pointwise convergence.

Definition. Nonuniform convergence of a series of functions. A series $\sum u_n(x)$ *converges pointwise but nonuniformly* on a set X to its sum $f(x)$ if there exists $\epsilon > 0$ such that for any N there are $n_N > N$ and $x_N \in X$ for which $|r_{n_N}(x_N)| = |f(x_N) - f_{n_N}(x_N)| \geq \epsilon$.

Remark. In some cases, the following more simple condition may be applied to show that the convergence is nonuniform: there exists $\epsilon > 0$ such that for any n there exists $x_n \in X$ for which $|r_n(x_n)| = |f(x_n) - f_n(x_n)| \geq \epsilon$. Evidently, this condition implies the condition given in the last Definition.

I.2.2.2 Conditions of the Uniform Convergence

Criterion of uniform convergence of a series. A series $\sum u_n(x)$ is uniformly convergent on X to its sum $f(x)$ if and only if $\sup_{x \in X}|f(x) - f_n(x)| = \sup_{x \in X}\left|\sum_{k=n+1}^{\infty} u_k(x)\right| \to 0$ as $n \to \infty$.

Cauchy criterion for uniform convergence of a series. A series of functions $\sum u_n(x)$ is uniformly convergent on X if and only if for every $\epsilon > 0$ there exists N (independent of x) such that for all $n > N$ and all $p > 0$ the inequality $|f_{n+p}(x) - f_n(x)| = \left|\sum_{k=n+1}^{n+p} u_k(x)\right| < \epsilon$ is fulfilled simultaneously for all $x \in X$.

Necessary condition for uniform convergence of a series. If a series of functions $\sum u_n(x)$ is uniformly convergent on X, then the sequence $u_n(x)$ converges uniformly to 0 on X.

Definition. Majorant (dominant) series. A numerical series $\sum M_n$ with nonnegative terms is said to be a *majorant (or dominant) series* for a series of functions $\sum u_n(x)$ on X if the inequalities $\sup_{x\in X}|u_n(x)| \le M_n$ hold for all sufficiently large indices $n \in \mathbb{N}$.

Weierstrass M-test. If for a series of functions $\sum u_n(x)$ defined on X there exists a convergent majorant series $\sum M_n$, then the series $\sum u_n(x)$ converges uniformly on X.

Definition. Uniform boundedness. A sequence $u_n(x)$ defined on X is *uniformly bounded* on X if there exists a constant C (independent of x) such that $|u_n(x)| \le C$ for $\forall n$ and $\forall x \in X$.

Dirichlet's theorem. Let a series $\sum u_n(x)v_n(x)$ be defined on X. If the partial sums of the series $\sum u_n(x)$ are uniformly bounded on X, and the sequence $v_n(x)$ is monotone with respect to n (for each fixed $x \in X$) and converges uniformly to 0 as $n \to +\infty$, then $\sum u_n(x)v_n(x)$ converges uniformly on X.

Abel's theorem. Let a series $\sum u_n(x)v_n(x)$ be defined on X. If the series $\sum u_n(x)$ converges uniformly on X, and the sequence $v_n(x)$ is uniformly bounded on X and monotone with respect to n (for each fixed $x \in X$), then $\sum u_n(x)v_n(x)$ converges uniformly on X.

Dini's theorem on uniform convergence. Let $\sum u_n(x)$ be a series of continuous functions on a compact set X. If $u_n(x) \ge 0$ on X for $\forall n$, and if the series converges to a continuous function $f(x)$ on X, then $\sum u_n(x)$ converges uniformly on X.

I.2.2.3 Properties of the Uniformly Convergent Series
Arithmetic properties of uniform convergence.

1) If $\sum u_n(x)$ and $\sum v_n(x)$ converge uniformly on X, then $\sum(u_n(x) + v_n(x))$ also converges uniformly on X.
2) If $\sum u_n(x)$ converges uniformly on X and $v(x)$ is a bounded function on X, then $\sum v(x)u_n(x)$ converges uniformly on the same set.

Theorem. Passage to limit term by term in series. If a series $\sum u_n(x)$ converges uniformly on X and for every n there exists a finite limit $\lim_{x\to x_0} u_n(x) = a_n$, then the sum of series $f(x) = \sum u_n(x)$ also has limit at x_0, the series $\sum a_n$ is also convergent, and $\lim_{x\to x_0} \sum u_n(x) = \sum \lim_{x\to x_0} u_n(x)$. It means that in a uniformly convergent series it is allowable to use *term-by-term passage to the limit*.

Theorem. Continuity and uniform convergence. If all the terms of a series $\sum u_n(x)$ are continuous functions on X and the series is uniformly convergent on X, then its sum is also continuous on X.

Theorem. Uniform continuity and uniform convergence. If a series $\sum u_n(x)$ of uniformly continuous on X functions $u_n(x)$ converges uniformly on X, then its sum is also uniformly continuous on X.

Theorem. Term-by-term integration of series. Let $u_n(x)$ be integrable (by Riemann) on $[a, b]$ functions. If a series $\sum u_n(x)$ is uniformly convergent on $[a, b]$ to a function $f(x)$, then $f(x)$ is (Riemann) integrable on $[a, b]$ and

$$\int_a^b \sum u_n(x)dx = \sum \int_a^b u_n(x)dx,$$

that is, the uniformly convergent series can be *integrated term by term*.

Theorem. Term-by-term differentiation of series. Let $u_n(x)$ be differentiable on a finite interval I functions, and assume $\sum u_n'(x)$ converges uniformly to $g(x)$ on I. If there exists a point $x_0 \in I$ where $\sum u_n(x_0)$ converges, then the series $\sum u_n(x)$ converges uniformly on I to a differentiable function $f(x)$ satisfying $f'(x) = g(x)$, that is, $(\sum u_n(x))' = \sum u_n'(x)$, $\forall x \in I$. In other words, the series can be *differentiated term by term*.

I.2.3 Families of Functions

I.2.3.1 Basic Definitions
Definition. A family of functions. A function $f(x, y)$ defined on $X \times Y \subset \mathbb{R}^2$ is called a *family of functions* depending on a parameter y if one of the variables, say y, is set apart for some reason.

Remark. Frequently, a family of functions depending on a parameter is also called a function depending on a parameter.

Definition. Convergence of a family of functions at a point. A family of functions $f(x, y)$ is said to be *convergent/divergent at a point* $x_0 \in X$ as y approaches y_0 if the function $f(x_0, y)$ converges/diverges as y approaches y_0.

Definition. Convergence of a family of functions on a set. A family of functions $f(x, y)$ is said to be *convergent pointwise (or simply convergent)* on X to $\varphi(x)$ if for each fixed $x \in X$ the function $f(x, y)$ converges to $\varphi(x)$ as y approaches y_0. Restated in more detailed $\epsilon - \delta$ form, it means that for each given $x \in X$ and for any $\epsilon > 0$ there exists a number $\delta > 0$ (dependent on ϵ and, perhaps, on x) such that $|f(x, y) - \varphi(x)| < \epsilon$ whenever $0 < |y - y_0| < \delta$.

The function $\varphi(x)$ is called the *limit function* of the family $f(x,y)$ on X.

Definition. Uniform convergence of a family of functions. A family of functions $f(x,y)$ is said to be *uniformly convergent* on X to $\varphi(x)$ if for any $\epsilon > 0$, there exists a number $\delta > 0$ (dependent on ϵ, but independent of x) such that if $0 < |y - y_0| < \delta$ then $|f(x,y) - \varphi(x)| < \epsilon$ for all $x \in X$ simultaneously.
The function $\varphi(x)$ is called the uniform limit of $f(x,y)$ on X.
Evidently, the uniform convergence implies the pointwise convergence.

Definition. Nonuniform convergence of a family of functions. A family $f(x,y)$ *converges pointwise but nonuniformly* on a set X to a limit function $\varphi(x)$ if there exists $\epsilon > 0$ such that for any $\delta > 0$ there are $x_\delta \in X$ and $y_\delta \in Y$, $0 < |y_\delta - y_0| < \delta$ such that $|f(x_\delta, y_\delta) - \varphi(x_\delta)| \geq \epsilon$.

Remark. In some cases, the following more simple condition may be applied to show that the convergence is nonuniform: there exists $\epsilon > 0$ such that for $\forall y \in Y$ there exists $x_y \in X$ for which $|f(x_y, y) - \varphi(x_y)| \geq \epsilon$. Evidently, this condition implies the condition given in the definition of nonuniform convergence.

I.2.3.2 Conditions of the Uniform Convergence
Criterion of uniform convergence of a family. A family $f(x,y)$ converges uniformly on X to a function $\varphi(x)$ as y approaches y_0 if and only if $\lim\limits_{y \to y_0} \sup\limits_{x \in X} |f(x,y) - \varphi(x)| = 0$.

Cauchy criterion for uniform convergence of a family. A family $f(x,y)$ converges uniformly on X as y approaches y_0 if and only if for every $\epsilon > 0$, there exists $\delta > 0$ (independent of x) such that $\forall y_1, y_2 \in Y$ the conditions $0 < |y_1 - y_0| < \delta$ and $0 < |y_2 - y_0| < \delta$ imply $|f(x,y_1) - f(x,y_2)| < \epsilon$ for all $x \in X$ simultaneously.

Dini's Theorem on uniform convergence. Let $f(x,y)$ be continuous on a compact set X for each fixed $y \in Y$. If $f(x,y)$ converges on X to a continuous function $\varphi(x)$ as $y \to y_0$, and if this convergence is monotone with respect to y for every fixed $x \in X$, then $f(x,y)$ converges uniformly to $\varphi(x)$ on X.

I.2.3.3 Properties of the Uniformly Convergent Families
Arithmetic properties of uniform convergence.

1) If $f(x,y)$ and $g(x,y)$ converge uniformly on X, then $f(x,y) + g(x,y)$ also converges uniformly on X.
2) If $f(x,y)$ converges uniformly on X and $g(x)$ is a bounded function on X, then $g(x)f(x,y)$ converges uniformly on the same set.

Theorem. Passage to limit. Let a family of functions $f(x,y)$ be defined on $X \times Y$ and x_0, y_0 be the limit points of X and Y, respectively. If $f(x,y)$ converges uniformly on X to $\varphi(x)$ as y approaches y_0 and for each $y \in Y$ there exists the limit $\lim_{x \to x_0} f(x, y) = \psi(y)$, then the limits $\lim_{x \to x_0} \varphi(x)$ and $\lim_{y \to y_0} \psi(y)$ exist and are equal, that is, $\lim_{x \to x_0} \lim_{y \to y_0} f(x, y) = \lim_{y \to y_0} \lim_{x \to x_0} f(x, y)$.

Theorem. Continuity and uniform convergence. If $f(x, y)$ converges uniformly on X to $\varphi(x)$ as y approaches y_0 and if each $f(x, y)$ is continuous in x on X for any fixed $y \in Y$, then the limit function $\varphi(x)$ is continuous on X.

Theorem. Interchange of the limit and differentiation signs. Let $f(x, y)$ be a family of functions defined on $I \times Y$, where I is a finite interval, and differentiable in $x \in I$ at any fixed $y \in Y$. If $f_x(x, y)$ converges uniformly on I to $\psi(x)$ and $f(x, y)$ converges at least at one point $x_0 \in I$, as y approaches y_0, then $f(x, y)$ converges uniformly on I to a differentiable function $\varphi(x)$ and $\varphi'(x) = \psi(x)$, $\forall x \in I$, that is, $\frac{d}{dx} \lim_{y \to y_0} f(x, y) = \lim_{y \to y_0} f_x(x, y)$.

I.2.3.4 Integrals Depending on a Parameter

Definition. Integral depending on a parameter. Let $f(x, y)$ be a family of functions defined on $[a, b] \times Y \subset \mathbb{R}^2$. *Integral depending on a parameter* has the form $\int_a^b f(x, y)dx$.

Theorem. Interchange of the limit and integration signs. Let $f(x, y)$ be a family of integrable on $x \in [a, b]$ functions at any fixed $y \in Y$, and assume $f(x, y)$ converges uniformly on $[a, b]$ to $\varphi(x)$ as y approaches y_0. Then the limit function $\varphi(x)$ is also integrable on $[a, b]$ and

$$\int_a^b \lim_{y \to y_0} f(x, y)dx = \int_a^b \varphi(x)dx = \lim_{y \to y_0} \int_a^b f(x, y)dx.$$

Corollary. Let $f(x, y)$ be a family of functions defined on $[a, b] \times Y$. Suppose for any fixed $y \in Y$ the function $f(x, y)$ is continuous in $x \in [a, b]$, and for any fixed $x \in [a, b]$ the function $f(x, y)$ is monotone in $y \in Y$. Assume also $f(x, y)$ converges pointwise on $[a, b]$ to a continuous function $\varphi(x)$ as y approaches y_0. Then

$$\int_a^b \lim_{y \to y_0} f(x, y)dx = \int_a^b \varphi(x)dx = \lim_{y \to y_0} \int_a^b f(x, y)dx.$$

Theorem. On continuity of integral. If $f(x, y)$ is continuous on $[a, b] \times [c, d]$, then $F(y) = \int_a^b f(x, y)dx$ is continuous on $[c, d]$.

Theorem. On differentiability under the integral sign. If $f(x, y)$ is continuous in $x \in [a, b]$ for any fixed $y \in [c, d]$, and if $f_y(x, y)$ is continuous on $[a, b] \times [c, d]$, then $F(y) = \int_a^b f(x, y)dx$ is differentiable on $[c, d]$ and $F'(y) = \int_a^b f_y(x, y)dx$.

Theorem. On integrability of the integral function. If $f(x, y)$ is continuous on $[a, b] \times [c, d]$, then $F(y) = \int_a^b f(x, y) dx$ is integrable on $[c, d]$ and $\int_c^d dy \int_a^b f(x, y) dx = \int_a^b dx \int_c^d f(x, y) dy$.

I.2.3.5 Improper Integrals Depending on a Parameter

Definition. Improper integral depending on a parameter. Let $f(x, y)$ be a family of functions defined on $[a, +\infty) \times Y \subset \mathbb{R}^2$. *Improper integral depending on a parameter* has the form $F(y) = \int_a^{+\infty} f(x, y) dx$. We assume that the improper integral $F(y)$ is convergent for each $y \in Y$.

Definition. Uniformly convergent improper integral. Consider an auxiliary function $G(y, A) = \int_a^A f(x, y) dx$ for $\forall A > a$. The improper integral $F(y)$ is *uniformly convergent* on Y if the function $G(y, A)$ converges uniformly on Y as $A \to +\infty$, that is, for $\forall \epsilon > 0$ there exists $A_\epsilon > a$ (independent of y) such that $|G(y, A) - F(y)| < \epsilon$ for any $A > A_\epsilon$ and for all $y \in Y$ simultaneously.

Using the *residual of the improper integral* $R(y, A) = F(y) - G(y, A) = \int_A^{+\infty} f(x, y) dx$ for each $y \in Y$, the last definition can be restated in the form: the improper integral $F(y)$ is uniformly convergent on Y if the function $R(y, A)$ converges uniformly on Y to 0 as $A \to +\infty$, that is, for $\forall \epsilon > 0$ there exists $A_\epsilon > a$ (independent of y) such that $|R(y, A| = | \int_A^{+\infty} f(x, y) dx| < \epsilon$ for any $A > A_\epsilon$ and for all $y \in Y$ simultaneously.

I.2.3.6 Conditions of the Uniform Convergence of Improper Integral

Cauchy criterion for uniform convergence. The integral $F(y) = \int_a^{+\infty} f(x, y) dx$ converges uniformly on Y if and only if for $\forall \epsilon > 0$, there exists $A_\epsilon > a$ (independent of y) such that $| \int_{A_1}^{A_2} f(x, y) dx| < \epsilon$ for any $A_1, A_2 > A_\epsilon$ and for all $y \in Y$ simultaneously.

Definition. Majorant (dominant) function. A nonnegative function $\varphi(x)$ defined on $[a, +\infty)$ is called a *majorant (or dominant) function* on Y for a family of functions $f(x, y)$ if $|f(x, y)| \leq \varphi(x)$, $\forall x \in [a, +\infty)$ and $\forall y \in Y$.

Weierstrass test. If for a family of functions $f(x, y)$ there exists a majorant function with convergent improper integral $\int_a^{+\infty} \varphi(x) dx$, then the integral $F(y) = \int_a^{+\infty} f(x, y) dx$ converges uniformly on Y.

Definition. Uniform boundedness. A family of functions $f(x, y)$ is *uniformly bounded* on $X \times Y$ if there exists $M > 0$ such that $|f(x, y)| \leq M$ for $\forall x \in X$ and $\forall y \in Y$.

An integral $G(y, A) = \int_a^A f(x, y) dx$ is uniformly bounded on Y if there exists $M > 0$ such that $|G(y, A)| = | \int_a^A f(x, y) dx| \leq M$ for $\forall A > a$ and $\forall y \in Y$.

Dirichlet's theorem. Let an integral $G(y, A) = \int_a^A f(x, y) dx$ be uniformly bounded on Y. If a family of functions $g(x, y)$ is monotone in $x \in [a, +\infty)$ for each fixed $y \in Y$, and $g(x, y)$ converges uniformly to 0 on Y as $x \to +\infty$, then the improper integral $\int_a^{+\infty} f(x, y)g(x, y)dx$ converges uniformly on Y.

Abel's theorem. Let an integral $F(y) = \int_a^{+\infty} f(x, y)dx$ be uniformly convergent on Y. If a family of functions $g(x, y)$ is uniformly bounded on $[a, +\infty) \times Y$ and monotone in $x \in [a, +\infty)$ for each fixed $y \in Y$, then the improper integral $\int_a^{+\infty} f(x, y)g(x, y)dx$ converges uniformly on Y.

I.2.3.7 Properties of the Uniformly Convergent Improper Integrals
Arithmetic properties of uniform convergence.

1) If $F(y) = \int_a^{+\infty} f(x, y)dx$ and $G(y) = \int_a^{+\infty} g(x, y)dx$ converge uniformly on Y, then $\int_a^{+\infty} f(x, y) + g(x, y)dx$ also converges uniformly on Y.
2) If $F(y) = \int_a^{+\infty} f(x, y)dx$ converges uniformly on Y and $g(y)$ is a bounded function on Y, then $\int_a^{+\infty} g(y)f(x, y)dx$ converges uniformly on the same set.

Theorem. Passage to limit. Let $f(x, y)$ be continuous in $x \in [a, +\infty)$ for each fixed $y \in Y$, and assume $f(x, y)$ converges uniformly on $[a, A]$, $\forall A > a$ to a function $\varphi(x)$ as $y \to y_0$. If $F(y) = \int_a^{+\infty} f(x, y)dx$ converges uniformly on Y, then the integral $\int_a^{+\infty} \varphi(x)dx$ converges and $\lim_{y \to y_0} \int_a^{+\infty} f(x, y)dx = \int_a^{+\infty} \lim_{y \to y_0} f(x, y)dx$.

Theorem. Continuity with respect to parameter. Let $f(x, y)$ be continuous on $[a, +\infty) \times [c, d]$. If $F(y) = \int_a^{+\infty} f(x, y)dx$ converges uniformly on $[c, d]$, then the function $F(y)$ is continuous on $[c, d]$.

Remark. Instead of $[c, d]$, one can use any type of interval.

Theorem. Differentiation with respect to parameter. Let $f(x, y)$ and $f_y(x, y)$ be continuous functions on $[a, +\infty) \times [c, d]$. If $F(y) = \int_a^{+\infty} f(x, y)dx$ converges (pointwise) on $[c, d]$ and $G(y) = \int_a^{+\infty} f_y(x, y)dx$ converges uniformly on $[c, d]$, then $F(y)$ is differentiable on $[c, d]$ and $F'(y) = G(y)$, $\forall y \in [c, d]$, that is, $\frac{d}{dy} \int_a^{+\infty} f(x, y)dx = \int_a^{+\infty} f_y(x, y)dx$.

Remark. Instead of $[c, d]$, one can use any type of interval.

Theorem. Integration with respect to parameter on a finite interval. Let $f(x, y)$ be continuous function on $[a, +\infty) \times [c, d]$. If $F(y) = \int_a^{+\infty} f(x, y)dx$ converges uniformly on $[c, d]$, then $F(y)$ is continuous on $[c, d]$ and $\int_c^d dy \int_a^{+\infty} f(x, y)dx = \int_a^{+\infty} dx \int_c^d f(x, y)dy$.

Corollary. Let $f(x, y)$ be a continuous and sign-preserving function on $[a, +\infty) \times [c, d]$. If $F(y) = \int_a^{+\infty} f(x, y)dx$ is continuous function on $[c, d]$, then $\int_c^d dy \int_a^{+\infty} f(x, y)dx = \int_a^{+\infty} dx \int_c^d f(x, y)dy$.

Theorem 1. Integration with respect to parameter on an infinite interval. Let $f(x, y)$ be continuous and nonnegative on $[a, +\infty) \times [c, +\infty)$, and assume that $F(y) = \int_a^{+\infty} f(x, y) dx$ is continuous on $[c, +\infty)$ and $G(x) = \int_c^{+\infty} f(x, y) dy$ is continuous on $[a, +\infty)$. If, additionally, one of the integrals $\int_c^{+\infty} F(y) dy$ or $\int_a^{+\infty} G(x) dx$ converges, then another integral also converges and these two integrals have the same value.

Theorem 2. Integration with respect to parameter on an infinite interval. Let $f(x, y)$ be a continuous function on $[a, +\infty) \times [c, +\infty)$, and assume that $F(y) = \int_a^{+\infty} f(x, y) dx$ converges uniformly on $[c, d]$, $\forall d > c$ and $G(x) = \int_c^{+\infty} f(x, y) dy$ converges uniformly on $[a, b]$, $\forall b > a$. If, additionally, at least one of the integrals $\int_c^{+\infty} dy \int_a^{+\infty} |f(x, y)| dx$ or $\int_a^{+\infty} dx \int_c^{+\infty} |f(x, y)| dy$ converges, then both integrals $\int_c^{+\infty} dy \int_a^{+\infty} f(x, y) dx$ and $\int_a^{+\infty} dx \int_c^{+\infty} f(x, y) dy$ are convergent and equal to each other.

CHAPTER 1

Conditions of Uniform Convergence

1.1 Pointwise, Absolute, and Uniform Convergence. Convergence on a Set and Subset

Example 1. A function $f(x, y)$, defined on $X \times Y$, has a limit for any fixed $x \in X$ as y approaches y_0, that is, $f(x, y)$ converges pointwise to a limit function $\varphi(x)$ as y approaches y_0, but the convergence of $f(x, y)$ to $\varphi(x)$ is nonuniform on X.

Solution
Let us consider $f(x, y) = \frac{xy}{x^2 + y^2}$ defined on $[0, 1] \times (0, 1]$ and choose $y_0 = 0$. If $x = 0$, then $f(0, y) = 0$ and consequently $\lim_{y \to 0} f(0, y) = \lim_{y \to 0} 0 = 0$. If $x \neq 0$, then $\lim_{y \to 0} f(x, y) = \lim_{y \to 0} \frac{xy}{x^2 + y^2} = 0$. Therefore, the limit function is defined for any $x \in [0, 1]$ and it is zero: $\varphi(x) = \lim_{y \to 0} f(x, y) = 0$. However, the convergence to $\varphi(x)$ is not uniform on $X = [0, 1]$. Indeed for $\forall y \in Y = (0, 1]$, there exists $x_y = y \in (0, 1]$ such that

$$|f(x_y, y) - \varphi(x_y)| = \frac{y^2}{2y^2} = \frac{1}{2} \nrightarrow_{y \to 0} 0,$$

that is, for $\varepsilon_0 = \frac{1}{2}$ whatever radius δ is chosen, there exists the point $x_y = y \in (0, 1]$ such that $|f(x_y, y) - \varphi(x_y)| = \frac{y^2}{2y^2} \geq \varepsilon_0$ although $|y| < \delta$. It means that the convergence is not uniform.

Remark 1. In the case of $Y = \mathbb{N}$, a similar example can be formulated as follows: a sequence of functions $f_n(x)$ converges (pointwise) on a set X, but this convergence is nonuniform. One of the counterexamples is $f_n(x) = x^n$, $X = (-1, 1)$. Since $|x| < 1$, one gets $\lim_{n \to \infty} f_n(x) = \lim_{n \to \infty} x^n = 0 = f(x)$, $\forall x \in X$. To show that this convergence is nonuniform, let us pick up $x_n = \left(1 - \frac{1}{n}\right) \in X$, for $\forall n \in \mathbb{N}, n \geq 2$; and for these points, we obtain

$$|f_n(x_n) - f(x_n)| = \left(1 - \frac{1}{n}\right)^n \xrightarrow[n \to \infty]{} e^{-1} \neq 0.$$

Counterexamples on Uniform Convergence: Sequences, Series, Functions, and Integrals, First Edition.
Andrei Bourchtein and Ludmila Bourchtein.

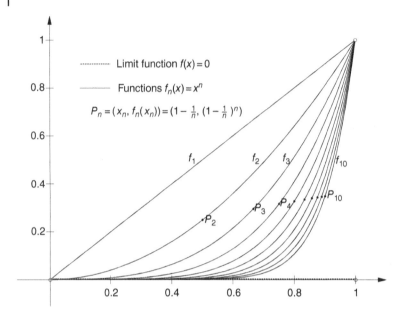

Figure 1.1 Examples 1, 4, and 28, sequence $f_n(x) = x^n$.

In other words, for $\varepsilon_0 = \frac{1}{4}$ whatever natural number N is chosen, for some (in this case, actually, for any) $n \in \mathbb{N}$, $n \geq 2$, there exists the point $x_n = 1 - \frac{1}{n}$ such that $|f_n(x_n) - f(x_n)| = \left(1 - \frac{1}{n}\right)^n \geq \frac{1}{4}$, that is, the convergence is not uniform. (In the last inequality, we have used the fact that the sequence $\left(1 - \frac{1}{n}\right)^n$ is increasing.)

Remark 2. A similar formulation can be made in the case of series: a series of functions converges (pointwise) on a set, but this convergence is nonuniform. The respective counterexample can be given with the series $\sum_{n=0}^{\infty} x^n$, $x \in X = (-1, 1)$. It is well known that the geometric series is convergent for $|x| < 1$ and $\sum_{n=0}^{\infty} x^n = \frac{1}{1-x} = f(x)$. To analyze the character of this convergence, first let us find the partial sums $f_n(x) = \sum_{k=0}^{n} x^k = \frac{1-x^{n+1}}{1-x}$ and the corresponding remainders $r_n(x) = f(x) - f_n(x) = \frac{x^{n+1}}{1-x}$. Choosing now $x_n = 1 - \frac{1}{n+1}$, $\forall n \in \mathbb{N}$, we obtain

$$r_n(x_n) = \frac{\left(1 - \frac{1}{n+1}\right)^{n+1}}{1 - 1 + \frac{1}{n+1}} = (n+1)\left(1 - \frac{1}{n+1}\right)^{n+1} \xrightarrow[n\to\infty]{} \infty.$$

Therefore, the convergence is nonuniform.

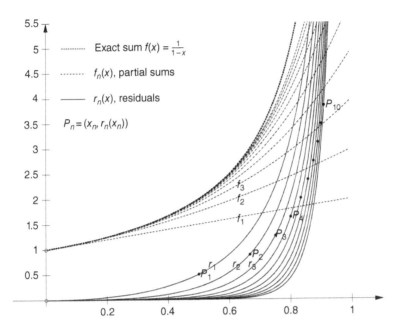

Figure 1.2 Examples 1 and 4, series $\sum_{n=0}^{\infty} x^n$.

Example 2. A series of functions converges on X and a general term of the series converges to zero uniformly on X, but the series converges nonuniformly on X.

Solution
Let us consider the series $\sum_{n=1}^{\infty} \frac{x^n}{n}$ on $X = [0, 1)$. This series converges for $\forall x \in X$, because $0 \le \frac{x^n}{n} \le x^n, \forall n$, and the geometric series $\sum_{n=1}^{\infty} x^n$ is convergent for $|x| < 1$. We can even find the sum of the series if we recall that the function $\ln(1 + x)$ has expansion in Taylor's series $\ln(1 + x) = \sum_{n=1}^{\infty} (-1)^{n-1} \frac{x^n}{n}$ convergent on $(-1, 1]$. Then, replacing x by $-x$, we obtain $\ln(1 - x) = -\sum_{n=1}^{\infty} \frac{x^n}{n}$ with convergence on $[-1, 1)$ and, in particular, on $X = [0, 1)$. Further, the general term $u_n(x) = \frac{x^n}{n}$ converges to 0 uniformly on $X = [0, 1)$, because $\lim_{n \to \infty} \frac{1}{n} = 0$ and evaluation $|u_n(x)| = \frac{|x|^n}{n} < \frac{1}{n}$ holds for $\forall x \in X$. Hence, the conditions of the statement are satisfied. However, the series is not convergent uniformly on X that can be shown by verifying the Cauchy criterion of the uniform convergence. In fact, for $\forall x \in X$ and for $\forall n, p \in \mathbb{N}$ we have the following evaluation:

$$\left| \sum_{k=n+1}^{n+p} \frac{x^k}{k} \right| = \frac{x^{n+1}}{n+1} + \cdots + \frac{x^{n+p}}{n+p} > p\frac{x^{n+p}}{n+p}.$$

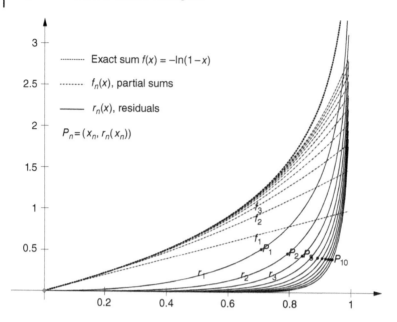

Figure 1.3 Examples 2, 26, 27, and 30, series $\sum_{n=1}^{\infty} \frac{x^n}{n}$.

Now, for $\forall n \in \mathbb{N}$, choosing $p_n = n$ and $x_n = \frac{1}{\sqrt[n]{2}} \in X$, we get

$$\left| \sum_{k=n+1}^{n+p_n} \frac{x_n^k}{k} \right| > n \frac{\left(1/\sqrt[n]{2}\right)^{2n}}{2n} = \frac{1}{8} \not\to 0,$$

which means that the Cauchy criterion is not satisfied and, therefore, the series does not converge uniformly on X.

Remark 1. The series $\sum_{n=1}^{\infty} \frac{x^n}{\sqrt{n}}$ considered on $X = (-1, 1)$ provides a similar counterexample. First, it converges for $\forall x \in X$, because $0 \leq \left| \frac{x^n}{\sqrt{n}} \right| \leq |x|^n$, $\forall n$, and the geometric series $\sum_{n=1}^{\infty} |x|^n$ is convergent for $|x| < 1$. Second, the inequality $|u_n(x)| = \left| \frac{x^n}{\sqrt{n}} \right| < \frac{1}{\sqrt{n}}$ holds for $\forall x \in X$; since $\lim_{n \to \infty} \frac{1}{\sqrt{n}} = 0$, it implies the uniform convergence of $\frac{x^n}{\sqrt{n}}$ to 0 on X. Hence, the statement conditions hold. To analyze the nature of the convergence of the series, let us evaluate the sum $\sum_{k=n+1}^{n+p} \frac{x^k}{\sqrt{k}}$ for $\forall n \in \mathbb{N}$, $p_n = n$, and $x_n = \frac{1}{\sqrt[n]{3}} \in X$:

$$\left| \sum_{k=n+1}^{n+p_n} \frac{x_n^k}{\sqrt{k}} \right| > n \frac{\left(1/\sqrt[n]{3}\right)^{2n}}{\sqrt{2n}} = \frac{1}{9\sqrt{2}} \sqrt{n} \xrightarrow[n \to \infty]{} \infty.$$

This means that the Cauchy criterion is not satisfied and, therefore, the series does not converge uniformly on X.

Remark 2. The converse general statement is true: if a series $\sum u_n(x)$ converges uniformly on X, then its general term converges to zero uniformly on X.

Example 3. A sequence of functions converges on X and there exists its subsequence that converges uniformly on X, but the original sequence does not converge uniformly on X.

Solution

Let us consider the sequence $f_n(x) = \begin{cases} \frac{x}{n}, n = 2k - 1 \\ \frac{1}{n}, n = 2k \end{cases}$, $\forall k \in \mathbb{N}, \; X = \mathbb{R}$.

For any fixed $x \in \mathbb{R}$, we have two partial limits: if $n = 2k - 1$, then $\lim_{k\to\infty} f_{2k-1}(x) = \lim_{k\to\infty} \frac{x}{2k-1} = 0$; and if $n = 2k$, then $\lim_{k\to\infty} f_{2k}(x) = \lim_{k\to\infty} \frac{1}{2k} = 0$. Therefore, this sequence converges to 0 on $\mathbb{R}$: $\lim_{n\to\infty} f_n(x) = 0 = f(x)$. Note also that the subsequence $f_{2k}(x)$ converges uniformly on $\mathbb{R}$, since the same evaluation $|f_{2k}(x) - f(x)| < \varepsilon$ holds simultaneously for all $x \in \mathbb{R}$. Or equivalently, for $\forall \varepsilon > 0$ there exists $K_\varepsilon = \left[\frac{1}{2\varepsilon}\right]$ such that for $\forall k > K_\varepsilon$ and simultaneously for all $x \in \mathbb{R}$ it follows that $|f_{2k}(x) - f(x)| < \varepsilon$. That is, the definition of the uniform convergence is satisfied for $f_{2k}(x)$. Nevertheless, the sequence $f_n(x)$ does not converge uniformly on $\mathbb{R}$. Indeed, whatever large index N we choose, there exists the index $n_N = 2N - 1 > N$ and the real point $x_N = 2N - 1$ such that

$$|f_{2N-1}(x_N) - f(x_N)| = \frac{2N-1}{2N-1} = 1 \nrightarrow_{N\to\infty} 0.$$

Hence, the convergence of $f_n(x)$ is not uniform on $\mathbb{R}$.

Example 4. A function $f(x, y)$ defined on $(a, b) \times Y$ converges to a limit function $\varphi(x)$ as y approaches y_0, and this convergence is uniform on any interval $[c, d] \subset (a, b)$, but the convergence is nonuniform on (a, b).

Solution

We can employ here the same functions used in Example 1. First, we consider $f(x, y) = \frac{xy}{x^2+y^2}$ defined on $(0, 1) \times (0, 1)$ and choose $y_0 = 0$. As in Example 1, the limit function is zero: $\varphi(x) = \lim_{y\to 0} f(x, y) = 0, \forall x \in (0, 1)$. However, the convergence to $\varphi(x)$ is not uniform on $(0, 1)$, because for $\forall y \in (0, 1)$ there exists $x_y = y$ such that

$$|f(x_y, y) - \varphi(x_y)| = \frac{y^2}{2y^2} = \frac{1}{2} \nrightarrow_{y\to 0} 0.$$

On the other hand, for any interval $[c, d] \subset (0, 1)$, the convergence is uniform. Indeed, for all $x \in [c, d]$ and for any $y > 0$, it follows that

$$|f(x, y) - \varphi(x)| = \frac{xy}{x^2 + y^2} \leq \frac{xy}{x^2} \leq \frac{1}{c}y.$$

Therefore, for any $\varepsilon > 0$, there exists $\delta_\varepsilon = c\varepsilon > 0$ (which is the same for all points in $[c, d]$) such that if $0 < y < \delta$, then $|f(x, y) - \varphi(x)| \leq \frac{1}{c}y < \varepsilon$ for all $x \in [c, d]$ simultaneously. It means that the convergence is uniform on $[c, d]$.

Remark 1. The sequence $f_n(x) = x^n$ on $x \in (-1, 1)$ from Example 1 provides the following counterexample: a sequence of functions $f_n(x)$, defined on (a, b), converges uniformly on any interval $[c, d] \subset (a, b)$, but the convergence is nonuniform on (a, b). The fact that the convergence to the limit function $f(x) = 0$ is not uniform on $(-1, 1)$ was already proved in Example 1. Let us show that the convergence is uniform on any $[c, d] \subset (-1, 1)$. Since we can always construct the interval $[-q, q]$, where $q = \max\{|c|, |d|\}$, such that $[c, d] \subset [-q, q] \subset (-1, 1)$, it is sufficient to prove the uniform convergence on $[-q, q]$. For this interval, we get $|f_n(x) - f(x)| = |x^n| \leq q^n$, for all $x \in [-q, q]$ at the same time. Since $\lim_{n \to \infty} q^n = 0$, that is, for any $\varepsilon > 0$ (it is sufficient to consider $\varepsilon < 1$), there exists $N_\varepsilon = \left[\frac{\ln \varepsilon}{\ln q}\right]$ such that $q^n < \varepsilon$ if $n > N_\varepsilon$, we can conclude that for any $\varepsilon > 0$ there exists exactly the same $N_\varepsilon = \left[\frac{\ln \varepsilon}{\ln q}\right]$ such that when $n > N_\varepsilon$, then $|x^n| \leq q^n < \varepsilon$ for all $x \in [-q, q]$ simultaneously. The last sentence is the definition of the uniform convergence on $[-q, q]$, and consequently, on $[c, d]$.

Remark 2. Finally, the series of Example 1 $\sum_{n=0}^{\infty} x^n$, $x \in (-1, 1)$ is an example of the situation when a series of functions converges uniformly on any interval $[c, d] \subset (a, b)$, but the convergence is nonuniform on (a, b). It was already shown in Example 1 that the convergence of the given series is not uniform on $(-1, 1)$. Let us consider an interval $[-q, q] \subset (-1, 1)$, $q > 0$ and show that the convergence is uniform on such an interval (this will imply the uniform convergence on any interval $[c, d] \subset (-1, 1)$). Since $|x^n| \leq q^n$ for any $x \in [-q, q]$ and the numerical series $\sum_{n=0}^{\infty} q^n$ is convergent (the geometrical series with $|q| < 1$), according to the Weierstrass test the series $\sum_{n=0}^{\infty} x^n$ converges uniformly on $[-q, q]$.

Remark 3. The nearly converse situation also takes place, as it is shown in Example 5.

Example 5. A sequence $f_n(x)$ converges on X, but this convergence is nonuniform on a closed interval $[a, b] \subset X$.

Solution

One of the counterexamples is $f_n(x) = nxe^{-n^2x^2}$ on $X = \mathbb{R}$. It is easy to show that this sequence approaches $f(x) \equiv 0$ on $\mathbb{R}$. In fact, for $x = 0$ one has $f_n(0) = 0$, $\forall n$ and, consequently, $\lim_{n\to\infty} f_n(0) = 0$. For $x \neq 0$, one can use the change of variable $t = nx$ and apply l'Hospital's rule:

$$\lim_{n\to\infty} f_n(x) = \lim_{t\to\pm\infty} \frac{t}{e^{t^2}} = \lim_{t\to\pm\infty} \frac{1}{2te^{t^2}} = 0.$$

Consider now $[a, b] \subset \mathbb{R}$ such that $a \leq 0 < b$. Choosing $N > \frac{1}{b}$ and $x_n = \frac{1}{n}$, one obtains the following evaluation for $\forall n > N$:

$$|f_n(x_n) - f(x_n)| = n|x_n|e^{-n^2x_n^2} = e^{-1} \not\to_{k\to\infty} 0,$$

which means that the convergence of $f_n(x)$ to 0 is nonuniform on such a closed interval.

Remark 1. A similar example for a series goes as follows: a series $\sum u_n(x)$ converges on a set X, but this series does not converge uniformly on a closed subinterval $[a, b] \subset X$. The series $\sum_{n=1}^\infty \frac{\sin nx}{n}$ provides an example. First, we show that it is convergent on $\mathbb{R}$. If $x_k = k\pi$, $\forall k \in \mathbb{Z}$, then $\sum_{n=1}^\infty \frac{\sin nx_k}{n} = \sum_{n=1}^\infty 0 = 0$. For $x \neq k\pi$, we can apply Dirichlet's theorem. For the partial sums $B_n(x) = \sum_{k=1}^n \sin kx$, the following evaluation holds:

$$|B_n(x)| = \left| \frac{1}{2\sin\frac{x}{2}} \sum_{k=1}^n 2\sin kx \sin\frac{x}{2} \right|$$

$$= \left| \frac{1}{2\sin\frac{x}{2}} \sum_{k=1}^n \left(\cos\left(k - \frac{1}{2}\right)x - \cos\left(k + \frac{1}{2}\right)x \right) \right|$$

$$= \frac{1}{2\left|\sin\frac{x}{2}\right|} \left| \cos\frac{x}{2} - \cos\left(n + \frac{1}{2}\right)x \right|$$

$$= \frac{1}{2\left|\sin\frac{x}{2}\right|} \left| 2\sin\frac{n+1}{2}x \cdot \sin\frac{n}{2}x \right| \leq \frac{1}{\left|\sin\frac{x}{2}\right|}$$

(note that the division by $\sin\frac{x}{2}$ is possible, since $x \neq k\pi$). Therefore, the sums $B_n(x)$ are bounded for any fixed $x \neq k\pi$. Besides, the numerical sequence $c_n = \frac{1}{n}$ is decreasing and approaches 0 as $n \to \infty$. Hence, all the conditions of Dirichlet's theorem are satisfied, and therefore the series is convergent on $\mathbb{R}$.

Let us show that this convergence is nonuniform on $(0, 2\pi)$ (and consequently on $[0, 2\pi]$ or any other interval containing $(0, 2\pi)$). To this end, we evaluate the sum $\sum_{k=n+1}^{n+p} u_k(x_n) = \sum_{k=n+1}^{n+p} \frac{\sin kx_n}{k}$ in the Cauchy criterion of uniform convergence. Choosing in this sum $p_n = n$ and $x_n = \frac{\pi}{6n}$, noting that $\frac{\pi}{6} < kx_n \leq \frac{\pi}{3}$

for any k such that $n < k \le n + p_n = 2n$, and recalling that $\sin t$ is positive and strictly increasing on $\left(0, \frac{\pi}{2}\right)$, we obtain

$$
\left| \sum_{k=n+1}^{n+p_n} u_k(x_n) \right| = \left| \sum_{k=n+1}^{2n} \frac{\sin kx_n}{k} \right|
$$

$$
= \frac{\sin\left(\frac{\pi}{6} + \frac{\pi}{6n}\right)}{n+1} + \frac{\sin\left(\frac{\pi}{6} + \frac{2\pi}{6n}\right)}{n+2} + \cdots + \frac{\sin\frac{\pi}{3}}{2n}
$$

$$
> \frac{\sin\frac{\pi}{6}}{n+1} + \frac{\sin\frac{\pi}{6}}{n+2} + \cdots + \frac{\sin\frac{\pi}{6}}{2n}
$$

$$
= \frac{1}{2}\left(\frac{1}{n+1} + \cdots + \frac{1}{2n}\right) > \frac{1}{2}\frac{n}{2n} = \frac{1}{4}.
$$

Hence, there exists $\varepsilon_0 = \frac{1}{4}$ such that for $\forall n$ there are $p_n = n$ and $x_n = \frac{\pi}{6n} \in (0, 2\pi)$ such that $\left| \sum_{k=n+1}^{n+p_n} u_k(x_n) \right| > \varepsilon_0$. This means that the Cauchy criterion is not satisfied on $(0, 2\pi)$ and, consequently, the series does not converge uniformly on this interval.

At the same time, the application of Dirichlet's theorem of uniform convergence reveals that the series converges uniformly on the interval $[a, 2\pi - a]$ for any $a \in (0, \pi)$. Indeed, since $\sin\frac{x}{2} > 0$, $\forall x \in [a, 2\pi - a]$, we can apply the same evaluations as above for the partial sums $B_n(x) = \sum_{k=1}^{n} \sin kx$ to obtain

$$
|B_n(x)| \le \frac{1}{|\sin x/2|} = \frac{1}{\sin x/2} \le \frac{1}{\sin a/2}, \quad \forall x \in [a, 2\pi - a],
$$

that is, the sums $B_n(x)$ are uniformly bounded on $[a, 2\pi - a]$. Since $c_n = \frac{1}{n} \underset{n \to \infty}{\to} 0$ and c_n is strictly decreasing, all the conditions of Dirichlet's theorem of uniform convergence are satisfied. Hence, the series converges uniformly on any interval $[a, 2\pi - a]$, $a \in (0, \pi)$.

Remark 2. For functions depending on a parameter, the corresponding formulation is as follows: a function $f(x, y)$ defined on $X \times Y$ has a limit $\lim_{y \to y_0} f(x, y) = \varphi(x)$ for $\forall x \in X$, but $f(x, y)$ converges to $\varphi(x)$ nonuniformly on a subinterval $[a, b] \subset X$. The function $f(x, y) = \frac{x^2 y^2}{x^4 + y^4}$ considered on $\mathbb{R} \times (0, +\infty)$ with the limit point $y_0 = 0$ provides the counterexample. This function converges to $\varphi(x) \equiv 0$ on $\mathbb{R}$ as $y \to 0$: for $x = 0$, one has $f(0, y) = 0$, $\forall y \in (0, +\infty)$ which implies $\lim_{y \to 0} f(0, y) = 0$; and for $x \ne 0$, one obtains by the arithmetic rules of the limits $\lim_{y \to 0} \frac{x^2 y^2}{x^4 + y^4} = \frac{0}{x^4} = 0$. Choose now $[a, b] \subset \mathbb{R}$ such that $a \le 0 < b$ and evaluate the difference $|f(x, y) - \varphi(x)|$ for $\forall y \in (0, b)$ and $x_y = y \in [a, b]$:

$$
|f(x_y, y) - \varphi(x_y)| = \frac{y^4}{2y^4} = \frac{1}{2} \underset{y \to 0}{\nrightarrow} 0.
$$

This result shows that the convergence is nonuniform on a chosen closed interval.

Remark 3. A strengthened versions of these statements are presented in Example 6.

Example 6. A sequence $f_n(x)$ converges on a set X, but it does not converge uniformly on any subinterval of X.

Solution
To construct a counterexample, let us place all the rational numbers of the interval $[0, 1]$ in a specific order of a numerical sequence r_n, $n = 1, 2, \cdots$ (this can be done, since the set of all the rational numbers of any interval is countable). Define now the functions $f_n(x)$ on $[0, 1]$ as follows:
$$f_n(x) = \begin{cases} 1, x = r_1, r_2, \cdots, r_n \\ 0, \text{otherwise} \end{cases}$$
. This sequence is monotone in n for any fixed $x \in [0, 1]$ (since $f_n(r_{n+1}) = 0 < 1 = f_{n+1}(r_{n+1})$ and $f_n(x) = f_{n+1}(x)$, $\forall x \neq r_{n+1}$) and bounded (since $0 \le f_n(x) \le 1$, $\forall n \in \mathbb{N}$, $\forall x \in [0, 1]$). Therefore, this sequence is convergent at any fixed $x \in [0, 1]$ and $f(x) = \lim_{n \to \infty} f_n(x) = \begin{cases} 1, x \in \mathbb{Q} \\ 0, x \in \mathbb{I} \end{cases} = D(x)$.
The convergence is nonuniform on $[0, 1]$ since for $\forall n$ there exist $x_n = r_{n+1}$ such that
$$|f_n(x_n) - f(x_n)| = |f_n(r_{n+1}) - f(r_{n+1})| = 1 \underset{n \to \infty}{\nrightarrow} 0.$$

Let us show that the convergence is also nonuniform on any interval $[a, b] \subset [0, 1]$, which will imply that the convergence is nonuniform on any interval in $[0, 1]$. In fact, since any interval contains infinitely many rational points, in $[a, b]$ there are infinitely many points of the sequence $r_1, r_2, r_3, \cdots$, which form a subsequence $r_{n_1}, r_{n_2}, r_{n_3}, \cdots, r_{n_k} \in [a, b]$, $\forall k \in \mathbb{N}$. Then for any $k \in \mathbb{N}$, there exist $n_k > k$ and $x_k = r_{n_{k+1}} \in [a, b]$ such that
$$|f_{n_k}(x_k) - f(x_k)| = |f_{n_k}(r_{n_{k+1}}) - f(r_{n_{k+1}})| = 1 \underset{k \to \infty}{\nrightarrow} 0,$$

which means that the convergence is nonuniform on $[a, b]$.

Remark 1. The corresponding example for a series that converges on X, but does not converge uniformly on any subinterval of X, can be easily constructed using the sequence of the given counterexample as partial sums of the series. For instance, the series $\sum u_n(x)$ with the terms $u_n(x) = \begin{cases} 1, x = r_n \\ 0, x \neq r_n \end{cases}$ defined on $[0, 1]$ has the partial sums $f_n(x)$ of the above counterexample and, consequently, this series converges on $[0, 1]$, but does not converge uniformly on any subinterval of $[0, 1]$.

Remark 2. Another example of this type, albeit for a sequence of continuous on X functions $f_n(x)$, is given in Example 26 of Chapter 2.

Example 7. A series $\sum u_n(x)$ converges uniformly on an interval, but it does not converge absolutely on the same interval.

Solution

The series $\sum_{n=1}^{\infty} (-1)^n \frac{x^2+n}{n^2}$ converges uniformly on the interval $[-a, a]$, $\forall a > 0$. In fact, for any fixed $x \in \mathbb{R}$, this is an alternating series, which converges by Leibniz's test: $\lim_{n\to\infty} \frac{x^2+n}{n^2} = 0$ and $\frac{x^2+n}{n^2}$ is strictly decreasing in n for any fixed $x \in \mathbb{R}$. For alternating series, the remainder can be evaluated through its first term: $|r_n(x)| \leq |u_{n+1}(x)| = \frac{x^2+n+1}{(n+1)^2}$. Therefore, for all $x \in [-a, a]$, we get

$$|r_n(x)| \leq \frac{x^2+n+1}{(n+1)^2} \leq \frac{a^2}{(n+1)^2} + \frac{1}{n+1} \xrightarrow[n\to\infty]{} 0,$$

which implies the uniform convergence of the series on $[-a, a]$. However, the series of the absolute values $\sum_{n=1}^{\infty} |u_n(x)| = \sum_{n=1}^{\infty} \frac{x^2+n}{n^2}$ diverges for any x: for $x = 0$, the series $\sum_{n=1}^{\infty} \frac{1}{n}$ is harmonic and divergent; for $\forall x \neq 0$, the series $\sum_{n=1}^{\infty} \left(\frac{x^2}{n^2} + \frac{1}{n} \right)$ is the sum of the two series—$\sum_{n=1}^{\infty} \frac{x^2}{n^2}$ and $\sum_{n=1}^{\infty} \frac{1}{n}$—where the former is convergent (*p*-series with $p = 2$) and the latter is divergent (the harmonic series), which implies the divergence of the sum. Hence, the given series is convergent on $\mathbb{R}$, uniformly convergent on $[-a, a]$, $\forall a > 0$, but it does not possess absolute convergence at any point.

Remark. The converse situation is considered in Example 8.

Example 8. A series $\sum u_n(x)$ converges absolutely on an interval, but it does not converge uniformly on the same interval.

Solution

The series $\sum_{n=0}^{\infty} (-1)^n x^n$ converges absolutely on the interval $X = [0, 1)$, because for $\forall x \in [0, 1)$ the series $\sum_{n=0}^{\infty} |u_n(x)| = \sum_{n=0}^{\infty} x^n$ is the geometric series with the nonnegative ratio less than 1. However, the convergence is not uniform on $[0, 1)$, because for $\forall n$ we can choose $x_n = 1 - \frac{1}{n+1} \in [0, 1)$ that gives the following evaluation:

$$|r_n(x_n)| = \left| \frac{(-1)^{n+1} x_n^{n+1}}{1 + x_n} \right| = \frac{\left(1 - \frac{1}{n+1} \right)^{n+1}}{1 + 1 - \frac{1}{n+1}} \xrightarrow[n\to\infty]{} \frac{1}{2} e^{-1} \neq 0.$$

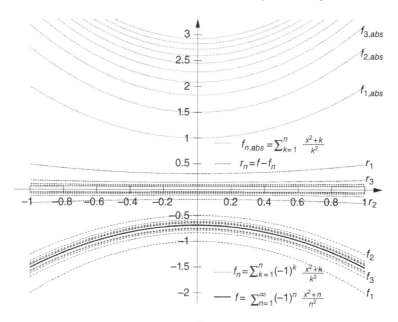

Figure 1.4 Example 7, series $\sum_{n=1}^{\infty} (-1)^n \frac{x^2+n}{n^2}$.

Example 9. A series $\sum u_n(x)$ converges absolutely and uniformly on $[a, b]$, but the series $\sum |u_n(x)|$ does not converge uniformly on $[a, b]$.

Solution

Let us consider the series $\sum_{n=0}^{\infty} u_n(x) = \sum_{n=0}^{\infty} (-1)^n (1-x)x^n$ on $[0, 1]$. If $x = 1$, then $u_n(1) = 0$ and the series converges at this point. If $x \neq 1$, then for each fixed x we have the geometric series with the ratio $q = -x$, and since $|q| = |x| < 1$, the series is convergent. The series of the absolute values $\sum_{n=0}^{\infty} |u_n(x)| = \sum_{n=0}^{\infty} (1-x)x^n$ is convergent on $[0, 1]$ for the very same reasons.

Let us analyze the uniform convergence of the original series. Since this series is alternating, we have the following evaluation for its remainder $r_n(x)$ for any n:

$$|r_n(x)| \leq |u_{n+1}(x)| = (1-x)x^{n+1}, \forall x \in [0, 1].$$

Note that the continuous function $h(x) = (1-x)x^{n+1}$ is positive on $(0, 1)$ and at the end points $h(0) = h(1) = 0$. Therefore, $h(x)$ achieves its global maximum in some interior point of $[0, 1]$. Solving the critical point equation

$$h'(x) = (n+1)x^n - (n+2)x^{n+1} = x^n[(n+1) - (n+2)x] = 0,$$

we find the only critical point $x_n = \frac{n+1}{n+2}$ on $(0, 1)$, which is the global maximum point of $h(x)$ on $[0, 1]$. Thus, for $\forall x \in [0, 1]$,

$$|r_n(x)| \le (1 - x)x^{n+1} \le (1 - x_n)x_n^{n+1}$$

$$= \frac{1}{n+2}\left(1 - \frac{1}{n+2}\right)^{n+1} \xrightarrow[n\to\infty]{} 0 \cdot e^{-1} = 0,$$

that is, the convergences is uniform on $[0, 1]$.

Finally, let us show that the series $\sum_{n=0}^{\infty} |u_n(x)| = \sum_{n=0}^{\infty} (1 - x)x^n$ does not converge uniformly on $[0, 1]$. It is sufficient to show that the convergence is nonuniform on $[0, 1)$, so let us evaluate the remainder $\tilde{r}_n(x)$ for $x \in [0, 1)$:

$$\tilde{r}_n(x) = \sum_{k=n+1}^{\infty} |u_k(x)| = \frac{(1 - x)x^{n+1}}{1 - x} = x^{n+1}.$$

Choosing now the points $x_n = \frac{1}{\sqrt[n+1]{2}} \in [0, 1)$ for each n, we obtain $\tilde{r}_n(x_n) = \left(\frac{1}{\sqrt[n+1]{2}}\right)^{n+1} = \frac{1}{2} \nrightarrow 0$, which shows that the convergence is nonuniform on $[0, 1)$ and, consequently, on $[0, 1]$.

Remark. The converse general statement is true: if a series $\sum_{n=0}^{\infty} |u_n(x)|$ converges uniformly on $[a, b]$, then the series $\sum_{n=0}^{\infty} u_n(x)$ converges absolutely and uniformly on $[a, b]$.

Example 10. A series $\sum u_n(x)$ converges absolutely and uniformly on X, but there is no bound of the general term $u_n(x)$ on X in the form $|u_n(x)| \le a_n$, $\forall n$ such that the series $\sum a_n$ converges.

Solution

One of the counterexamples is the series $\sum_{n=1}^{\infty} u_n(x)$ with the general term

$$u_n(x) = \begin{cases} 0, x \in [0, 2^{-n-1}] \cup [2^{-n}, 1] \\ \frac{1}{n}\sin^2(2^{n+1}\pi x), x \in (2^{-n-1}, 2^{-n}) \end{cases}$$ defined on $X = [0, 1]$. Note that $u_n(x) \ge 0$ for $\forall x \in [0, 1]$, so the convergence and absolute convergence is the same thing for this series. At the points $x = 0$, $x = 2^{-n}$, $\forall n \in \mathbb{N}$, and for $\forall x \in \left[\frac{1}{2}, 1\right]$, we get $u_n(x) = 0$, $\forall n \in \mathbb{N}$ and the series converges to zero. If $x \in \left(0, \frac{1}{2}\right)$, $x \ne 2^{-n}$, $n \ge 2$, then each of such points lies in only one of the intervals $(2^{-n-1}, 2^{-n})$, because these intervals have no common points: $(2^{-n-1}, 2^{-n}) \cap (2^{-n-2}, 2^{-n-1}) = \emptyset$, $\forall n \in \mathbb{N}$. Therefore, there is only one n_x such that $x \in (2^{-n_x-1}, 2^{-n_x})$. Then $u_{n_x}(x) = \frac{1}{n_x}\sin^2(2^{n_x+1}\pi x)$ and $u_n(x) = 0$, $\forall n \ne n_x$ and, consequently, $\sum_{n=1}^{\infty} u_n(x) = u_{n_x}(x)$, which shows the (absolute) convergence of this series on $[0, 1]$.

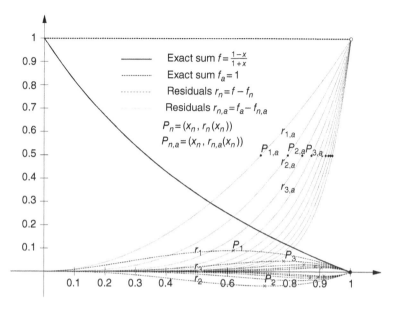

Figure 1.5 Examples 9 and 10 (second counterexample), series $\sum_{n=0}^{\infty}(-1)^n(1-x)x^n$.

Applying the Cauchy criterion and employing similar reasoning, we can also prove that the convergence is uniform. Indeed, since for any fixed $x \in [0,1]$ at most only one term in the entire series is nonzero and this term satisfies the inequality $|u_{n_x}(x)| = \left|\frac{1}{n_x}\sin^2(2^{n_x+1}\pi x)\right| \le \frac{1}{n_x}$, we obtain the following evaluation $\left|\sum_{k=n+1}^{n+p} u_k(x)\right| \le \frac{1}{n+1} < \frac{1}{n}$, which holds for $\forall n, p \in \mathbb{N}$ and simultaneously for $\forall x \in [0,1]$. Hence, for $\forall \varepsilon > 0$, there exists $N_\varepsilon = \left[\frac{1}{\varepsilon}\right]$ such that for $\forall n > N_\varepsilon$, $\forall p \in \mathbb{N}$ and simultaneously for all $x \in [0,1]$, it follows that $\left|\sum_{k=n+1}^{n+p} u_k(x)\right| < \frac{1}{n} < \varepsilon$, that is, the series converges uniformly on $[0,1]$ according to the Cauchy criterion of the uniform convergence.

Nevertheless, the functions $u_n(x)$ do not admit majoration on $[0,1]$ by the constants a_n such that the series $\sum_{n=1}^{\infty} a_n$ converges. Indeed, for $\forall n \in \mathbb{N}$, the inequality $|u_n(x)| \le \frac{1}{n}$ is exact (in the sense that $\frac{1}{n}$ is the lowest upper bound for $|u_n(x)|$) on $[0,1]$, because there exists the point $x_n = 3 \cdot 2^{-n-2} \in (2^{-n-1}, 2^{-n})$ such that $u_n(x_n) = \frac{1}{n}\sin^2\left(\frac{3}{2}\pi\right) = \frac{1}{n}$, and the series $\sum_{n=1}^{\infty} \frac{1}{n}$ diverges.

Another interesting counterexample is the series of Example 9: $\sum_{n=0}^{\infty} u_n(x) = \sum_{n=0}^{\infty}(-1)^n(1-x)x^n$ on $[0,1]$. It was shown in Example 9 that this series converges absolutely and uniformly on $[0,1]$. For each fixed n, the function $|u_n(x)| > 0$, $\forall x \in (0,1)$, and $|u_n(0)| = |u_n(1)| = 0$. Therefore, the continuous function $|u_n(x)|$ achieves its global maximum in an interior point of $[0,1]$,

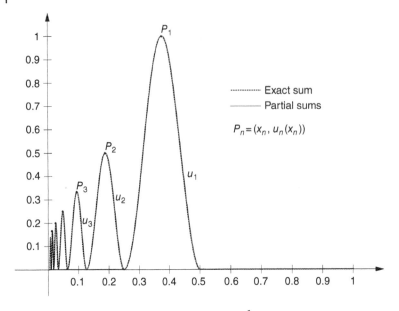

Figure 1.6 Example 10, series $\sum_{n=1}^{\infty} u_n(x)$, $u_n(x) = \begin{cases} 0, x \in [0, 2^{-n-1}] \cup [2^{-n}, 1] \\ \frac{1}{n}\sin^2(2^{n+1}\pi x), x \in (2^{-n-1}, 2^{-n}) \end{cases}$.

which can be found by solving the critical point equation:

$$|u_n(x)|' = (x^n - x^{n+1})' = nx^{n-1} - (n+1)x^n$$
$$= (n+1)x^{n-1}\left(\frac{n}{n+1} - x\right) = 0.$$

The unique solution on $(0, 1)$ is $x_n = \frac{n}{n+1}$ and, consequently,

$$|u_n(x)| \le \max_{[0,1]}|u_n(x)| = |u_n(x_n)| = \frac{1}{n+1}\left(1 - \frac{1}{n+1}\right)^n.$$

Since $\lim_{n\to\infty}\left(1 - \frac{1}{n+1}\right)^n = e^{-1}$ and the series $\sum_{n=0}^{\infty}\frac{1}{n+1}$ diverges, according to the comparison test, the series $\sum_{n=0}^{\infty}\frac{1}{n+1}\left(1 - \frac{1}{n+1}\right)^n$ also diverges. Note that for each n, the majorant term $a_n = \frac{1}{n+1}\left(1 - \frac{1}{n+1}\right)^n$ is exact (i.e., the minimum possible) for $|u_n(x)|$ on $[0, 1]$. Therefore, there is no convergent majorant series $\sum_{n=0}^{\infty}a_n$ such that $|u_n(x)| \le a_n$.

Remark. The converse general statement is true and represents the famous Weierstrass M-test.

1.2   Uniform Convergence of Sequences and Series of Squares and Products

Example 11. A sequence $f_n(x)$ converges uniformly on X to a function $f(x)$, but $f_n^2(x)$ does not converge uniformly on X to $f^2(x)$.

Solution

The sequence $f_n(x) = \ln \frac{nx}{n+1}$ converges on $X = (0, +\infty)$ to $f(x) = \ln x$, because $\lim_{n \to \infty} \ln \frac{nx}{n+1} = \ln \left(\lim_{n \to \infty} \frac{n}{n+1} \right) x = \ln x$. The following evaluation shows that this convergence is uniform:

$$|f_n(x) - f(x)| = \left| \ln \frac{nx}{n+1} - \ln x \right| = \left| \ln \frac{n}{n+1} \right| < \varepsilon.$$

So for $\forall \varepsilon > 0$, there exists $N_\varepsilon = \left[\frac{1}{e^\varepsilon - 1} \right]$, which depends only on ε, such that for $\forall n > N_\varepsilon$ and simultaneously for all $x \in X$ we have $|f_n(x) - f(x)| < \varepsilon$.

Due to arithmetic properties of the limits, the sequence $f_n^2(x)$ also converges to $f^2(x) = \ln^2 x$ for any fixed $x \in X$ (it can also be shown directly: $\lim_{n \to \infty} \ln^2 \frac{nx}{n+1} =$ $\left(\ln \left(\lim_{n \to \infty} \frac{n}{n+1} \right) x \right)^2 = \ln^2 x$). However, this convergence is not uniform. In fact, for each $x \in X$, we get

$$|f_n^2(x) - f^2(x)| = \left| \ln^2 \frac{nx}{n+1} - \ln^2 x \right|$$

$$= \left| \ln \frac{nx}{n+1} - \ln x \right| \cdot \left| \ln \frac{nx}{n+1} + \ln x \right|$$

$$= \ln \frac{n+1}{n} \cdot \left| \ln \frac{nx^2}{n+1} \right|.$$

Choosing now $x_n = \frac{1}{ne^n} \in X$, we obtain

$$|f_n^2(x_n) - f^2(x_n)| = \ln \frac{n+1}{n} \cdot \left| \ln \frac{n}{(n+1)n^2 e^{2n}} \right|$$

$$= \ln \frac{n+1}{n} \cdot \ln(n(n+1)e^{2n})$$

$$= \ln \frac{n+1}{n} \cdot \ln n(n+1) + 2n \ln \frac{n+1}{n} > 1$$

for sufficiently large n (since the first term is positive and the limit of the second equals two: $\lim_{n \to \infty} 2n \ln \frac{n+1}{n} = \lim_{n \to \infty} 2 \ln \left(1 + \frac{1}{n} \right)^n = 2 \ln e = 2$, that is, $2n \ln \frac{n+1}{n} >$ 1 for large n). Therefore, the convergence is not uniform: for $\varepsilon_0 = 1$ whatever N is chosen, it can be found that $\tilde{n} > N$ and corresponding $x_{\tilde{n}} \in X$ such that $|f_{\tilde{n}}^2(x_{\tilde{n}}) - f^2(x_{\tilde{n}})| > \varepsilon_0 = 1$.

Remark 1. Naturally, the following example can also be constructed: sequences $f_n(x)$ and $g_n(x)$ converge uniformly on X to $f(x)$ and $g(x)$, respectively, but $f_n(x)g_n(x)$ does not converge uniformly on X to $f(x)g(x)$. In the case $f_n(x) = g_n(x)$, we have the original example with the square of function. For different sequences, we can use the same $f_n(x) = \ln \frac{nx}{n+1}$ and slightly different $g_n(x) = \ln \frac{nx}{2n+5}$. The sequence $g_n(x)$ converges to $g(x) = \ln \frac{x}{2}$, and this convergence is uniform on $X = (0, +\infty)$ due to the evaluation

$$|g_n(x) - g(x)| = \left| \ln \frac{nx}{2n+5} - \ln \frac{x}{2} \right|$$

$$= \left| \ln \frac{2n}{2n+5} \right| = \ln \left(1 + \frac{5}{2n} \right) \underset{n \to \infty}{\to} 0.$$

Consequently, $f_n(x)g_n(x)$ converges to $f(x)g(x) = \ln x \ln \frac{x}{2}$ for each fixed $x \in X$ due to arithmetic rules of the limits. However, this convergence is not uniform on X, as it is shown below: for each $x \in X$, we have

$$|f_n(x)g_n(x) - f(x)g(x)| = \left| \ln \frac{nx}{n+1} \ln \frac{nx}{2n+5} - \ln x \ln \frac{x}{2} \right|$$

$$= \left| \ln \frac{nx}{n+1} \ln \frac{nx}{2n+5} - \ln \frac{nx}{n+1} \ln \frac{x}{2} \right.$$

$$\left. + \ln \frac{nx}{n+1} \ln \frac{x}{2} - \ln x \ln \frac{x}{2} \right|$$

$$= \left| \ln \frac{nx}{n+1} \ln \frac{2n}{2n+5} + \ln \frac{x}{2} \ln \frac{n}{n+1} \right|,$$

and for the special choice of the points $x_n = \frac{1}{ne^n} \in X$, we obtain

$$|f_n(x_n)g_n(x_n) - f(x_n)g(x_n)| = \left| \ln \frac{1}{(n+1)e^n} \ln \frac{2n}{2n+5} + \ln \frac{1}{2ne^n} \ln \frac{n}{n+1} \right|$$

$$= \left| (n + \ln(n+1)) \ln \left(1 + \frac{5}{2n} \right) + (n + \ln 2n) \ln \left(1 + \frac{1}{n} \right) \right|$$

$$= \frac{5}{2} \cdot \frac{2n}{5} \ln \left(1 + \frac{5}{2n} \right) + \frac{\ln(n+1)}{2n/5} \frac{2n}{5} \ln \left(1 + \frac{5}{2n} \right)$$

$$+ n \ln \left(1 + \frac{1}{n} \right) + \frac{\ln 2n}{n} n \ln \left(1 + \frac{1}{n} \right) \underset{n \to \infty}{\to} \frac{5}{2} \cdot 1$$

$$+ 0 \cdot 1 + 1 + 0 \cdot 1 = \frac{7}{2}.$$

In the evaluation of the last limit, we have used the following auxiliary limits:

$$\lim_{n \to \infty} \frac{n}{\alpha} \ln \left(1 + \frac{\alpha}{n} \right) = \lim_{n \to \infty} \ln \left(1 + \frac{\alpha}{n} \right)^{n/\alpha} = \ln e = 1, \quad \forall \alpha \neq 0$$

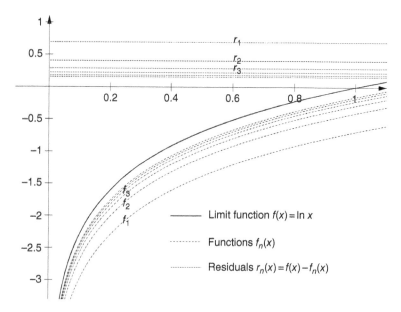

Figure 1.7 Examples 11 and 17, sequence $f_n(x) = \ln \frac{nx}{n+1}$.

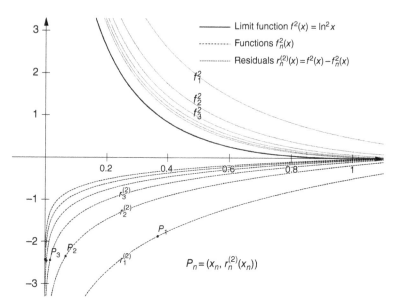

Figure 1.8 Example 11, sequence of squares $f_n^2(x) = \ln^2 \frac{nx}{n+1}$.

according to the second remarkable limit, and

$$\lim_{t \to \infty} \frac{\ln(\alpha t + \beta)}{t} = \lim_{t \to \infty} \frac{\alpha/(\alpha t + \beta)}{1} = 0, \quad \forall \alpha > 0, \forall \beta$$

due to l'Hospital's rule. Therefore, for $\varepsilon_0 = 1$ whatever N is chosen, there is $\tilde{n} > N$ and corresponding $x_{\tilde{n}} \in X$ such that $|f_{\tilde{n}}(x_{\tilde{n}})g_{\tilde{n}}(x_{\tilde{n}}) - f(x_{\tilde{n}})g(x_{\tilde{n}})| > \varepsilon_0 = 1$, that is, the convergence is nonuniform.

Remark 2. The following general statement is true for the sum and difference: if $f_n(x)$ and $g_n(x)$ converge uniformly on X to $f(x)$ and $g(x)$, respectively, then $f_n(x) \pm g_n(x)$ converges uniformly on X to $f(x) \pm g(x)$.

Remark 3. The following general statement is true for the product: if $f_n(x)$ and $g_n(x)$ converge uniformly on X to $f(x)$ and $g(x)$, respectively, and $f(x)$ and $g(x)$ are bounded on X, then $f_n(x) \cdot g_n(x)$ converges uniformly on X to $f(x) \cdot g(x)$. (Note the requirement of boundedness of the limit functions in this formulation.)

Example 12. Sequences $f_n(x)$ and $g_n(x)$ converge nonuniformly on X to $f(x)$ and $g(x)$, respectively, but $f_n(x) \cdot g_n(x)$ converges to $f(x) \cdot g(x)$ uniformly on X.

Solution
Consider the sequences $f_n(x) = \frac{1}{n\sqrt{x}}$ and $g_n(x) = nxe^{-nx}$ on $X = (0, +\infty)$. Both sequences converge to 0 for any fixed $x \in X$:

$$\lim_{n \to \infty} f_n(x) = \lim_{n \to \infty} \frac{1}{n\sqrt{x}} = 0 = f(x);$$

$$\lim_{n \to \infty} g_n(x) = \lim_{n \to \infty} nxe^{-nx} = \lim_{t \to +\infty} \frac{t}{e^t} = \lim_{t \to +\infty} \frac{1}{e^t} = 0 = g(x).$$

Therefore, $\lim_{n \to \infty} f_n(x) \cdot g_n(x) = 0$.

Let us investigate the nature of the convergence of these sequences. For $f_n(x)$, choosing $x_n = \frac{1}{n^2} \in X$, we obtain

$$|f_n(x_n) - f(x_n)| = \frac{1}{n\sqrt{x_n}} = \frac{n}{n} = 1 \not\to_{n \to \infty} 0,$$

that is, this convergence is nonuniform on X. Similarly, choosing $x_n = \frac{1}{n} \in X$, we can show the nonuniform convergence of $g_n(x)$ on X:

$$|g_n(x_n) - g(x_n)| = nx_n e^{-nx_n} = 1 \cdot e^{-1} \not\to_{n \to \infty} 0.$$

Finally, for $f_n(x) \cdot g_n(x)$, we have $|f_n(x)g_n(x) - f(x)g(x)| = \sqrt{x}e^{-nx}$. The derivative of the right-hand side is

$$(\sqrt{x}e^{-nx})_x = \left(\frac{1}{2\sqrt{x}} - n\sqrt{x}\right)e^{-nx} = \frac{n}{\sqrt{x}}e^{-nx}\left(\frac{1}{2n} - x\right).$$

Therefore, the point $x_n = \frac{1}{2n} \in X$ is the only local (and global) maximum of $\sqrt{x}e^{-nx}$ on X. Consequently,

$$|f_n(x)g_n(x) - f(x)g(x)| \leq \max_{(0,+\infty)} \sqrt{x}e^{-nx} = \sqrt{x_n}e^{-nx_n} = \frac{1}{\sqrt{2n}}e^{-1/2} \underset{n\to\infty}{\to} 0,$$

that is, $f_n(x) \cdot g_n(x)$ converges uniformly on X to 0.

Example 13. A sequence $f_n^2(x)$ converges uniformly on X, but $f_n(x)$ diverges on X.

Solution
Consider the sequence $f_n(x) = (-1)^n \frac{n+1}{n} x$ on $X = (0, 1]$. The sequence of squares $f_n^2(x) = \frac{(n+1)^2}{n^2} x^2$ converges uniformly on $(0, 1]$ to x^2, because

$$|f_n^2(x) - x^2| = \left|\frac{(n+1)^2}{n^2}x^2 - x^2\right| = x^2 \frac{2n+1}{n^2} \leq \frac{2}{n} + \frac{1}{n^2} \underset{n\to\infty}{\to} 0.$$

However, there is no limit of $f_n(x)$ for any fixed $x \in (0, 1]$, since two partial limits give different results: $f_{2n}(x) = \frac{2n+1}{2n}x \underset{n\to\infty}{\to} x$ and $f_{2n+1}(x) = -\frac{2n+2}{2n+1}x \underset{n\to\infty}{\to} -x$.

Remark 1. The same sequence can be used to exemplify the following situation: a sequence $|f_n(x)|$ converges uniformly on X, but $f_n(x)$ diverges on X. Indeed, although $f_n(x) = (-1)^n \frac{n+1}{n} x$ does not converge on $X = (0, 1]$, the sequence $|f_n(x)| = \frac{n+1}{n} x$ converges uniformly to x on $X = (0, 1]$:

$$||f_n(x)| - x| = \left|\frac{n+1}{n}x - x\right| = \frac{1}{n}|x| \leq \frac{1}{n} \underset{n\to\infty}{\to} 0.$$

Remark 2. Note that the inequality $||f_n(x)| - |f(x)|| \leq |f_n(x) - f(x)|$ ensures the validity of the converse general statement: if $f_n(x)$ converges uniformly on X to $f(x)$, then $|f_n(x)|$ converges uniformly on X to $|f(x)|$.

Remark 3. The following example also takes place: a sequence $f_n^2(x)$ converges uniformly on X and $f_n(x)$ converges on X, but the convergence of $f_n(x)$ is nonuniform. Consider the sequence $f_n(x)$ on $[0, 1]$ similar to that analyzed in Example 6: $f_n(x) = \begin{cases} 1, x = r_1, r_2, \cdots, r_n \\ -1, \text{otherwise} \end{cases}$, where r_n is the sequence of all the rational points in $[0, 1]$ ordered in some way. This sequence is monotone in n for any fixed $x \in [0, 1]$ (since $f_n(r_{n+1}) = -1 < 1 = f_{n+1}(r_{n+1})$ and $f_n(x) = f_{n+1}(x)$, $\forall x \neq r_{n+1}$) and bounded (since $-1 \leq f_n(x) \leq 1$, $\forall n \in \mathbb{N}$, $\forall x \in [0, 1]$). Therefore, this sequence is convergent at any fixed $x \in [0, 1]$ and $f(x) = \lim_{n \to \infty} f_n(x) = \begin{cases} 1, x \in \mathbb{Q} \\ -1, x \in \mathbb{I} \end{cases}$. The sequence of the squares consists of the same constant function $f_n^2(x) = 1$, $\forall n \in \mathbb{N}$ and, therefore, it converges uniformly on $[0, 1]$ to $f^2(x) = 1$. At the same time, using the same reasoning as in Example 6, one can show that the convergence of $f_n(x)$ is nonuniform on $[0, 1]$ and on any subinterval of $[0, 1]$.

Example 14. A sequence $f_n(x) \cdot g_n(x)$ converges uniformly on X to 0, but neither $f_n(x)$ nor $g_n(x)$ converges to 0 on X.

Solution
The sequences $f_n(x) = nx + (-1)^n nx$ and $g_n(x) = nx - (-1)^n nx$ are divergent for every fixed $x \in X = (0, +\infty)$:

$$f_{2n}(x) = 4nx \underset{n \to \infty}{\to} +\infty, f_{2n+1}(x) = 0 \underset{n \to \infty}{\to} 0;$$

$$g_{2n}(x) = 0 \underset{n \to \infty}{\to} 0, g_{2n+1}(x) = (4n + 2)x \underset{n \to \infty}{\to} +\infty.$$

However, $f_n(x) \cdot g_n(x) = n^2 x^2 - n^2 x^2 = 0$ converges uniformly to 0 on X.

Another interesting counterexample includes the sequences $f_n(x) = \begin{cases} \frac{nx}{n+1}, x \in \mathbb{Q} \cap X \\ \frac{x}{n}, x \in \mathbb{I} \cap X \end{cases}$ and $g_n(x) = \begin{cases} \frac{x}{n}, x \in \mathbb{Q} \cap X \\ \frac{nx}{n+1}, x \in \mathbb{I} \cap X \end{cases}$ defined on $X = (0, 1]$. Both sequences converge on X to nonzero functions $f(x) = \lim_{n \to \infty} f_n(x) = \begin{cases} x, x \in \mathbb{Q} \cap X \\ 0, x \in \mathbb{I} \cap X \end{cases}$ and $g(x) = \lim_{n \to \infty} g_n(x) = \begin{cases} 0, x \in \mathbb{Q} \cap X \\ x, x \in \mathbb{I} \cap X \end{cases}$, respectively. At the same time, $\lim_{n \to \infty} f_n(x)g_n(x) = \lim_{n \to \infty} \frac{x^2}{n+1} = 0$, for $\forall x \in (0, 1]$ and this convergence is uniform on X, since the estimate

$$|f_n(x)g_n(x) - 0| = \frac{x^2}{n+1} \leq \frac{1}{n+1} \underset{n \to \infty}{\to} 0$$

holds simultaneously for all $x \in (0, 1]$.

Example 15. A sequence $f_n(x)$ converges uniformly on X to a function $f(x)$, $f_n(x) \neq 0, f(x) \neq 0$, $\forall x \in X$, but $\frac{1}{f_n(x)}$ does not converge uniformly on X to $\frac{1}{f(x)}$.

Solution
The sequence $f_n(x) = \frac{nx}{n+2}$ converges uniformly on $X = (0, 1)$ to $f(x) = x$:

$$|f_n(x) - f(x)| = \left| \frac{nx}{n+2} - x \right| = \frac{2}{n+2}|x| < \frac{2}{n+2} < \varepsilon$$

and the last inequality holds for $\forall \varepsilon > 0$ and simultaneously for all $x \in X$ if we choose $\forall n > N_\varepsilon = \left[\frac{2}{\varepsilon}\right]$. On the other hand,

$$\left|\frac{1}{f_n(x)} - \frac{1}{f(x)}\right| = \left|\frac{n+2}{nx} - \frac{1}{x}\right| = \frac{2}{n}\frac{1}{|x|},$$

and for $x_n = \frac{1}{n} \in X$, it follows that $\left|\frac{1}{f_n(x_n)} - \frac{1}{f(x_n)}\right| = n\frac{2}{n} = 2$, which means that the convergence is nonuniform.

Example 16. A sequence $f_n(x)$ is bounded uniformly on $\mathbb{R}$ and converges uniformly on $[-a, a]$, $\forall a > 0$, to a function $f(x)$, but the numerical sequence $\sup_{x \in \mathbb{R}} f_n(x)$ does not converge to $\sup_{x \in \mathbb{R}} f(x)$.

Solution
Consider the sequence $f_n(x) = e^{-(x-n)^2}$, which is defined and uniformly bounded on $\mathbb{R}$: $0 < e^{-(x-n)^2} \leq 1$, $\forall n$, $\forall x \in \mathbb{R}$. This sequence converges to zero on $\mathbb{R}$, since for any fixed $x \in \mathbb{R}$ one has $(x-n)^2 \underset{n\to\infty}{\to} +\infty$ and, consequently, $\lim_{n\to\infty} e^{-(x-n)^2} = \lim_{t\to+\infty} e^{-t} = 0$. Hence, $f(x) \equiv 0$ on $\mathbb{R}$ and, consequently, $\sup_{x\in\mathbb{R}} f(x) = 0$. On the other hand, $\sup_{x\in\mathbb{R}} f_n(x) = 1$, $\forall n \in \mathbb{N}$, since $f_n(x) \leq 1$ and $f_n(n) = 1$. This means that $\sup_{x\in\mathbb{R}} f_n(x) = 1$ does not converge to $\sup_{x\in\mathbb{R}} f(x) = 0$. It just remains to prove the uniform convergence of $f_n(x)$ on $[-a, a]$, $\forall a > 0$. For any fixed $a > 0$, there exists the natural number $N_a > a$. Then for $\forall n > N_a$, one gets $(x-n)^2 \geq (a-n)^2$ for each $x \in [-a, a]$. Therefore,

$$|f_n(x) - f(x)| = e^{-(x-n)^2} \leq e^{-(a-n)^2}$$

for all $x \in [-a, a]$. Since $\exp(-(a-n)^2) \underset{n\to\infty}{\to} 0$, the last inequality guarantees the uniformity of the convergence on $[-a, a]$, where $a > 0$ is arbitrary. Note, however, that the convergence of $f_n(x)$ is not uniform on $\mathbb{R}$, which is evident if one chooses $x_n = n$ leading to

$$|f_n(x_n) - f(x_n)| = e^{-(n-n)^2} = 1.$$

Remark 1. Two other interesting counterexamples are $f_n(x) = \arctan\frac{x}{n}$ and $f_n(x) = \frac{2nx}{n^2+x^2}$. For instance, for the first function the reasoning can be as follows. First, note that the sequence is uniformly bounded on $\mathbb{R}$ ($\left|\arctan\frac{x}{n}\right| < \frac{\pi}{2}$, $\forall n \in \mathbb{N}$ and $\forall x \in \mathbb{R}$). Second, it converges to zero for any fixed $x \in \mathbb{R}$ ($\lim_{n\to\infty} \arctan\frac{x}{n} = 0$). Further, this convergence is uniform on $[-a, a]$, $\forall a > 0$ due to the evaluation

$$|f_n(x) - f(x)| = \left|\arctan\frac{x}{n}\right| = \arctan\frac{|x|}{n} \leq \arctan\frac{a}{n},$$

that holds for all $x \in [-a, a]$ (here we used the properties that $\arctan t$ is an odd and a strictly increasing function). Since $\lim\limits_{n\to\infty} \arctan \frac{a}{n} = 0$, the last evaluation implies the uniform convergence. Hence, all the conditions of the example are satisfied, but still the sequence $\sup\limits_{x\in\mathbb{R}} f_n(x)$ does not converge to $\sup\limits_{x\in\mathbb{R}} f(x)$, because $\sup\limits_{x\in\mathbb{R}} \arctan \frac{x}{n} = \frac{\pi}{2}$ for any n, while $\sup\limits_{x\in\mathbb{R}} f(x) = \sup\limits_{x\in\mathbb{R}} 0 = 0$. Note, that just like in the first counterexample, the convergence of $f_n(x)$ is not uniform on $\mathbb{R}$: for any n one can choose $x_n = n$ to obtain

$$|f_n(x_n) - f(x_n)| = \arctan \frac{n}{n} = \arctan 1 = \frac{\pi}{4} \neq 0.$$

Remark 2. The following general statement is true: if a sequence $f_n(x)$ is bounded uniformly on $\mathbb{R}$ and converges uniformly on $\mathbb{R}$ to a function $f(x)$, then the numerical sequence $\sup\limits_{x\in\mathbb{R}} f_n(x)$ converges to $\sup\limits_{x\in\mathbb{R}} f(x)$. (Note the requirement of uniform convergence on $\mathbb{R}$ to the limit function in this formulation.)

Remark 3. The condition of uniform convergence of a sequence $f_n(x)$ to a function $f(x)$ on X is equivalent to the condition $\lim\limits_{n\to\infty} \sup\limits_{x\in X} |f_n(x) - f(x)| = 0$.

Example 17. Suppose each function $f_n(x)$ maps X on Y and function $g(y)$ is continuous on Y; the sequence $f_n(x)$ converges uniformly on X, but the sequence $g_n(x) = g(f_n(x))$ does not converge uniformly on X.

Solution
Let us consider the sequence $f_n(x) = \ln \frac{nx}{n+1}$ on $X = (0, +\infty)$ and function $g(y) = e^y$. Each of the functions $f_n(x)$ maps $(0, +\infty)$ on the entire real line and the function $g(y)$ is continuous on $\mathbb{R}$. In Example 11, it was shown that the sequence $f_n(x)$ converges uniformly on X to the function $f(x) = \ln x$. The corresponding sequence $g_n(x) = g(f_n(x)) = \frac{nx}{n+1}$ converges for any fixed $x \in (0, +\infty)$:

$$\lim_{n\to\infty} g_n(x) = \lim_{n\to\infty} \frac{nx}{n+1} = x = g(f(x)),$$

but this convergence is not uniform. In fact,

$$|g_n(x) - g(f(x))| = \left| \frac{nx}{n+1} - x \right| = \frac{1}{n+1} x$$

and choosing $x_n = (n+1) \in X$, one gets

$$|g_n(x_n) - g(f(x_n))| = \frac{x_n}{n+1} = 1.$$

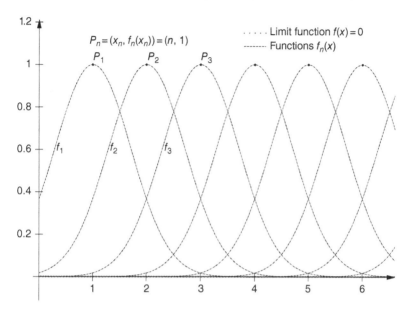

Figure 1.9 Example 16, sequence $f_n(x) = e^{-(x-n)^2}$.

Remark. The following example also takes place: suppose functions $f_n(x)$ map X on Y and function $g(y)$ is continuous on Y; the sequence $f_n(x)$ converges nonuniformly on X, but the sequence $g_n(x) = g(f_n(x))$ converges uniformly on X. In the trivial case, one can use an arbitrary nonuniformly convergent sequence $f_n(x)$ and the constant function $g(y) \equiv 1$. For a nonconstant function $g(y)$, one can use the above sequence $f_n(x) = \frac{nx}{n+1}$ defined on $X = (0, +\infty)$ and the function $g(y) = \ln y$.

Example 18. A series $\sum u_n^2(x)$ converges uniformly on X, but the series $\sum u_n(x)$ does not converge uniformly on X.

Solution
In Remark 1 to Example 5, it was shown that the series $\sum_{n=1}^{\infty} \frac{\sin nx}{n}$ converges nonuniformly on $\mathbb{R}$. However, the series $\sum_{n=1}^{\infty} \frac{\sin^2 nx}{n^2}$ converges uniformly on $\mathbb{R}$ according to the Weierstrass test: $\frac{\sin^2 nx}{n^2} \leq \frac{1}{n^2}$, $\forall x \in \mathbb{R}$ and the majorant series $\sum_{n=1}^{\infty} \frac{1}{n^2}$ converges.

Example 19. A series $\sum u_n^2(x)$ converges uniformly on X, but the series $\sum u_n(x)$ does not converge (even pointwise) on X.

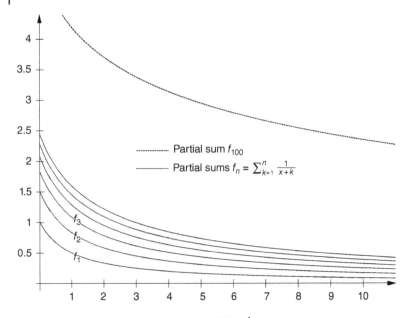

Figure 1.10 Examples 19, 20, and 21, series $\sum_{n=1}^{\infty} \frac{1}{x+n}$.

Solution

Consider the series $\sum_{n=1}^{\infty} \frac{1}{x+n}$ on $X = [0, +\infty)$. This series is divergent at each point $x \in X$, since $\lim_{n\to\infty} \frac{1/n}{1/(x+n)} = 1$ and the harmonic series $\sum_{n=1}^{\infty} \frac{1}{n}$ is divergent. At the same time, the series $\sum_{n=1}^{\infty} \frac{1}{(x+n)^2}$ converges uniformly on $X = [0, +\infty)$ due to the Weierstrass test: $\frac{1}{(x+n)^2} \le \frac{1}{n^2}, \forall x \in X$ and the majorant series $\sum_{n=1}^{\infty} \frac{1}{n^2}$ converges.

Remark. If $\sum u_n(x)$ diverges, it may happen that $\sum_{n=1}^{\infty} u_n^2(x)$ also diverges. For instance, the series $\sum_{n=1}^{\infty} \frac{1}{\sqrt{x}+\sqrt{n}}$ diverges at each point $x \in X = [0, 1]$ according to the comparison test: $\lim_{n\to\infty} \frac{1/\sqrt{n}}{1/(\sqrt{x}+\sqrt{n})} = 1$ and the series $\sum_{n=1}^{\infty} \frac{1}{\sqrt{n}}$ is divergent. Due to the very same arguments, the series of squares also diverges on $X = [0, 1]$: $\lim_{n\to\infty} \frac{1/n}{1/(\sqrt{x}+\sqrt{n})^2} = 1$ and the series $\sum_{n=1}^{\infty} \frac{1}{n}$ is divergent.

Example 20. A series $\sum u_n(x)v_n(x)$ converges uniformly on X, but at least one of the series $\sum u_n(x)$ or $\sum v_n(x)$ does not converge uniformly on X.

Solution

Consider $u_n(x) = \frac{1}{x+n}$ and $v_n(x) = \frac{1}{x^2+n^2}$ on $X = [0, +\infty)$. The series $\sum_{n=1}^{\infty} u_n(x)v_n(x)$ converges uniformly on X due to the Weierstrass test:

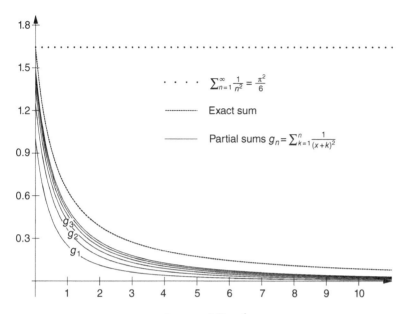

Figure 1.11 Example 19, series of squares $\sum_{n=1}^{\infty} \frac{1}{(x+n)^2}$.

$$|u_n(x)v_n(x)| = \frac{1}{x+n} \frac{1}{x^2+n^2} \leq \frac{1}{n^3}, \quad \forall x \in [0, +\infty)$$

and $\sum \frac{1}{n^3}$ is a convergent series. The same reasoning shows the uniform convergence of the series $\sum_{n=1}^{\infty} v_n(x)$ on X:

$$|v_n(x)| = \frac{1}{x^2+n^2} \leq \frac{1}{n^2}, \quad \forall x \in [0, +\infty)$$

and $\sum \frac{1}{n^2}$ is a convergent series. However, the series $\sum_{n=1}^{\infty} u_n(x)$ diverges on X, since $\lim\limits_{n\to\infty} \frac{1/n}{1/(x+n)} = 1$ and the series $\sum \frac{1}{n}$ is divergent.

Example 21. A series $\sum u_n(x)v_n(x)$ converges uniformly on X, but neither $\sum u_n(x)$ nor $\sum v_n(x)$ converges (even pointwise) on X.

Solution
Consider $u_n(x) = \frac{1}{x+n}$ and $v_n(x) = \frac{1}{\sqrt{x+\sqrt{n}}}$ on $X = [0, +\infty)$. The series $\sum_{n=1}^{\infty} u_n(x)v_n(x)$ converges uniformly on X due to the Weierstrass test:

$$|u_n(x)v_n(x)| = \frac{1}{x+n} \frac{1}{\sqrt{x}+\sqrt{n}} \leq \frac{1}{n^{3/2}}, \quad \forall x \in [0, +\infty)$$

and $\sum \frac{1}{n^{3/2}}$ is a convergent series. However, both $\sum_{n=1}^{\infty} u_n(x)$ and $\sum_{n=1}^{\infty} v_n(x)$ diverge on X according to the comparison test: $\lim_{n\to\infty} \frac{1/n}{1/(x+n)} = 1$ and the series $\sum \frac{1}{n}$ diverges; $\lim_{n\to\infty} \frac{1/\sqrt{n}}{1/(\sqrt{x}+\sqrt{n})} = 1$ and the series $\sum \frac{1}{\sqrt{n}}$ diverges.

Example 22. Series $\sum u_n(x)$ and $\sum v_n(x)$ converge nonuniformly on X, but $\sum u_n(x)v_n(x)$ converges uniformly on X.

Solution
The series $\sum_{n=1}^{\infty} \frac{\sin nx}{n}$ converges nonuniformly on $\mathbb{R}$ (see Remark 1 to Example 5), and so does the series $\sum_{n=1}^{\infty} \frac{\sin nx}{\sqrt{n}}$ (applying for the latter series the same reasoning as for the former in Remark 1 to Example 5). However, the series $\sum_{n=1}^{\infty} \frac{\sin^2 nx}{n^{3/2}}$ converges uniformly on $\mathbb{R}$ according to the Weierstrass test: $\frac{\sin^2 nx}{n^{3/2}} \leq \frac{1}{n^{3/2}}$, $\forall x \in \mathbb{R}$ and the majorant series $\sum_{n=1}^{\infty} \frac{1}{n^{3/2}}$ converges.

Example 23. A series $\sum u_n(x)$ converges uniformly on X, but $\sum u_n^2(x)$ does not converge uniformly on X.

Solution
The uniform convergence of the series $\sum_{n=1}^{\infty} u_n(x) = \sum_{n=1}^{\infty} (-1)^n \frac{x^n}{\sqrt[3]{n}}$ on $X = (0,1)$ can be proved by applying Abel's theorem. In fact, the series $\sum (-1)^n \frac{1}{\sqrt[3]{n}}$ converges by Leibniz's test of alternating series (and this convergence is uniform on X, since the series does not depend on x), and the sequence x^n is monotone in n for each $x \in (0,1)$ and is uniformly bounded, since $x^n < 1$, $\forall x \in (0,1)$, $\forall n \in \mathbb{N}$.

On the other hand, the convergence of the series $\sum_{n=1}^{\infty} u_n^2(x) = \sum_{n=1}^{\infty} \frac{x^{2n}}{n^{2/3}}$ is nonuniform on $X = (0,1)$. In fact, this series converges on $X = (0,1)$, since $u_n^2(x) = \frac{x^{2n}}{n^{2/3}} \leq x^{2n}$, $\forall n \in \mathbb{N}$, and the series $\sum x^{2n}$ is convergent by being a geometric series with the ratio in $(0,1)$. At the same time, applying the Cauchy criterion of the uniform convergence with $p_n = n$ and $x_n = \left(1 - \frac{1}{4n}\right) \in X$, one obtains

$$\left| \sum_{k=n+1}^{n+p_n} u_k^2(x_n) \right| = \sum_{k=n+1}^{n+p_n} \frac{x_n^{2k}}{k^{2/3}} > n \frac{x_n^{4n}}{(2n)^{2/3}} = \frac{n^{1/3}}{4^{1/3}} \left(1 - \frac{1}{4n}\right)^{4n} \xrightarrow[n\to\infty]{} +\infty,$$

that is, the series $\sum u_n^2(x)$ converges nonuniformly on $X = (0,1)$.

Remark 1. If, in the given counterexample, one changes the set X to $(0,1]$, then $\sum_{n=1}^{\infty} u_n(x) = \sum_{n=1}^{\infty} (-1)^n \frac{x^n}{\sqrt[3]{n}}$ converges uniformly on $X = (0,1]$ (due to the

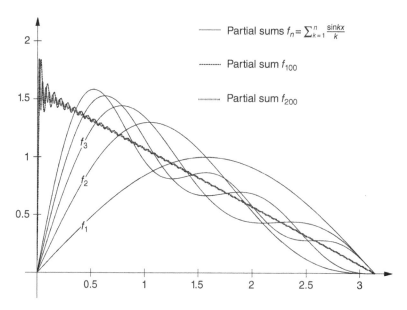

Figure 1.12 Examples 22, 5, and 18, series $\sum_{n=1}^{\infty} \frac{\sin nx}{n}$.

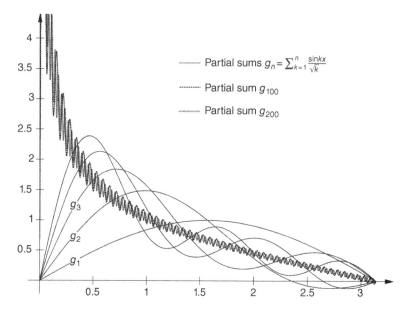

Figure 1.13 Examples 22 and 24, series $\sum_{n=1}^{\infty} \frac{\sin nx}{\sqrt{n}}$.

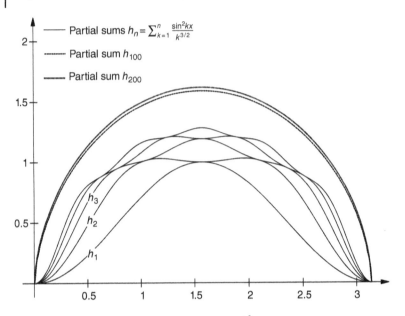

Figure 1.14 Example 22, series of products $\sum_{n=1}^{\infty} \frac{\sin^2 nx}{n^{3/2}}$.

same reasoning as before), but $\sum_{n=1}^{\infty} u_n^2(x) = \sum_{n=1}^{\infty} \frac{x^{2n}}{n^{2/3}}$ diverges at $x = 1$ since $\sum_{n=1}^{\infty} u_n^2(1) = \sum_{n=1}^{\infty} \frac{1}{n^{2/3}}$ is a divergent p-series.

Remark 2. Evidently, the following more general example can also be constructed: both series $\sum u_n(x)$ and $\sum v_n(x)$ converge uniformly on X, but the series $\sum u_n(x)v_n(x)$ does not converge uniformly on X. In the particular case $u_n(x) = v_n(x)$, the counterexample is already provided above. Let us consider the case when $u_n(x) \neq v_n(x)$. For instance, using the same arguments as before, one can prove that both $\sum_{n=1}^{\infty} u_n(x) = \sum_{n=1}^{\infty} (-1)^n \frac{x^n}{\sqrt[3]{n}}$ and $\sum_{n=1}^{\infty} v_n(x) = \sum_{n=1}^{\infty} (-1)^n \frac{x^n}{\sqrt{n}}$ converge uniformly on $X = (0,1)$, but the series $\sum_{n=1}^{\infty} u_n(x)v_n(x) = \sum_{n=1}^{\infty} \frac{x^{2n}}{n^{5/6}}$ converges nonuniformly on $X = (0,1)$. For the last series, its convergence follows from the evaluation $\frac{x^{2n}}{n^{5/6}} \leq x^{2n}$, $\forall n \in \mathbb{N}$ and the convergence of the geometric series $\sum_{n=1}^{\infty} x^{2n}$ for $\forall x \in (0,1)$, while the nonuniformity can be shown by the Cauchy criterion, choosing as above $p_n = n$ and $x_n = \left(1 - \frac{1}{4n}\right) \in X$:

$$\left| \sum_{k=n+1}^{n+p_n} u_k(x_n)v_k(x_n) \right| = \sum_{k=n+1}^{2n} \frac{x_n^{2k}}{k^{5/6}} > n \frac{x_n^{4n}}{(2n)^{5/6}} = \frac{n^{1/6}}{2^{5/6}} \left(1 - \frac{1}{4n}\right)^{4n} \underset{n \to +\infty}{\to} \infty.$$

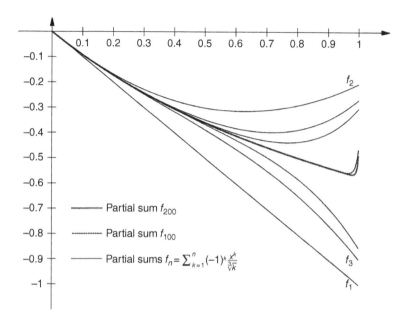

Figure 1.15 Example 23, series $\sum_{n=1}^{\infty} (-1)^n \frac{x^n}{\sqrt[3]{n}}$.

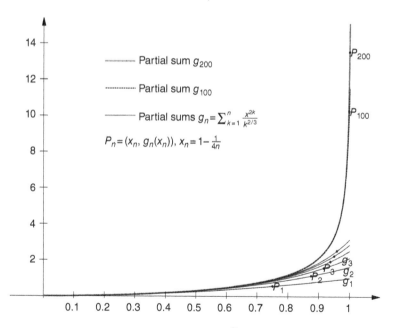

Figure 1.16 Example 23, series of squares $\sum_{n=1}^{\infty} \frac{x^{2n}}{n^{2/3}}$.

Example 24. A series $\sum u_n(x)$ converges uniformly on X, but $\sum u_n^2(x)$ does not converge (even pointwise) on X.

Solution

The series $\sum_{n=1}^{\infty} u_n(x) = \sum_{n=1}^{\infty} \frac{\sin nx}{\sqrt{n}}$ is uniformly convergent on $X = [a, \pi - a]$, $\forall a \in \left(0, \frac{\pi}{2}\right)$, which can be shown using the same considerations as for $\sum_{n=1}^{\infty} \frac{\sin nx}{n}$ in Remark 1 to Example 5. Let us prove that the series $\sum_{n=1}^{\infty} u_n^2(x) = \sum_{n=1}^{\infty} \frac{\sin^2 nx}{n}$ is divergent on X. Note that the series $\sum_{n=1}^{\infty} \frac{\cos 2nx}{n}$ is uniformly convergent on X just like the series in Remark 1 to Example 5. The series of squares can be rewritten in the form $\sum_{n=1}^{\infty} u_n^2(x) = \sum_{n=1}^{\infty} \frac{1 - \cos 2nx}{2n}$, that is, the general term $u_n^2(x)$ is the difference of the general term of the divergent harmonic series and uniformly convergent series. This implies the divergence of the series $\sum_{n=1}^{\infty} \frac{\sin^2 nx}{n}$ at each point of X.

Remark. Naturally, the following more general situation also takes place: both series $\sum u_n(x)$ and $\sum v_n(x)$ converge uniformly on X, but the series $\sum u_n(x)v_n(x)$ does not converge (even pointwise) on X. In the particular case $u_n(x) = v_n(x)$, the counterexample is given above. Let us consider the case when $u_n(x) \neq v_n(x)$. For instance, using the same arguments as before, one can prove that both $\sum_{n=1}^{\infty} \frac{\cos nx}{\sqrt[3]{n}}$ and $\sum_{n=1}^{\infty} \frac{\cos nx}{\sqrt[4]{n}}$ converge uniformly on $X = [a, \pi - a]$, $\forall a \in \left(0, \frac{\pi}{2}\right)$, but the series $\sum_{n=1}^{\infty} u_n(x)v_n(x) = \sum_{n=1}^{\infty} \frac{\cos^2 nx}{n^{7/12}}$ diverges at each point of X, because its general term can be represented in the form $\frac{\cos^2 nx}{n^{7/12}} = \frac{1}{2n^{7/12}} + \frac{\cos 2nx}{2n^{7/12}}$, where the first summand is a general term of the divergent p-series, while the second is a general term of the uniformly convergent series.

Example 25. Both series $\sum u_n(x)$ and $\sum v_n(x)$ are nonnegative for $\forall x \in X$, $\lim_{n \to \infty} \frac{u_n(x)}{v_n(x)} = 1$ and one of these series converges uniformly on X, but another series does not converge uniformly on X.

Solution

Consider the two nonnegative series $\sum_{n=1}^{\infty} u_n(x) = \sum_{n=1}^{\infty} \frac{x^2}{n^4 + x^4}$ and $\sum_{n=1}^{\infty} v_n(x) = \sum_{n=1}^{\infty} \frac{x^2}{n^4 + x^2}$ on $X = \mathbb{R}$. The limit of the general terms equals 1: $\lim_{n \to \infty} \frac{u_n(x)}{v_n(x)} = \lim_{n \to \infty} \frac{n^4 + x^2}{n^4 + x^4} = 1$ for $\forall x \in \mathbb{R}$. Also, the series $\sum_{n=1}^{\infty} \frac{x^2}{n^4 + x^4}$ converges uniformly on $X = \mathbb{R}$ by the Weierstrass test, since $u_n(x) = \frac{1}{2n^2} \frac{2n^2 x^2}{n^4 + x^4} \leq \frac{1}{2n^2}$, $\forall x \in \mathbb{R}$ and the series $\sum \frac{1}{2n^2}$ converges. Therefore, all the statement conditions hold. At the same time, the second series $\sum_{n=1}^{\infty} \frac{x^2}{n^4 + x^2}$ converges on $X = \mathbb{R}$ since $\frac{x^2}{n^4 + x^2} \leq \frac{x^2}{n^4}$, $\forall x \in \mathbb{R}$ and the series $\sum \frac{1}{n^4}$ converges. However, the convergence of the second

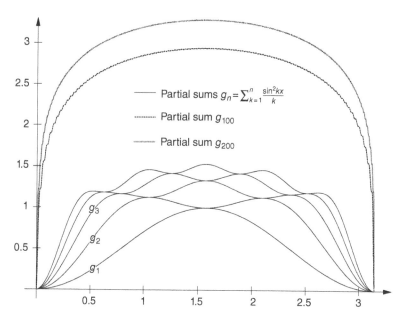

Figure 1.17 Example 24, series of squares $\sum_{n=1}^{\infty} \frac{\sin^2 nx}{n}$.

series is nonuniform since its general term does not converge to 0 uniformly: for $x_n = n^2$, one gets $v_n(x_n) = \frac{x_n^2}{n^4+x_n^2} = \frac{1}{2} \nrightarrow 0$.

Remark. For numerical series and for the pointwise convergence of series of functions, the corresponding general statement is true and represents a particular case of the Comparison test for nonnegative series: if both series $\sum u_n(x)$ and $\sum v_n(x)$ are nonnegative for $\forall x \in X$, $\lim_{n\to\infty} \frac{u_n(x)}{v_n(x)} = const > 0$ and one of these series converges on X, then another series also converges on X.

1.3 Dirichlet's and Abel's Theorems

Remark to Examples 26–29. In the following four examples, the conditions of Dirichlet's theorem, which provides sufficient conditions for the uniform convergence of the series $\sum u_n(x)v_n(x)$, are analyzed. It is shown that none of the three conditions stated in the theorem can be dropped. At the same time, these conditions are not necessary: all of them can be violated for an uniformly convergent series.

Example 26. The partial sums of $\sum u_n(x)$ are bounded for $\forall x \in X$, and the sequence $v_n(x)$ is monotone in n for each fixed $x \in X$ and converges uniformly on X to 0, but the series $\sum u_n(x)v_n(x)$ does not converge uniformly on X.

Solution

Let $u_n(x) = x^n$ and $v_n(x) = \frac{1}{n}$ be defined on $X = (0, 1)$. The series $\sum_{n=1}^{\infty} u_n(x) = \sum_{n=1}^{\infty} x^n$ converges on X, since this is a geometric series with the ratio in $(0, 1)$. Therefore, the partial sums of this series are bounded for each fixed $x \in X$. However, the boundedness is not uniform as is seen from the evaluation of the partial sums at the points $x_n = 1 - \frac{1}{n}$, $n > 1$ lying in X:

$$\sum_{k=1}^{n} x_n^k = \frac{x_n}{1 - x_n}(1 - x_n^n) = \frac{1 - 1/n}{1/n}\left(1 - \left(1 - \frac{1}{n}\right)^n\right)$$

$$= (n - 1)\left(1 - \left(1 - \frac{1}{n}\right)^n\right) \underset{n \to \infty}{\to} +\infty,$$

that is, the first condition in Dirichlet's theorem is weakened. The remaining two conditions hold: $v_n(x) = \frac{1}{n}$ is monotone and $v_n(x) = \frac{1}{n} \underset{n \to \infty}{\to} 0$ (and the last convergence is uniform, because v_n does not depend on x).

The series of the products $\sum_{n=1}^{\infty} u_n(x)v_n(x) = \sum_{n=1}^{\infty} \frac{x^n}{n}$ converges on $(0, 1)$, since $0 < \frac{x^n}{n} \le x^n, \forall n \in \mathbb{N}$ and the geometric series converges for $\forall x \in (0, 1)$. However, this convergence is nonuniform, which can be shown by the Cauchy criterion: choosing $p_n = n$ and $\tilde{x}_n = 1 - \frac{1}{2n} \in X, \forall n \in \mathbb{N}$, one obtains:

$$\left|\sum_{k=n+1}^{n+p_n} u_k(\tilde{x}_n)v_k(\tilde{x}_n)\right| = \sum_{k=n+1}^{2n} \frac{\tilde{x}_n^k}{k} > n\frac{\tilde{x}_n^{2n}}{2n} = \frac{1}{2}\left(1 - \frac{1}{2n}\right)^{2n} \underset{n \to +\infty}{\to} \frac{1}{2}e^{-1} \ne 0.$$

Example 27. The partial sums of $\sum u_n(x)$ are uniformly bounded on X, and the sequence $v_n(x)$ converges uniformly on X to 0, but the series $\sum u_n(x)v_n(x)$ does not converge uniformly on X.

Solution

Let $u_n(x) = \frac{(-1)^n}{\sqrt{n}}$ and $v_n(x) = (-1)^n \frac{x^n}{\sqrt{n}}$ be defined on $X = (0, 1)$. The series $\sum_{n=1}^{\infty} u_n(x) = \sum_{n=1}^{\infty} \frac{(-1)^n}{\sqrt{n}}$ converges by Leibniz's test and, consequently, its partial sums are bounded (and this boundedness is uniform on X, because the general term does not depend on x). The sequence $v_n(x) = (-1)^n \frac{x^n}{\sqrt{n}}$ converges uniformly on X to 0 due to the following evaluation: $\left|(-1)^n \frac{x^n}{\sqrt{n}}\right| \le \frac{1}{\sqrt{n}}$, $\forall x \in (0, 1)$ and $\frac{1}{\sqrt{n}} \underset{n \to +\infty}{\to} 0$. Thus, both conditions of the statement hold, but the series of the products $\sum_{n=1}^{\infty} u_n(x)v_n(x) = \sum_{n=1}^{\infty} \frac{x^n}{n}$ converges nonuniformly on $(0, 1)$ as was shown in Example 26. Note that the sequence $v_n(x) = (-1)^n \frac{x^n}{\sqrt{n}}$

is not monotone in n, that is, the condition of the monotonicity of $v_n(x)$ in Dirihlet's theorem is violated.

Example 28. The partial sums of $\sum u_n(x)$ are uniformly bounded on X, and the sequence $v_n(x)$ is monotone in n for each fixed $x \in X$ and converges on X to 0, but the series $\sum u_n(x)v_n(x)$ does not converge uniformly on X.

Solution
Consider $u_n(x) = (-1)^n$ and $v_n(x) = x^n$ on $X = (0,1)$. The partial sums $\sum_{k=1}^{n} u_k(x)$ are uniformly bounded on X: $\left|\sum_{k=1}^{n} u_k(x)\right| = \left|\sum_{k=1}^{n} (-1)^k\right| \le 1$, for $\forall n \in \mathbb{N}$ and $\forall x \in (0,1)$. The sequence $v_n(x) = x^n$ is decreasing in n and $v_n(x) = x^n \xrightarrow[n \to +\infty]{} 0$ for each fixed $x \in (0,1)$. Thus, the conditions of the statement are satisfied. However, the series of the products converges nonuniformly. In fact, $\sum_{n=1}^{\infty} u_n(x)v_n(x) = \sum_{n=1}^{\infty} (-1)^n x^n$ is a convergent geometric series on $(0,1)$ (the ratio $-x \in (-1,0)$), but the evaluation of its residual shows that this convergence is nonuniform: choosing $x_n = 1 - \frac{1}{n+1} \in (0,1)$, one obtains

$$\left|\sum_{k=n+1}^{\infty} (-1)^k x_n^k\right| = \left|\frac{(-1)^{n+1} x_n^{n+1}}{1 + x_n}\right|$$

$$= \frac{1}{2 - \frac{1}{n+1}}\left(1 - \frac{1}{n+1}\right)^{n+1} \xrightarrow[n \to +\infty]{} \frac{1}{2}e^{-1} \ne 0.$$

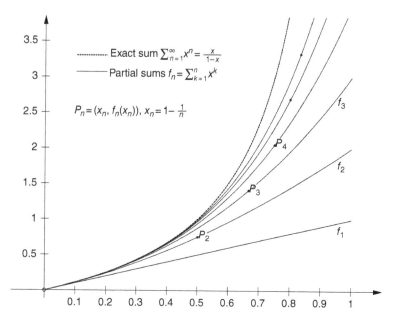

Figure 1.18 Examples 26, 29, and 33, series $\sum_{n=1}^{\infty} u_n(x) = \sum_{n=1}^{\infty} x^n$.

Note that the third condition of Dirichlet's theorem (the uniform convergence of $v_n(x)$) is weakened in the above statement, and the chosen sequence $v_n(x) = x^n$ converges nonuniformly to 0 on $X = (0, 1)$, since for $x_n = 1 - \frac{1}{n} \in (0, 1)$, $\forall n \in \mathbb{N}$ it follows $|v_n(x_n)| = \left(1 - \frac{1}{n}\right)^n \underset{n \to +\infty}{\to} e^{-1} \neq 0$.

Example 29. The partial sums of $\sum u_n(x)$ are not uniformly bounded on X, and the sequence $v_n(x)$ is not monotone in n and does not converge uniformly on X to 0, but still the series $\sum u_n(x)v_n(x)$ converges uniformly on X.

Solution
Consider $u_n(x) = x^n$ and $v_n(x) = \frac{(-1)^n}{xn^2}$ on $X = (0, 1)$. Let us check the conditions of the statement. First, the partial sums of $\sum_{n=1}^{\infty} u_n(x) = \sum_{n=1}^{\infty} x^n$ are not uniformly bounded on X (see for details Example 26). Second, the sequence $v_n(x)$ is not monotone in n (it is alternating for each fixed $x \in (0, 1)$). Finally, $v_n(x) = \frac{(-1)^n}{xn^2}$ converges to 0 for each fixed $x \in (0, 1)$, but this convergence is not uniform, because choosing $x_n = \frac{1}{n^2}$, $\forall n \in \mathbb{N}$ one obtains $\left|\frac{(-1)^n}{x_n n^2}\right| = 1 \not\to 0$. In this way, all the conditions in Dirichlet's theorem are violated. Nevertheless, the series $\sum_{n=1}^{\infty} u_n(x)v_n(x) = \sum_{n=1}^{\infty} (-1)^n \frac{x^{n-1}}{n^2}$ converges uniformly on $(0, 1)$

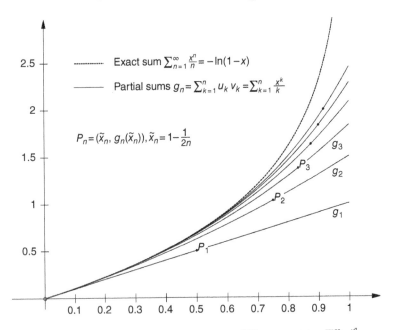

Figure 1.19 Examples 26, 27, 30, 31, and 32, series $\sum_{n=1}^{\infty} u_n(x)v_n(x) = \sum_{n=1}^{\infty} \frac{x^n}{n}$.

according to the Weierstrass test: $\left|(-1)^n \frac{x^{n-1}}{n^2}\right| \le \frac{1}{n^2}$, for $\forall n \in \mathbb{N}$ and $\forall x \in (0,1)$, and the majorant series $\sum \frac{1}{n^2}$ converges.

Remark. The functions $u_n(x) = \frac{x}{n}$ and $v_n(x) = (-1)^n x$ considered on $X = (0,10]$ exhibit even "wilder" behavior. In fact, the partial sums of the series $\sum_{n=1}^{\infty} \frac{x}{n}$ are not bounded at any point $x \in (0,10]$ since this series is positive and divergent at each $x \in (0,10]$. The sequence $(-1)^n x$ is not monotone in n and diverges at each $x \in (0,10]$. Nevertheless, the series $\sum_{n=1}^{\infty} u_n(x)v_n(x) = \sum_{n=1}^{\infty} (-1)^n \frac{x^2}{n}$ converges uniformly on $(0,10]$ since the following evaluation of the residual (resulting from Leibniz's test for alternating series)

$$|r_n(x)| = \left| \sum_{k=n+1}^{\infty} (-1)^k \frac{x^2}{k} \right| \le \left| (-1)^{n+1} \frac{x^2}{n+1} \right| \le \frac{100}{n+1} \xrightarrow[n\to+\infty]{} 0$$

is true for all $x \in (0,10]$ simultaneously.

Remark to Examples 30–33. In the next four examples, we analyze the sufficient conditions of Abel's theorem for the uniform convergence of the series $\sum u_n(x)v_n(x)$. The situation here is quite similar to that for Dirichlet's

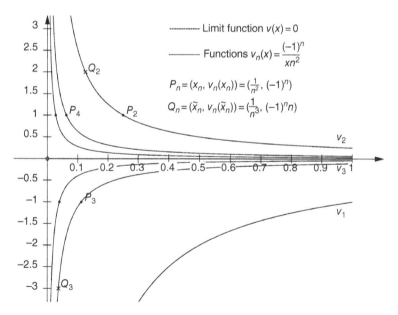

Figure 1.20 Examples 29 and 33, sequence $v_n(x) = \frac{(-1)^n}{xn^2}$.

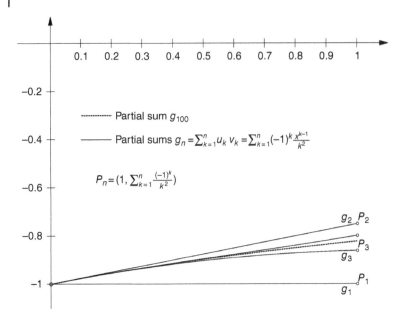

Figure 1.21 Examples 29 and 33, series $\sum_{n=1}^{\infty} u_n(x) v_n(x) = \sum_{n=1}^{\infty} (-1)^n \frac{x^{n-1}}{n^2}$.

theorem: none of the three conditions can be dropped, but, at the same time, all of them can be violated for an uniformly convergent series.

Example 30. A series $\sum u_n(x)$ converges on X, and a sequence $v_n(x)$ is monotone in n for each fixed $x \in X$ and uniformly bounded on X, but the series $\sum u_n(x) v_n(x)$ does not converge uniformly on X.

Solution
For $u_n(x) = x^{n-1}$ and $v_n(x) = \frac{x}{n}$ on $X = (0, 1)$, the series $\sum_{n=1}^{\infty} u_n(x) = \sum_{n=1}^{\infty} x^{n-1}$ is convergent at each point $x \in (0, 1)$, since this is a geometric series with the ratio in $(0, 1)$, and the sequence $v_n(x) = \frac{x}{n}$ is monotone in n and uniformly bounded on X: $\left|\frac{x}{n}\right| \leq \frac{1}{n} \leq 1$, $\forall n \in \mathbb{N}$, $\forall x \in X$. Thus, all the statement conditions are satisfied. However, the series $\sum_{n=1}^{\infty} u_n(x) v_n(x) = \sum_{n=1}^{\infty} \frac{x^n}{n}$ converges nonuniformly on $(0, 1)$ (see Example 26 for details). Note, that in the statement conditions, the first condition of Abel's theorem (the uniform convergence of $\sum u_n(x)$) is weakened, and the chosen series $\sum_{n=1}^{\infty} u_n(x) = \sum_{n=1}^{\infty} x^{n-1}$ converges nonuniformly on $X = (0, 1)$, which can be seen from the evaluation of the series residual for $x_n = 1 - \frac{1}{n} \in X$, $\forall n > 1$:

$$\left|\sum_{k=n+1}^{\infty} x_n^{k-1}\right| = \frac{x_n^n}{1 - x_n} = n \cdot \left(1 - \frac{1}{n}\right)^n \xrightarrow{n \to +\infty} \infty.$$

Example 31. A series $\sum u_n(x)$ converges uniformly on X, and a sequence $v_n(x)$ is uniformly bounded on X, but the series $\sum u_n(x)v_n(x)$ does not converge uniformly on X.

Solution

Consider $u_n(x) = \frac{(-1)^n}{n}$ and $v_n(x) = (-1)^n x^n$ on $X = (0,1)$. The series $\sum_{n=1}^{\infty} u_n(x) = \sum_{n=1}^{\infty} \frac{(-1)^n}{n}$ converges by Leibniz's test, and this convergence is uniform since $u_n(x)$ does not depend on x. The uniform boundedness of $v_n(x) = (-1)^n x^n$ is also easily verified: $|(-1)^n x^n| \leq 1$, for $\forall n \in \mathbb{N}$, $\forall x \in (0,1)$. Thus, all the statement conditions hold. However, as was shown in Example 26, the series $\sum_{n=1}^{\infty} u_n(x)v_n(x) = \sum_{n=1}^{\infty} \frac{x^n}{n}$ converges nonuniformly on $(0,1)$. Note that the condition of monotonicity of $v_n(x)$ in Abel's theorem is omitted in this example and, consequently, the choice of the nonmonotone sequence $v_n(x) = (-1)^n x^n$ resulted in nonuniform convergence of the series of the products.

Example 32. A series $\sum u_n(x)$ converges uniformly on X, and a sequence $v_n(x)$ is monotone in n for each fixed $x \in X$, but the series $\sum u_n(x)v_n(x)$ does not converge uniformly on X.

Solution

For $u_n(x) = \frac{x^{n-1}}{n^2}$ and $v_n(x) = nx$ on $X = (0,1)$, all the statement conditions are satisfied. In fact, the sequence $v_n(x) = nx$ is monotone in n for each fixed $x \in (0,1)$, and the series $\sum_{n=1}^{\infty} u_n(x) = \sum_{n=1}^{\infty} \frac{x^{n-1}}{n^2}$ converges uniformly on $(0,1)$ according to the Weierstrass test: $\left| \frac{x^{n-1}}{n^2} \right| \leq \frac{1}{n^2}$, $\forall x \in (0,1)$ and the series $\sum \frac{1}{n^2}$ is a convergent p-series. However, the series $\sum_{n=1}^{\infty} u_n(x)v_n(x) = \sum_{n=1}^{\infty} \frac{x^n}{n}$ converges nonuniformly on $(0,1)$ (see details in Example 26). Note that the condition of uniform boundedness of $v_n(x)$ in Abel's theorem is omitted in the statement. The chosen sequence $v_n(x) = nx$ is not bounded for any $x \in (0,1)$, and this led to nonuniform convergence of the series $\sum u_n(x)v_n(x)$.

Remark. The following strengthened version of this example can also be constructed: a series $\sum u_n(x)$ converges uniformly on X, and a sequence $v_n(x)$ is monotone and bounded in n for each fixed $x \in X$, but the series $\sum u_n(x)v_n(x)$ does not converge uniformly on X. The counterexample can be provided by $u_n(x) = \frac{x^2}{(1+x)^n}$ and $v_n(x) = \frac{n^2}{(3n^2+2)x}$ on $X = (0,+\infty)$. The series $\sum_{n=0}^{\infty} u_n(x) = \sum_{n=0}^{\infty} \frac{x^2}{(1+x)^n}$ converges on X as a geometric series with the ratio $\frac{1}{1+x} \in (0,1)$, $\forall x \in X$. To show the uniformity of this convergence on X, let us consider the residual

$$r_n(x) = \sum_{k=n+1}^{\infty} \frac{x^2}{(1+x)^k} = \frac{\frac{x^2}{(1+x)^{n+1}}}{1 - \frac{1}{1+x}} = \frac{x}{(1+x)^n}$$

and solve the critical point equation for each $n > 1$ fixed:

$$r_n'(x) = \frac{1 - (n-1)x}{(1+x)^{n+1}} = 0$$

that gives $x_n = \frac{1}{n-1}$. Since the derivative $r_n'(x)$ is positive at the left of x_n and negative at the right, the critical point x_n is the maximum on X and, consequently, for each fixed n one gets the following evaluation of the residual:

$$r_n(x) \leq \max_{(0,+\infty)} |r_n(x)| = r_n(x_n) = \frac{1}{n-1}\left(1 + \frac{1}{n-1}\right)^{-n} \underset{n\to\infty}{\longrightarrow} 0 \cdot e^{-1} = 0,$$

that is, the series $\sum_{n=0}^{\infty} u_n(x)$ converges uniformly on X.

As for the sequence $v_n(x) = \frac{n^2}{(3n^2+2)x}$, for each fixed $x \in X$, its terms are monotonic ($v_{n+1}(x) > v_n(x)$) and bounded ($|v_n(x)| = \frac{n^2}{(3n^2+2)x} < \frac{1}{3x}$). Thus, all the conditions of the statement are satisfied.

Nevertheless, the series $\sum_{n=0}^{\infty} u_n(x)v_n(x) = \sum_{n=0}^{\infty} \frac{n^2}{3n^2+2}\frac{x}{(1+x)^n}$ converges nonuniformly on X. Indeed, the convergence on X follows from the inequality $0 < \frac{n^2}{3n^2+2}\frac{x}{(1+x)^n} < \frac{1}{3}\frac{x}{(1+x)^n}$ and the convergence of the geometric series $\sum \frac{x}{(1+x)^n}$ for each fixed $x \in X$. Applying now the Cauchy criterion with $p_n = n$ and $x_n = \frac{1}{n}$, one obtains

$$\left|\sum_{k=n+1}^{n+p_n} u_k(x_n)v_k(x_n)\right| = \sum_{k=n+1}^{2n} \frac{k^2}{3k^2+2}\frac{x_n}{(1+x_n)^k} > \frac{n}{4}\frac{x_n}{(1+x_n)^{2n}}$$

$$= \frac{n}{4}\frac{1}{n}\left(1 + \frac{1}{n}\right)^{-2n} \underset{n\to\infty}{\longrightarrow} \frac{1}{4}e^{-2} \neq 0,$$

which means that the convergence is nonuniform on $X = (0,+\infty)$. Note that although $v_n(x)$ is bounded for each fixed $x \in X$, it is not uniformly bounded on X, since for $x_n = \frac{1}{3n^2+2} \in X$ one gets $v_n(x_n) = n^2 \underset{n\to\infty}{\longrightarrow} +\infty$.

Example 33. A series $\sum u_n(x)$ does not converge uniformly on X, and a sequence $v_n(x)$ is not monotone in n and is not uniformly bounded on X, but still the series $\sum u_n(x)v_n(x)$ converges uniformly on X.

Solution

Consider $u_n(x) = x^n$ and $v_n(x) = \frac{(-1)^n}{xn^2}$ on $X = (0,1)$. Let us check the conditions of the statement. First, using the same reasoning as in Example 30, one can prove that the series $\sum_{n=1}^{\infty} x^n$ converges nonuniformly on $(0,1)$. Then, the sequence $v_n(x)$ is not monotone in n (it is alternating for each fixed $x \in (0,1)$). Finally, $v_n(x) = \frac{(-1)^n}{xn^2}$ converges to 0 for each fixed $x \in (0,1)$, but this sequence does not bounded uniformly on $(0,1)$, since for $x_n = \frac{1}{n^3} \in (0,1)$ one has $|v_n(x_n)| = n \underset{n\to+\infty}{\longrightarrow} \infty$. Thus, all the conditions of Abel's theorem are violated. Nevertheless, the series $\sum_{n=1}^{\infty} u_n(x)v_n(x) = \sum_{n=1}^{\infty}(-1)^n \frac{x^{n-1}}{n^2}$ converges uniformly on $(0,1)$ according to the Weierstrass test as was shown in Example 29.

Remark. The conditions of Abel's theorem are violated even stronger for the functions $u_n(x) = \frac{x}{n}$ and $v_n(x) = (-1)^n \sqrt{nx}$ considered on $X = (0,2]$. In fact, the series $\sum_{n=1}^{\infty} \frac{x}{n}$ diverges at each $x \in (0,2]$. The sequence $(-1)^n \sqrt{nx}$ is not monotone in n and is unbounded at each $x \in (0,2]$ because $|v_n(x)| = |(-1)^n \sqrt{nx}| = \sqrt{nx} \underset{n \to +\infty}{\to} +\infty, \ \forall x \in (0,2]$. Nevertheless, the series $\sum_{n=1}^{\infty} u_n(x)v_n(x) = \sum_{n=1}^{\infty} (-1)^n \frac{x^2}{\sqrt{n}}$ converges uniformly on $(0,2]$ as is seen from the evaluation of the residual (following from Leibniz's test for alternating series)

$$|r_n(x)| = \left| \sum_{k=n+1}^{\infty} (-1)^k \frac{x^2}{\sqrt{k}} \right| \leq \left| (-1)^{n+1} \frac{x^2}{\sqrt{n+1}} \right| \leq \frac{4}{\sqrt{n+1}} \underset{n \to +\infty}{\to} 0,$$

which is satisfied for all $x \in (0,2]$ simultaneously.

Exercises

1   Show that Example 1 can be illustrated by the sequence $f_n(x) = \frac{nx}{n^2 x^2 + 1}$ on $X = [0,1]$.

2   Use the points $x_n = \frac{1}{\sqrt[n]{2}}, \ \forall n \in \mathbb{N}$ to prove that the sequence $f_n(x) = x^n$ of Example 1 converges nonuniformly on $X = (-1,1)$.

3   Show that the series $\sum_{n=1}^{\infty} 2^n \sin \frac{1}{5^n x}$ converges on $X = (0,\infty)$, but the convergence is not uniform.

4   Check if the series $\sum_{n=1}^{\infty} \frac{x^n}{\sqrt[n]{n}}$ on $X = (-1,1)$ can be used for Example 2.

5   Use the sequence $f_n(x) = \frac{x+n+(-1)^n x}{n^2}$ on $X = \mathbb{R}$ to illustrate the statement in Example 3.

6   Construct a counterexample to the following false statement: "if $f(x,y)$ defined on $[a,b] \times Y$ converges to a limit function $\varphi(x)$ as y approaches y_0, and this convergence is uniform on any interval $[c,b], \ \forall c \in (a,b)$, then the convergence is also uniform on $[a,b]$." Compare with the statement in Example 4. Formulate similar false statements for sequences and series and disprove them by counterexamples. (Hint: for the functions depending on a parameter, try $f(x,y) = \frac{2xy^2}{x^2+y^4}$ on $[0,1] \times (0,1]$ with the limit point $y_0 = 0$; for the sequences—$f_n(x) = \frac{nx}{n^2 x^2 + 1}$ on $[0,1]$; and for the series—$\sum u_n(x) = \sum \frac{(1-x)^n}{n}$ on $(0,1]$.)

7 Verify that

a) the function $f(x, y) = \frac{x^2}{y^2} e^{-x/y}$ on $X \times Y = [0, +\infty) \times (0, 1]$ with the limit point $y_0 = 0$

b) the function $f(x, y) = \begin{cases} \frac{x}{y} \sin \frac{y}{x}, x \neq 0 \\ 1, x = 0 \end{cases}$ on $X \times Y = \mathbb{R} \times (0, 1]$ with the limit point $y_0 = 0$

c) the sequence $f_n(x) = \frac{2n^2 x}{1+n^4 x^2}$ on $X = \mathbb{R}$

d) the series $\sum_{n=1}^{\infty} u_n(x) = \sum_{n=1}^{\infty} \frac{\sin nx}{\sqrt{n}}$ on $X = \mathbb{R}$

provide counterexamples for Example 5.

8 Verify that the series $\sum_{n=1}^{\infty} (-1)^n \frac{x^2+n^2}{n^3}$ on $X = [-1, 1]$ is one more counterexample to the statement of Example 7.

9 Use the series $\sum_{n=1}^{\infty} (-1)^{n-1} \frac{x^n}{n}$, $X = (-1, 1)$ for Example 8.

10 Show the feasibility of Example 9 by using counterexamples with

a) the series $\sum_{n=1}^{\infty} (-1)^{n-1} \frac{x^2}{(1+x^2)^n}$, $X = \mathbb{R}$

b) the series $\sum_{n=0}^{\infty} (-1)^n x(1 - x)^n$, $X = [0, 1]$.

11 Use the series $\sum_{n=1}^{\infty} u_n(x)$ with

a) $u_n(x) = \begin{cases} 0, x \in [0, 3^{-n-1}] \cup [3^{-n}, 1] \\ \frac{1}{\sqrt{n}} \cos^2 \left(\frac{3^{n+1}}{2} \pi x \right), x \in (3^{-n-1}, 3^{-n}) \end{cases}$ on $X = [0, 1]$

b) $u_n(x) = (-1)^{n-1} \frac{x^2}{(1+x^2)^n}$ on $X = \mathbb{R}$

to show the feasibility of Example 10.

12 Verify that the sequence $f_n(x) = \frac{x^2 n^2}{2n^2+5}$ on $X = (0, 1]$ specifies Example 15.

13 Check the statement of Example 16 for the sequence $f_n(x) = \frac{2nx}{n^2+x^2}$ on $X = \mathbb{R}$.

14 Verify whether the sequences $f_n(x)$, $g_n(x)$ and $f_n(x) \cdot g_n(x)$ are convergent or divergent on X. In the case of the convergence, analyze its character:

a) $f_n(x) = \ln \frac{(n^2+1)x}{n^2}$, $g_n(x) = \ln \frac{3n+2}{n} x^2$ on $X = (0, +\infty)$

b) $f_n(x) = g_n(x) = \ln \frac{n^2 x^2}{n^2+1}$ on $X = (0, +\infty)$

c) $f_n(x) = \frac{x}{n}$, $g_n(x) = \frac{\sin nx}{nx}$ on $X = (0, +\infty)$

d) $f_n(x) = \frac{x}{n}$, $g_n(x) = \ln \frac{(n+1)x}{n}$ on $X = (0, +\infty)$

e) $f_n(x) = \frac{x}{n}$, $g_n(x) = \frac{\sin nx}{nx}$ on $X = (0, 1]$

f) $f_n(x) = g_n(x) = (-1)^n \frac{5n^2+2}{3n^2+1} x$ on $X = (0, 1]$.

Formulate false statements for which these sequences represent counterexamples.

15 Show that the sequences $f_n(x) = \begin{cases} \frac{n^2+1}{n^2}x^2, x \in \mathbb{Q} \\ \frac{x^2}{n^2}, x \in \mathbb{I} \end{cases}$ and $g_n(x) = \begin{cases} \frac{1}{n^2}x^2, x \in \mathbb{Q} \\ \frac{n^2+1}{n^2}x^2, x \in \mathbb{I} \end{cases}$ defined on $X = [0,1]$ provide a counterexample to Example 14.

16 Verify whether the series $\sum u_n(x)$, $\sum v_n(x)$ and $\sum u_n(x) \cdot v_n(x)$ are convergent or divergent on X. In the case of the convergence, analyze its character:

a) $u_n(x) = v_n(x) = \frac{\sin nx}{\sqrt{n}}$ on $X = \mathbb{R}$

b) $u_n(x) = v_n(x) = \frac{\sin nx}{n^{2/3}}$ on $X = \mathbb{R}$

c) $u_n(x) = v_n(x) = \frac{1}{x^{2/3}+n^{2/3}}$ on $X = [0, +\infty)$

d) $u_n(x) = \frac{1}{x+n}, v_n(x) = \frac{1}{x+\ln^2 n}$ on $X = (0, +\infty)$

e) $u_n(x) = v_n(x) = \frac{\cos nx}{\sqrt{n}}$ on $X = [a, \pi - a], \forall a \in \left(0, \frac{\pi}{2}\right)$

f) $u_n(x) = \frac{\sin nx}{\sqrt{n}}, v_n(x) = \frac{\sin nx}{\sqrt[4]{n}}$ on $X = [a, \pi - a], \forall a \in \left(0, \frac{\pi}{2}\right)$

g) $u_n(x) = v_n(x) = (-1)^n \frac{x^n}{\sqrt{n}}$ on $X = (0, 1)$

h) $u_n(x) = (-1)^n \frac{x^n}{\ln n}, v_n(x) = (-1)^n \frac{x^n}{\sqrt{n}}, n \geq 2$ on $X = (0, 1)$

i) $u_n(x) = \frac{\sin nx}{n^{2/3}}, v_n(x) = \frac{\sin nx}{n}$ on $X = \mathbb{R}$

j) $u_n(x) = v_n(x) = \frac{x^n}{n}$ on $X = [0, 1)$

k) $u_n(x) = \frac{x^n}{n}, v_n(x) = \frac{x^n}{n^{1/3}}$ on $X = [0, 1)$

l) $u_n(x) = v_n(x) = (-1)^n \frac{x^n}{\sqrt[4]{n}}$ on $X = [0, 1)$

m) $u_n(x) = v_n(x) = (-1)^n \frac{x}{\sqrt{n}}$ on $X = (0, 1)$.

Formulate false statements for which these series represent counterexamples.

17 Show that the series $\sum_{n=1}^{\infty} \frac{2x^3}{n^6+x^6}$ and $\sum_{n=1}^{\infty} \frac{2x^3}{n^6+x^3}$ on $X = (0, +\infty)$ exemplify the statement in Example 25.

18 For given $u_n(x)$ and $v_n(x)$ on the specified set X, verify the conditions of Dirichlet's theorem and investigate the character of the convergence of the series $\sum u_n(x) \cdot v_n(x)$:

a) $u_n(x) = x^{n-1}, v_n(x) = \frac{x}{n}$ on $X = (-1, 1)$

b) $u_n(x) = x^{n-1}, v_n(x) = (-1)^n \frac{x}{n}$ on $X = [0, 1)$

c) $u_n(x) = x^{n+1}, v_n(x) = (-1)^n \frac{1}{n^{3/2}x}$ on $X = (0, 1)$

d) $u_n(x) = (-1)^n x^{n-1}$, $v_n(x) = (-1)^n \frac{x}{n}$ on $X = [0, 1)$

e) $u_n(x) = \frac{1}{\sqrt{n}}$, $v_n(x) = \sin nx$ on $X = \left[\frac{\pi}{10}, \frac{19\pi}{10} \right]$

f) $u_n(x) = (-1)^n$, $v_n(x) = \left(\frac{x^2}{1+x^2} \right)^n$ on $X = \mathbb{R}$

g) $u_n(x) = x^2$, $v_n(x) = \frac{(-1)^n}{(1+x^2)^n}$ on $X = (0, +\infty)$

h) $u_n(x) = \frac{x}{n}$, $v_n(x) = \frac{\sin nx}{x}$ on $X = \left[\frac{\pi}{6}, \frac{11\pi}{6} \right]$

i) $u_n(x) = x^{2n}$, $v_n(x) = \frac{(-1)^n}{(1+x^2)^n}$ on $X = (-1, 1)$.

Formulate false statements for which these functions and series represent counterexamples.

19 For given $u_n(x)$ and $v_n(x)$ on the specified set X, verify the conditions of Abel's theorem and investigate the character of the convergence of the series $\sum u_n(x) \cdot v_n(x)$:

a) $u_n(x) = (-1)^n$, $v_n(x) = \left(\frac{x^2}{1+x^2} \right)^n$ on $X = \mathbb{R}$

b) $u_n(x) = x^{n-1}$, $v_n(x) = \frac{x}{n}$ on $X = (-1, 1)$

c) $u_n(x) = \frac{x^2}{n}$, $v_n(x) = \frac{\sin nx}{nx^2}$ on $X = (0, +\infty)$

d) $u_n(x) = \frac{1}{nx}$, $v_n(x) = \sqrt{nx} \sin nx$ on $X = \left[\frac{\pi}{10}, \frac{19\pi}{10} \right]$

e) $u_n(x) = \frac{(-1)^n}{n}$, $v_n(x) = (-1)^n \sin nx$ on $X = \mathbb{R}$

f) $u_n(x) = (-1)^n \frac{x^2}{(1+x^2)^n}$, $v_n(x) = \frac{2n+1}{(n+1)x^2}$ on $X = (0, +\infty)$

g) $u_n(x) = \frac{\sin nx}{\sqrt{n}}$, $v_n(x) = \frac{1}{\sqrt{n}}$ on $X = \mathbb{R}$

h) $u_n(x) = \frac{\sin nx}{n^2}$, $v_n(x) = n^{3/2}$ on $X = \mathbb{R}$

i) $u_n(x) = \frac{x^n}{n\sqrt{n}}$, $v_n(x) = x\sqrt{n}$ on $X = (0, 1)$

j) $u_n(x) = \frac{\sin nx}{n}$, $v_n(x) = (-1)^n \sqrt{n}$ on $X = \left(0, \frac{\pi}{2} \right)$.

Formulate false statements for which these functions and series represent counterexamples.

Further Reading

S. Abbott. *Understanding Analysis*, Springer, New York, 2002.

D. Bressoud, *A Radical Approach to Real Analysis*, MAA, Washington, DC, 2007.

T.J.I. Bromwich, *An Introduction to the Theory of Infinite Series*, AMS, Providence, RI, 2005.

B.M. Budak and S.V. Fomin, *Multiple Integrals, Field Theory and Series*, Mir Publisher, Moscow, 1978.

G.M. Fichtengolz, *Differential- und Integralrechnung, Vol.1–3*, V.E.B. Deutscher Verlag Wiss., Berlin, 1968.

V.A. Ilyin and E.G. Poznyak, *Fundamentals of Mathematical Analysis, Vol.1,2*, Mir Publisher, Moscow, 1982.

K. Knopp, *Theory and Applications of Infinite Series*, Dover Publication, Mineola, NY, 1990.

C.H.C. Little, K.L. Teo and B. Brunt, *Real Analysis via Sequences and Series*, Springer, New York, 2015.

W. Rudin, *Principles of Mathematical Analysis*, McGraw-Hill, New York, 1976.

V.A. Zorich, *Mathematical Analysis I, II*, Springer, Berlin, 2004.

CHAPTER 2

Properties of the Limit Function: Boundedness, Limits, Continuity

2.1 Convergence and Boundedness

Example 1. A sequence of bounded on X functions converges on X to a function, which is unbounded on X.

Solution

Consider the sequence of functions $f_n(x) = \begin{cases} \min\left(n, \frac{1}{x}\right), 0 < x \leq 1 \\ 0, x = 0 \end{cases}$, each

of which is bounded on $[0,1]$, because $0 \leq f_n(x) \leq n$, $\forall x \in [0,1]$. Since for any fixed $x \in (0,1]$ there exists $n_0 \in \mathbb{N}$ such that $n_0 > \frac{1}{x}$, for all $n > n_0$ we have $f_n(x) = \frac{1}{x}$ and, consequently, $\lim_{n \to \infty} f_n(x) = \frac{1}{x}$, $\forall x \in (0,1]$. For $x = 0$, we get $\lim_{n \to \infty} f_n(0) = \lim_{n \to \infty} 0 = 0$. Hence, the limit function is $f(x) = \begin{cases} \frac{1}{x}, 0 < x \leq 1 \\ 0, x = 0 \end{cases}$, which is unbounded on $[0,1]$. Note, that this convergence is not uniform, since for any $n \in \mathbb{N}$ one can choose the point $x_n = \frac{1}{2n} \in [0,1]$, which gives $|f_n(x_n) - f(x_n)| = |n - 2n| = n \underset{n \to \infty}{\to} \infty$.

Remark 1. Note that the functions in the counterexample are not uniformly bounded on $[0,1]$. In fact, for any $M > 0$ there exist $n > M$ and $x_n = \frac{1}{n} \in (0,1]$ such that $f_n(x_n) = n > M$. This is the cause of the unboundedness of the limit function.

Remark 2. The corresponding general statement will be true if the condition of the uniform boundedness of $f_n(x)$ on X would be added: a sequence of uniformly bounded on X functions converges on X to a bounded function $f(x)$.

Remark 3. For bounded functions, the uniform convergence on X implies the uniform boundedness. Therefore, the additional condition of the uniform convergence of $f_n(x)$ on X is sufficient for the general statement becomes true: if a

Counterexamples on Uniform Convergence: Sequences, Series, Functions, and Integrals, First Edition.
Andrei Bourchtein and Ludmila Bourchtein.
© 2017 John Wiley & Sons, Inc. Published 2017 by John Wiley & Sons, Inc.
Companion website: www.wiley.com/go/bourchtein/counterexamples_on_uniform_convergence

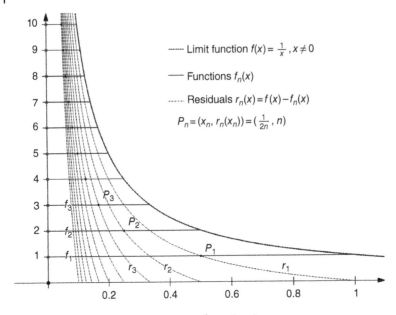

Figure 2.1 Example 1, sequence $f_n(x) = \begin{cases} \min\left(n, \frac{1}{x}\right), 0 < x \le 1 \\ 0, x = 0 \end{cases}$.

sequence of bounded on X functions converges uniformly on X, then the limit function $f(x)$ is bounded on X.

Remark 4. The general statement with the opposite conclusion—"if a sequence of nonuniformly bounded functions converges on X, then the limit function is not bounded on X"—is also false as shown in Example 2.

Example 2. A sequence of functions is not uniformly bounded on X, but it converges on X to a function, which is bounded on X.

Solution

Consider the sequence of functions $f_n(x) = \begin{cases} n^2x + n - n^3, x \in \left[n - \frac{1}{n}, n\right] \\ -n^2x + n + n^3, x \in \left[n, n + \frac{1}{n}\right] \\ 0, \text{otherwise} \end{cases}$

defined on $X = [0, +\infty)$. Like in Example 1, each of the functions is bounded, because $0 \le f_n(x) \le n$, $\forall x \in [0, +\infty)$ for any fixed n, but the sequence is not uniformly bounded on $[0, +\infty)$, since for any $M > 0$ there exist $n > M$ and $x_n = n \in [0, +\infty)$ such that $f_n(x_n) = n > M$. Let us find the limit function. For any fixed $x \in [0, +\infty)$ there exists $n_0 \in \mathbb{N}$ such that $n_0 > x + 1$, so for all $n > n_0$

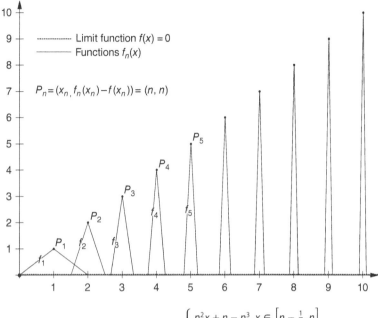

Figure 2.2 Example 2, sequence $f_n(x) = \begin{cases} n^2x + n - n^3, x \in \left[n - \frac{1}{n}, n\right] \\ -n^2x + n + n^3, x \in \left[n, n + \frac{1}{n}\right] \\ 0, \text{otherwise} \end{cases}$.

we have $n - \frac{1}{n} > n - 1 > x$, which implies that $\lim\limits_{n \to \infty} f_n(x) = 0, \ \forall x \in [0, +\infty)$. Therefore, the limit function is bounded on $[0, +\infty)$.

Note, that the convergence of this sequence is not uniform (just like in Example 1). Indeed, choosing for any $n \in \mathbb{N}$ the point $x_n = n \in [0, +\infty)$, we have $|f_n(x_n) - f(x_n)| = n \underset{n \to \infty}{\to} \infty$.

Remark to Examples 1 and 2. In these two examples, although the sequences of the functions are not bounded uniformly, each of the functions is bounded on X. Analogous examples can be constructed for sequences of unbounded on X functions.

Example 3. A sequence of unbounded and discontinuous on X functions converges on X to a function, which is bounded and continuous on X.

Solution

The sequence of functions $f_n(x) = \begin{cases} \frac{1}{nx}, 0 < x \leq 1 \\ 0, x = 0 \end{cases}$ considered on $X = [0, 1]$ gives a counterexample. Each of these functions is unbounded on $[0, 1]$, since for any $M > 0$ there exists $x_n = \frac{1}{n(M+1)} \in [0, 1]$ such that $f_n(x_n) = M + 1 > M$. Also, each of the functions has an essential discontinuity at $x = 0$, since $\lim_{x \to 0_+} f_n(x) = \lim_{x \to 0_+} \frac{1}{nx} = +\infty$. However, this sequence converges to the bounded and continuous function $f(x) \equiv 0$. Indeed, for $x = 0$, we get $\lim_{n \to \infty} f_n(0) = \lim_{n \to \infty} 0 = 0$, and for any fixed $x \in (0, 1]$, we obtain $\lim_{n \to \infty} f_n(x) = \lim_{n \to \infty} \frac{1}{nx} = 0$. Note, that the convergence is not uniform, since for any $n \in \mathbb{N}$ there is the point $x_n = \frac{1}{n} \in [0, 1]$ such that $|f_n(x_n) - f(x_n)| = 1 \nrightarrow_{n \to \infty} 0$.

Remark 1. If the condition of discontinuity is dropped in the statement, then the same sequence considered on $X = (0, 1]$ provides a counterexample.

Remark 2. The strengthened version of this example with the added condition of the uniform convergence also takes place. The corresponding counterexample, albeit for the case of bounded functions, is given in Example 11.

Remark 3. The general statement—"if a sequence of unbounded and discontinuous functions converges to an unbounded and discontinuous function, then the convergence is uniform"—is also false as shown in Example 4.

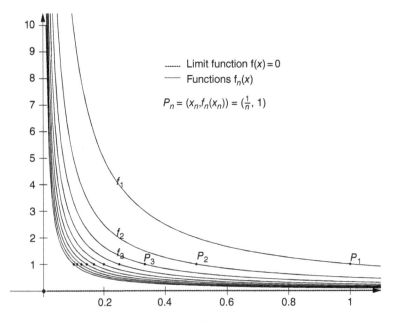

Figure 2.3 Example 3, sequence $f_n(x) = \begin{cases} \frac{1}{nx}, 0 < x \le 1 \\ 0, x = 0 \end{cases}$.

Example 4. A sequence of unbounded and discontinuous on X functions converges on X to an unbounded and discontinuous function, but the convergence is nonuniform on X.

Solution

The sequence of functions $f_n(x) = \begin{cases} \frac{n+1}{nx}, 0 < x \le 1 \\ 0, x = 0 \end{cases}$ converges on $X = [0,1]$

to the function $f(x) = \begin{cases} \frac{1}{x}, 0 < x \le 1 \\ 0, x = 0 \end{cases}$, since for $x = 0$, we have $\lim\limits_{n\to\infty} f_n(0) =$

$\lim\limits_{n\to\infty} 0 = 0$, and for any fixed $x \in (0,1]$, we obtain $\lim\limits_{n\to\infty} f_n(x) = \lim\limits_{n\to\infty} \frac{n+1}{nx} = \frac{1}{x}$.
Each $f_n(x)$, as well as $f(x)$, has essential discontinuity at the origin:
$\lim\limits_{x\to0_+} f_n(x) = \lim\limits_{x\to0_+} \frac{n+1}{nx} = +\infty$, $\lim\limits_{x\to0_+} f(x) = \lim\limits_{x\to0_+} \frac{1}{x} = +\infty$. It also shows that
$f_n(x)$ and $f(x)$ are unbounded on $[0,1]$. However, the convergence to $f(x)$ is
nonuniform, since for any $n \in \mathbb{N}$ there is the point $x_n = \frac{1}{n} \in [0,1]$ such that
$|f_n(x_n) - f(x_n)| = \left|\frac{n+1}{n}n - n\right| = 1 \underset{n\to\infty}{\nrightarrow} 0$.

Remark. A similar example can be considered for the continuous functions.
For instance, each of the functions $f_n(x) = \frac{n+1}{n}x$ considered on $X = \mathbb{R}$ is continuous but still unbounded (albeit bounded on any finite interval), and the
sequence converges on $\mathbb{R}$ to $f(x) = \lim\limits_{n\to\infty} \frac{n+1}{n}x = x$, which is also continuous and
unbounded function on $\mathbb{R}$. Choosing for each $n \in \mathbb{N}$ the points $x_n = n$, one
can prove that the convergence is not uniform: $|f_n(x_n) - f(x_n)| = \left|\frac{n+1}{n}n - n\right| = 1 \underset{n\to\infty}{\nrightarrow} 0$.

Remark to Examples 3 and 4. Based on these examples, it might be suggested
that a sequence of unbounded functions can converge only nonuniformly. The
next example shows that this is not the case.

Example 5. A sequence of unbounded and discontinuous on X functions converges on X to an unbounded and discontinuous function, but this convergence
is uniform on X.

Solution

The sequence of functions $f_n(x) = \begin{cases} \ln \frac{n+1}{n}x, x \in (0, +\infty) \\ 0, x = 0 \end{cases}$ converges on

$X = [0, +\infty)$ to the function $f(x) = \begin{cases} \ln x, x \in (0, +\infty) \\ 0, x = 0 \end{cases}$, since for $x = 0$,

we have $\lim\limits_{n\to\infty} f_n(0) = \lim\limits_{n\to\infty} 0 = 0$, and for any fixed $x \in (0, +\infty)$, we obtain
$\lim\limits_{n\to\infty} f_n(x) = \lim\limits_{n\to\infty} \ln \frac{n+1}{n}x = \ln x$. Each $f_n(x)$, as well as $f(x)$, is unbounded and has
essential discontinuity at 0, because $\lim\limits_{x\to0_+} f_n(x) = \lim\limits_{x\to0_+} \ln \frac{n+1}{n}x = -\infty$, $\forall n \in \mathbb{N}$,

and $\lim\limits_{x\to 0_+} f(x) = \lim\limits_{x\to 0_+} \ln x = -\infty$. At the same time, the convergence to $f(x)$ is uniform. Indeed, from the evaluation

$$|f_n(x) - f(x)| = |\ln\frac{n+1}{n}x - \ln x| = \ln\frac{n+1}{n} = \ln\left(1 + \frac{1}{n}\right) < \frac{1}{n}, \forall x > 0,$$

it follows that $|f_n(x) - f(x)| < \frac{1}{n}$ for $\forall x \in [0, +\infty)$, and since $\frac{1}{n} \xrightarrow[n\to\infty]{} 0$, this means that $f_n(x)$ converges uniformly to $f(x)$ on $[0, +\infty)$.

Remark. The condition of discontinuity in the statement can be dropped. The respective counterexample can be given with the same sequence considered on $X = (0, +\infty)$ or $X = (0, 1]$.

Example 6. A sequence of uniformly bounded on X functions converges on X, but the convergence is nonuniform on X.

Solution

The sequence $f_n(x) = \begin{cases} nx\sin\frac{1}{nx}, x \neq 0 \\ 1, x = 0 \end{cases}$ is uniformly bounded on $X = [0, 1]$

since $|f_n(x)| = \left|\frac{\sin\frac{1}{nx}}{\frac{1}{nx}}\right| \leq 1$, $\forall n \in \mathbb{N}$ and $\forall x \in [0, 1]$. Further, for $x = 0$ one has

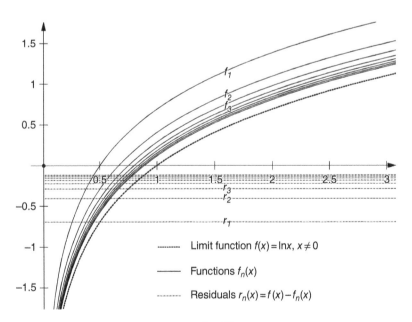

Figure 2.4 Example 5, sequence $f_n(x) = \begin{cases} \ln\frac{n+1}{n}x, x \in (0, +\infty) \\ 0, x = 0 \end{cases}$.

Within figure legend:
Limit function $f(x) = \ln x$, $x \neq 0$
Functions $f_n(x)$
Residuals $r_n(x) = f(x) - f_n(x)$

$\lim\limits_{n\to\infty} f_n(0) = \lim\limits_{n\to\infty} 1 = 1$ and for $x \neq 0$, using the first remarkable limit, one obtains

$$\lim_{n\to\infty} f_n(x) = \lim_{n\to\infty} \frac{\sin \frac{1}{nx}}{\frac{1}{nx}} = \lim_{t\to 0_+} \frac{\sin t}{t} = 1.$$

Therefore, the limit of $f_n(x)$ exists and $f(x) = \lim\limits_{n\to\infty} f_n(x) = 1, \forall x \in [0,1]$. Thus, the statement conditions are satisfied, but the convergence on $[0,1]$ is nonuniform: choosing $x_n = \frac{1}{n} \in [0,1]$ for $\forall n \in \mathbb{N}$, one gets

$$|f_n(x_n) - f(x_n)| = \left| n \cdot \frac{1}{n} \sin 1 - 1 \right| = 1 - \sin 1 \not\to_{n\to\infty} 0.$$

2.2 Limits and Continuity of Limit Functions

Example 7. A series $\sum u_n(x)$ converges on X and $\lim\limits_{x\to x_0} u_n(x)$ exists for each n, but the corresponding limit of the series cannot be calculated term by term, that is, one cannot interchange the order of the limit and infinite sum: $\lim\limits_{x\to x_0} \sum u_n(x) \neq \sum \lim\limits_{x\to x_0} u_n(x)$.

Solution
Let us consider the series $\sum_{n=1}^{\infty}(x^n - x^{n+1})$ on $X = (-1,1)$ with the limit point $x_0 = 1$. The partial sums are easily found:

$$f_n(x) = \sum_{k=1}^{n} u_k(x) = (x - x^2) + (x^2 - x^3) + \cdots + (x^n - x^{n+1}) = x - x^{n+1}.$$

Since $|x| < 1$, it follows that $\lim\limits_{n\to\infty} x^n = 0$, and therefore, the series is convergent on $(-1,1)$, with the sum of the series being $f(x) = \lim\limits_{n\to\infty} f_n(x) = \lim\limits_{n\to\infty}(x - x^{n+1}) = x$. All the required limits exist: $\lim\limits_{x\to 1_-} u_n(x) = \lim\limits_{x\to 1_-}(x^n - x^{n+1}) = 0, \forall n \in \mathbb{N}$. However,

$$\sum_{n=1}^{\infty} \lim_{x\to 1_-} u_n(x) = \sum_{n=1}^{\infty} \lim_{x\to 1_-}(x^n - x^{n+1}) = \sum_{n=1}^{\infty} 0 = 0$$

$$\neq 1 = \lim_{x\to 1_-} x = \lim_{x\to 1_-} f(x) = \lim_{x\to 1_-} \sum_{n=1}^{\infty} u_n(x).$$

Let us verify the type of convergence of this series. Evaluating the remainder

$$|r_n(x)| = \left| \sum_{k=n+1}^{\infty} u_k(x) \right| = |f_n(x) - f(x)| = |x - x^{n+1} - x| = |x|^{n+1}$$

and choosing for $\forall n \in \mathbb{N}$ the corresponding $x_n = \frac{1}{\sqrt[n+1]{2}} \in (-1, 1)$, we obtain $|r_n(x_n)| = \frac{1}{2} \nrightarrow_{n \to \infty} 0$, which means that the convergence is nonuniform on $(-1, 1)$.

Remark 1. A similar example can be given for a sequence of functions: a sequence $f_n(x)$ converges on X and $\lim_{x \to x_0} f_n(x)$ exists for each n, but $\lim_{x \to x_0} \lim_{n \to \infty} f_n(x) \neq \lim_{n \to \infty} \lim_{x \to x_0} f_n(x)$. It can be shown by the counterexample with the sequence $f_n(x) = \frac{n^2 x^2 - 1}{n^2 x^2 + 1}$ defined on $X = (0, 1)$ with the limit point $x_0 = 0$. This sequence converges to the limit function

$$f(x) = \lim_{n \to \infty} f_n(x) = \lim_{n \to \infty} \frac{n^2 x^2 - 1}{n^2 x^2 + 1} = 1, \forall x \in (0, 1)$$

and $\lim_{x \to 0_+} f(x) = \lim_{x \to 0_+} 1 = 1$. Also, each function has the limit at 0: $\lim_{x \to 0_+} f_n(x) = \lim_{x \to 0_+} \frac{n^2 x^2 - 1}{n^2 x^2 + 1} = -1$, $\forall n \in \mathbb{N}$. Therefore, $\lim_{x \to 0_+} \lim_{n \to \infty} f_n(x) = 1 \neq -1 = \lim_{n \to \infty} \lim_{x \to 0_+} f_n(x)$. To see that the convergence of $f_n(x)$ to $f(x)$ is not uniform, we can choose $x_n = \frac{1}{n} \in (0, 1)$ for any $n \in \mathbb{N}$ in order to obtain

$$|f_n(x_n) - f(x_n)| = \left| \frac{n^2 \cdot \frac{1}{n^2} - 1}{n^2 \cdot \frac{1}{n^2} + 1} - 1 \right| = 1 \nrightarrow_{n \to \infty} 0.$$

Remark 2. A similar example for a function depending on a continuous parameter is formulated as follows: a function $f(x, y)$ is defined on $X \times Y$, converges on X to a limit function $\varphi(x)$, as $y \to y_0$, and $\lim_{x \to x_0} f(x, y)$ exists for each $y \in Y$, but $\lim_{x \to x_0} \lim_{y \to y_0} f(x, y) \neq \lim_{y \to y_0} \lim_{x \to x_0} f(x, y)$. For a counterexample, one can use the function $f(x, y) = \frac{y^2}{x^2 + y^2}$ defined on $X \times Y = (0, 1] \times (0, 1]$ with the limit points $x_0 = 0$ for X and $y_0 = 0$ for Y. The limit function is

$$\varphi(x) = \lim_{y \to 0_+} f(x, y) = \lim_{y \to 0_+} \frac{y^2}{x^2 + y^2} = 0, \forall x \in (0, 1],$$

and therefore, $\lim_{x \to 0_+} \varphi(x) = \lim_{x \to 0_+} 0 = 0$. However, another order of limits gives a different result:

$$\lim_{y \to 0_+} \lim_{x \to 0_+} f(x, y) = \lim_{y \to 0_+} \lim_{x \to 0_+} \frac{y^2}{x^2 + y^2} = \lim_{y \to 0_+} 1 = 1.$$

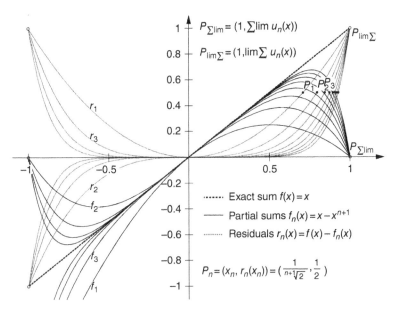

Figure 2.5 Example 7, series $\sum_{n=1}^{\infty}(x^n - x^{n+1})$.

Analyzing the convergence, we see that for $\forall y \in (0, 1]$ there exists $x_y = y \in (0, 1]$ that leads to $|f(x_y, y) - \varphi(x_y)| = \frac{y^2}{y^2+y^2} = \frac{1}{2} \nrightarrow 0$. This means that the convergence is nonuniform.

Another interesting counterexample of a similar type is $f(x, y) = \arctan \frac{x}{y}$ defined on $X \times Y = (0, +\infty) \times (0, 1)$ with the limit points $x_0 = 0$ for X and $y_0 = 0$ for Y.

Remark 3. In this example, both limits are finite, but assume different values. Example 8 shows that one of the limits can be infinite.

Example 8. A series $\sum u_n(x)$ converges on X and $\lim_{x \to x_0} u_n(x)$ exists for each n, but $\lim_{x \to x_0} \sum u_n(x) \neq \sum \lim_{x \to x_0} u_n(x)$ since the left-hand side limit is infinite.

Solution
For such a case, consider the series $\sum_{n=1}^{\infty} u_n(x)$ where $u_n(x) = \frac{(n-1)^2 x^2 + 2}{(n-1)^2 x^3 + 1} - \frac{n^2 x^2 + 2}{n^2 x^3 + 1}$ are defined on $X = (0, \infty)$. The partial sums of this telescoping series are found in the form $f_n(x) = \sum_{k=1}^{n} u_k(x) = 2 - \frac{n^2 x^2 + 2}{n^2 x^3 + 1}$, and its sum is $f(x) = \sum_{n=1}^{\infty} u_n(x) = $

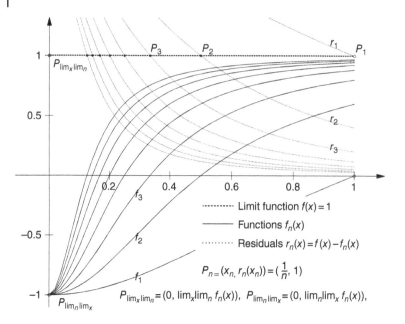

Figure 2.6 Example 7, sequence $f_n(x) = \frac{n^2 x^2 - 1}{n^2 x^2 + 1}$.

$\lim\limits_{n \to \infty} f_n(x) = 2 - \frac{1}{x}$, $\forall x \in X$. Therefore, $\lim\limits_{x \to 0_+} \sum_{n=1}^{\infty} u_n(x) = \lim\limits_{x \to 0_+} \left(2 - \frac{1}{x}\right) = -\infty$. Nevertheless, at the same limit point $x = 0$, the limit of each $u_n(x)$ exists: $\lim\limits_{x \to 0_+} u_1(x) = \lim\limits_{x \to 0_+} \left(2 - \frac{x^2 + 2}{x^3 + 1}\right) = 0$, $\lim\limits_{x \to 0_+} u_n(x) = 2 - 2 = 0$, $\forall n \neq 1$, and the right-hand side limit in the statement is finite: $\sum_{n=1}^{\infty} \lim\limits_{x \to 0_+} u_n(x) = \sum_{n=1}^{\infty} 0 = 0$.

Note that the above series converges nonuniformly on $(0, \infty)$, since for $\forall n \in \mathbb{N}$ choosing $x_n = \frac{1}{n} \in (0, \infty)$, one gets

$$|r_n(x_n)| = |f(x_n) - f_n(x_n)| = \left|2 - \frac{1}{x_n} - 2 + \frac{n^2 x_n^2 + 2}{n^2 x_n^3 + 1}\right| = \left|\frac{3}{\frac{1}{n} + 1} - n\right| \xrightarrow[n \to \infty]{} \infty.$$

Remark 1. Analogous example with infinite limit can be given for a sequence: a sequence $f_n(x)$ converges on X and $\lim\limits_{x \to x_0} f_n(x)$ exists for each n, but $\lim\limits_{x \to x_0} \lim\limits_{n \to \infty} f_n(x) \neq \lim\limits_{n \to \infty} \lim\limits_{x \to x_0} f_n(x)$ since the left-hand side limit is infinite. If one considers the functions $f_n(x) = \frac{n^2 x^2 + 2}{n^2 x^3 + 1}$ on $X = (0, \infty)$ (similar to the partial sums in the first counterexample), then $\lim\limits_{n \to \infty} f_n(x) = \frac{1}{x}$ and $\lim\limits_{x \to 0_+} \lim\limits_{n \to \infty} f_n(x) = \lim\limits_{x \to 0_+} \frac{1}{x} = \infty$, while at the same limit point $x = 0$, the limit of each $f_n(x)$ exists:

$\lim_{x \to 0_+} f_n(x) = 2$, $\forall n \in \mathbb{N}$, and the right-hand side limit in the example statement is finite: $\lim_{n \to \infty} \lim_{x \to 0_+} f_n(x) = \lim_{n \to \infty} 2 = 2$. Note that the convergence of $f_n(x)$ to $f(x)$ is nonuniform on X as shown by the following evaluation: choosing $x_n = \frac{1}{n}$, $\forall n \in \mathbb{N}$, one gets

$$|f_n(x_n) - f(x_n)| = \left| \frac{n^2 \cdot \frac{1}{n^2} + 2}{n^2 \cdot \frac{1}{n^3} + 1} - n \right| \underset{n \to \infty}{\to} \infty.$$

Remark 2. For a function depending on a parameter one gets: a function $f(x, y)$ is defined on $X \times Y$, converges on X to a limit function $\varphi(x)$, as $y \to y_0$, and $\lim_{x \to x_0} f(x, y)$ exists for each $y \in Y$, but $\lim_{x \to x_0} \lim_{y \to y_0} f(x, y) \neq \lim_{y \to y_0} \lim_{x \to x_0} f(x, y)$ since the left-hand side limit is infinite. For a counterexample, one can consider $f(x, y) = \frac{x^4 - 3y^4}{x^5 + y^4}$ on $X \times Y = (0, 1] \times (0, 1]$ using the limit points $x_0 = 0$ and $y_0 = 0$. On the one hand, the left-hand side limit in the statement is infinite, since

$$\varphi(x) = \lim_{y \to 0_+} f(x, y) = \lim_{y \to 0_+} \frac{x^4 - 3y^4}{x^5 + y^4} = \frac{1}{x}, \forall x \in (0, 1]$$

and

$$\lim_{x \to 0_+} \varphi(x) = \lim_{x \to 0_+} \frac{1}{x} = +\infty.$$

On the other hand, the right-hand side limit is finite:

$$\lim_{y \to 0_+} \lim_{x \to 0_+} f(x, y) = \lim_{y \to 0_+} \lim_{x \to 0_+} \frac{x^4 - 3y^4}{x^5 + y^4} = \lim_{y \to 0_+} (-3) = -3.$$

Thus, the limits not only cannot be interchanged, but one of them is infinite while another is finite. It is easy to show that the convergence of $f(x, y)$ to $\varphi(x)$ is nonuniform on $(0, 1]$: choosing $x_y = y \in (0, 1]$ for any $y \in (0, 1]$, one obtains

$$|f(x_y, y) - \varphi(x_y)| = \left| \frac{y^4 - 3y^4}{y^5 + y^4} - \frac{1}{y} \right| = \frac{1}{y} + \frac{2}{1 + y} \underset{y \to 0_+}{\to} +\infty.$$

Remark 3. Under the conditions of the statement, the limit in the left-hand side may even not exist (neither finite nor infinite), as shown in Example 9.

Example 9. A series $\sum u_n(x)$ converges on X, also $\lim_{x \to x_0} u_n(x)$ exists for each n and one of the two conditions is satisfied: either $\sum \lim_{x \to x_0} u_n(x)$ converges or $\lim_{x \to x_0} \sum u_n(x)$ exists, but nevertheless the remaining condition does not hold.

Solution

Let us consider the series $\sum_{n=1}^{\infty} u_n(x)$ with functions $u_n(x)$ defined on $X = (-1, 1)$ as follows: $u_n(x) = \begin{cases} x^n - x^{n+1}, x \in \mathbb{Q} \cap X \\ 0, x \in \mathbb{I} \cap X \end{cases}$. At the limit point $x_0 = 1$, one gets $\lim_{x \to 1_-, x \in \mathbb{Q}} u_n(x) = \lim_{x \to 1_-, x \in \mathbb{Q}} (x^n - x^{n+1}) = 0$ and $\lim_{x \to 1_-, x \in \mathbb{I}} u_n(x) = \lim_{x \to 1_-, x \in \mathbb{I}} 0 = 0$. Therefore, $\lim_{x \to 1_-} u_n(x) = 0$, $\forall n \in \mathbb{N}$ and $\sum_{n=1}^{\infty} \lim_{x \to 1_-} u_n(x) = \sum_{n=1}^{\infty} 0 = 0$. However,

$$f(x) = \sum_{n=1}^{\infty} u_n(x) = \begin{cases} \frac{(1-x)x}{1-x} = x, x \in \mathbb{Q} \cap X \\ 0, x \in \mathbb{I} \cap X \end{cases},$$

that gives $\lim_{x \to 1_-, x \in \mathbb{Q}} f(x) = \lim_{x \to 1_-, x \in \mathbb{Q}} x = 1$ and $\lim_{x \to 1_-, x \in \mathbb{I}} f(x) = \lim_{x \to 1_-, x \in \mathbb{I}} 0 = 0$, which means that $\lim_{x \to 1_-} f(x)$ does not exist. Note that choosing $x_n = 1 - \frac{1}{n+1} \in \mathbb{Q} \cap X, \forall n \in \mathbb{N}$, one obtains

$$|r_n(x_n)| = \left| \sum_{k=n+1}^{\infty} x_n^k(1 - x_n) \right| = \frac{(1 - x_n)x_n^{n+1}}{1 - x_n}$$

$$= \left(1 - \frac{1}{n+1} \right)^{n+1} \xrightarrow[n \to \infty]{} e^{-1} \neq 0,$$

which means that the convergence is nonuniform on $(-1, 1)$.

Another example for series is $\sum_{n=0}^{\infty} u_n(x)$ with $u_n(x) = (-1)^n x^n$ on $X = (0, 1)$. In this case, $\lim_{x \to 1_-} \sum_{n=0}^{\infty} (-1)^n x^n = \lim_{x \to 1_-} \frac{1}{1+x} = \frac{1}{2}$. However, $\lim_{x \to 1_-} u_n(x) = (-1)^n$, $\forall n \in \mathbb{N}$ and the series $\sum_{n=0}^{\infty} \lim_{x \to 1_-} u_n(x) = \sum_{n=0}^{\infty} (-1)^n$ diverges. This happens since the convergence of the series is nonuniform on X: if for an evaluation of $\sum_{k=n+1}^{n+p} u_k(x)$, $\forall n \in \mathbb{N}$ in the Cauchy criterion one chooses $p_n = 1$ and $x_n = \frac{1}{\sqrt[n+1]{2}} \in X$, then

$$\left| \sum_{k=n+1}^{n+p} u_k(x_n) \right| = |(-1)^{n+1} x_n^{n+1}| = \frac{1}{2} \xrightarrow[n \to \infty]{} 0.$$

Remark 1. Under similar conditions, any of the two iterated limits may not exist in the statement on sequences. Suppose $f_n(x)$ converges on X and $\lim_{x \to x_0} f_n(x)$ exists for each n. Even if one of the limits— $\lim_{n \to \infty} \lim_{x \to x_0} f_n(x)$ or $\lim_{x \to x_0} \lim_{n \to \infty} f_n(x)$—exists, another one may not exist.

In fact, consider the sequence $f_n(x) = \begin{cases} 1, x \in \mathbb{Q} \cap [0, 1) \\ x^n, x \in \mathbb{I} \cap [0, 1) \end{cases}$ on $X = [0, 1)$. Since $\lim_{x \to 1_-, x \in \mathbb{Q}} f_n(x) = \lim_{x \to 1_-} 1 = 1$ and $\lim_{x \to 1_-, x \in \mathbb{I}} f_n(x) = \lim_{x \to 1_-} x^n = 1$, it follows that

$\lim_{x \to 1_-} f_n(x) = 1$, $\forall x \in [0, 1)$ and, consequently, $\lim_{n \to \infty} \lim_{x \to 1_-} f_n(x) = 1$. On the other hand, $\lim_{n \to \infty, x \in \mathbb{Q}} f_n(x) = \lim_{n \to \infty} 1 = 1$ and $\lim_{n \to \infty, x \in \mathbb{I}} f_n(x) = \lim_{n \to \infty} x^n = 0$, which implies

$$\lim_{n \to \infty} f_n(x) = \begin{cases} 1, x \in \mathbb{Q} \\ 0, x \in \mathbb{I} \end{cases} = D(x), \forall x \in [0, 1). \text{ Therefore, the second iterated limit}$$

$\lim_{x \to 1_-} \lim_{n \to \infty} f_n(x) = \lim_{x \to 1_-} D(x)$ does not exist. It happens because the convergence of the sequence is not uniform on $X = [0, 1)$: $\forall n > 1$ choosing $x_n = \frac{1}{\sqrt[n]{2}} \in \mathbb{I} \cap X$, one gets

$$|f_n(x_n) - f(x_n)| = x_n^n = \frac{1}{2} \nrightarrow_{n \to \infty} 0.$$

Another counterexample is the sequence $f_n(x) = (-1)^n \frac{\sin nx}{nx}$ defined on $X = (0, \infty)$ with the limit point $x_0 = 0$. For any fixed $x > 0$ the sequence converges to zero: $f(x) = \lim_{n \to \infty} f_n(x) = \lim_{n \to \infty} (-1)^n \frac{\sin nx}{nx} = 0$, and therefore, the first iterated limit exists and equals zero: $\lim_{x \to 0_+} \lim_{n \to \infty} f_n(x) = \lim_{x \to 0_+} f(x) = 0$. Also, $\lim_{x \to 0_+} f_n(x) = (-1)^n$ for each $n \in \mathbb{N}$, but $\lim_{n \to \infty} \lim_{x \to 0_+} f_n(x) = \lim_{n \to \infty} (-1)^n$ does not exist. Such a behavior is related to nonuniform convergence of the sequence on X: for $\forall n$, there exists $x_n = \frac{1}{n} \in X$ such that

$$|f_n(x_n) - f(x_n)| = \left| (-1)^n \frac{\sin 1}{1} \right| = \sin 1 \nrightarrow_{n \to \infty} 0.$$

Remark 2. Similar situation may occur for functions depending on a parameter. Assume a function $f(x, y)$ is defined on $X \times Y$, converges on X to a limit function $\varphi(x)$, as $y \to y_0$, and $\lim_{x \to x_0} f(x, y)$ exists for each $y \in Y$. Even though one of the two iterated limits—$\lim_{y \to y_0} \lim_{x \to x_0} f(x, y)$ or $\lim_{x \to x_0} \lim_{y \to y_0} f(x, y)$—exists, another one may not exist.

First, consider the function $f(x, y) = \begin{cases} \frac{y^2}{x^2 + y^2}, x \in \mathbb{Q} \\ 1, x \in \mathbb{I} \end{cases}$ on $X \times Y = (0, 1] \times (0, 1]$. At the limit point $(0, 0)$, the first iterated limit exists:

$$\lim_{x \to 0_+, x \in \mathbb{Q}} f(x, y) = \lim_{x \to 0_+, x \in \mathbb{Q}} \frac{y^2}{x^2 + y^2} = 1, \lim_{x \to 0_+, x \in \mathbb{I}} f(x, y) = \lim_{x \to 0_+, x \in \mathbb{I}} 1 = 1,$$

and, consequently, $\lim_{y \to 0_+} \lim_{x \to 0_+} f(x, y) = \lim_{y \to 0_+} 1 = 1$. However, the second iterated limit does not exist, since

$$\lim_{y \to 0_+, x \in \mathbb{Q}} f(x, y) = \lim_{y \to 0_+, x \in \mathbb{Q}} \frac{y^2}{x^2 + y^2} = 0, \lim_{y \to 0_+, x \in \mathbb{I}} f(x, y) = \lim_{y \to 0_+, x \in \mathbb{I}} 1 = 1,$$

and the modified Dirichlet function $\tilde{D}(x) = 1 - D(x) = \lim_{y \to 0_+} f(x, y) =$

$\begin{cases} 0, x \in \mathbb{Q} \cap (0, 1] \\ 1, x \in \mathbb{I} \cap (0, 1] \end{cases}$ has no limit. Note that such a situation occurs because the convergence of $f(x, y)$ to $\tilde{D}(x)$ is nonuniform on $X = (0, 1]$: for $\forall y \in \mathbb{Q} \cap (0, 1]$, one can choose $x_y = y$ to obtain

$$|f(x_y, y) - \tilde{D}(x_y)| = \left| \frac{y^2}{x_y^2 + y^2} - 0 \right| = \left| \frac{1}{2} - 0 \right| = \frac{1}{2} \nrightarrow_{y \to 0_+} 0.$$

For a second counterexample, consider $f(x, y) = \frac{yD(y)}{x} \arctan \frac{x}{y}$ on $X \times Y = (0, +\infty) \times (0, +\infty)$ with the limit point $(0, 0)$. Since $D(y)$ and $\arctan \frac{x}{y}$ are bounded functions, one obtains $\varphi(x) = \lim_{y \to 0_+} f(x, y) = \lim_{y \to 0_+} \frac{y}{x} D(y) \arctan \frac{x}{y} = 0$ for any $x \in (0, +\infty)$. Therefore, one of the iterated limits exists: $\lim_{x \to 0_+} \lim_{y \to 0_+} f(x, y) = \lim_{x \to 0_+} \varphi(x) = 0$. Nevertheless, the second limit does not exist, since $\lim_{x \to 0_+} f(x, y) = \lim_{x \to 0_+} D(y) \frac{\arctan \frac{x}{y}}{\frac{x}{y}} = D(y)$, $\forall y \in (0, +\infty)$ and the limit of $D(y)$ does not exist at any point.

Note that the convergence of $f(x, y)$ to $\varphi(x)$ is nonuniform on $X = (0, +\infty)$: for $\forall y \in \mathbb{Q} \cap (0, +\infty)$, one can choose $x_y = y \in (0, +\infty)$ such that

$$|f(x_y, y) - \varphi(x_y)| = \arctan 1 = \frac{\pi}{4} \nrightarrow_{y \to 0_+} 0.$$

Remark to Examples 7–9. In all these counterexamples, the convergence is nonuniform on X. If the convergence would be uniform, the corresponding general statements will be true, that is, the calculation of the limit with respect to x can be interchanged with summation (for a series), or with the calculation of the limit with respect to discrete or continuous variable (for a sequence or a function depending on a parameter, respectively). Nevertheless, the following example shows that the uniform convergence is sufficient but not necessary condition for commutative properties of the operators.

Example 10. A series converges nonuniformly on X, but still a limit of this series can be calculated term by term.

Solution
Consider the series $\sum_{n=1}^{\infty} (nxe^{-nx} - (n - 1)xe^{-(n-1)x})$ on $X = [0, 1]$ with the limit point $x_0 = 0$. First, let us find out if this series is convergent. If $x = 0$, then $u_n(x) = 0$, $\forall n \in \mathbb{N}$, and $f(0) = \sum_{n=1}^{\infty} 0 = 0$. If $x \in (0, 1]$, then

$$f_n(x) = \sum_{k=1}^{n} u_k(x)$$

$$= xe^{-x} + 2xe^{-2x} - xe^{-x} + \cdots + nxe^{-nx} - (n - 1)xe^{-(n-1)x} = nxe^{-nx},$$

and consequently,

$$f(x) = \lim_{n\to\infty} f_n(x) = \lim_{n\to\infty} nxe^{-nx} = \lim_{t\to+\infty} \frac{t}{e^t} = \lim_{t\to+\infty} \frac{1}{e^t} = 0$$

(to calculate the last limit, we use the change of variables $nx = t$, which implies $t \xrightarrow[n\to\infty]{} +\infty$ for any $x > 0$, and then apply l'Hospital's rule). Hence, the series is convergent to $f(x) = 0$, $\forall x \in [0, 1]$. To show that this convergence is not uniform, for the series remainder $|r_n(x)| = \left|\sum_{k=n+1}^{\infty} u_k(x)\right| = |f_n(x) - f(x)| = nxe^{-nx}$, we can choose $x_n = \frac{1}{n} \in [0, 1]$ for any given $n \in \mathbb{N}$ and obtain $|r_n(x_n)| = n \cdot \frac{1}{n} e^{-n \cdot \frac{1}{n}} = e^{-1} \xrightarrow[n\to\infty]{} 0$. However, the limit, as x approaches 0, can be calculated term by term:

$$\lim_{x\to 0_+} u_n(x) = \lim_{x\to 0_+} (nxe^{-nx} - (n-1)xe^{-(n-1)x}) = 0,$$

and therefore,

$$\sum_{n=1}^{\infty} \lim_{x\to 0_+} u_n(x) = \sum_{n=1}^{\infty} 0 = 0 = \lim_{x\to 0_+} 0 = \lim_{x\to 0_+} f(x) = \lim_{x\to 0_+} \sum_{n=1}^{\infty} u_n(x).$$

Remark 1. The corresponding example for a sequence goes as follows: a sequence $f_n(x)$ converges nonuniformly on X, but still $\lim_{x\to x_0} \lim_{n\to\infty} f_n(x) = \lim_{n\to\infty} \lim_{x\to x_0} f_n(x)$. It can be exemplified by the sequence $f_n(x) = x^n - x^{2n}$ defined on $X = [0, 1]$ with the limit point $x_0 = 1$. For the limit function, we have: $f(x) = \lim_{n\to\infty} f_n(x) = \lim_{n\to\infty} (x^n - x^{2n}) = 0$ if $x \in [0, 1)$, and $f(1) = \lim_{n\to\infty} f_n(1) = \lim_{n\to\infty} (1 - 1) = 0$ if $x = 1$. Thus, the sequence converges to $f(x) \equiv 0$ on $[0, 1]$. Let us calculate the required limits at 1: $\lim_{x\to 1_-} f_n(x) = \lim_{x\to 1_-} (x^n - x^{2n}) = 1 - 1 = 0$, $\forall n \in \mathbb{N}$, and $\lim_{x\to 1_-} f(x) = \lim_{x\to 1_-} 0 = 0$. Therefore,

$$\lim_{x\to 1_-} \lim_{n\to\infty} f_n(x) = \lim_{x\to 1_-} f(x) = 0 = \lim_{n\to\infty} 0 = \lim_{n\to\infty} \lim_{x\to 1_-} f_n(x).$$

At the same time, the convergence is not uniform on $[0, 1]$ due to the following argument: for any $n \in \mathbb{N}$, we can choose $x_n = \frac{1}{\sqrt[n]{2}} \in [0, 1]$, which gets

$$|f_n(x_n) - f(x_n)| = \left|\left(\frac{1}{\sqrt[n]{2}}\right)^n - \left(\frac{1}{\sqrt[n]{2}}\right)^{2n}\right| = \left|\frac{1}{2} - \frac{1}{4}\right| = \frac{1}{4} \xrightarrow[n\to\infty]{} 0.$$

Remark 2. A similar example for a function depending on a continuous parameter can be formulated as follows: a function $f(x, y)$ is defined on $X \times Y$ and converges nonuniformly on X to a limit function $\varphi(x)$, as $y \to y_0$, but $\lim_{x\to x_0} \lim_{y\to y_0} f(x, y) = \lim_{y\to y_0} \lim_{x\to x_0} f(x, y)$. A counterexample is provided by the function

$f(x, y) = \frac{xy}{x^2+y^2}$ defined on $X \times Y = (0, 1] \times (0, 1]$ with the limit points $x_0 = 0$ for X and $y_0 = 0$ for Y. The limit function is $\varphi(x) = \lim_{y \to 0_+} f(x, y) = \lim_{y \to 0_+} \frac{xy}{x^2+y^2} = 0$, $\forall x \in (0, 1]$ and also $\lim_{x \to 0_+} f(x, y) = \lim_{x \to 0_+} \frac{xy}{x^2+y^2} = 0$, $\forall y \in (0, 1]$. Therefore, the limits can be interchanged:

$$\lim_{y \to 0_+} \lim_{x \to 0_+} f(x, y) = \lim_{y \to 0_+} 0 = 0 = \lim_{x \to 0_+} \varphi(x) = \lim_{x \to 0_+} \lim_{y \to 0_+} f(x, y).$$

Nevertheless, the convergence is not uniform on $X = (0, 1]$, since for $\forall y \in (0, 1]$ there exists $x_y = y \in (0, 1]$ such that

$$|f(x_y, y) - \varphi(x_y)| = \frac{y^2}{y^2 + y^2} = \frac{1}{2} \nrightarrow_{y \to 0_+} 0.$$

Example 11. A sequence of discontinuous functions converges uniformly on X, but the limit function is continuous on X.

Solution

Consider the sequence $f_n(x) = \begin{cases} \frac{1}{n}, 0 < x < \frac{1}{n} \\ 0, x = 0, \frac{1}{n} \leq x \leq 1 \end{cases}$ defined on $X = [0, 1]$. It converges to $f(x) \equiv 0$ on $[0, 1]$ because

$$f(0) = \lim_{n \to \infty} f_n(0) = \lim_{n \to \infty} 0 = 0, \; f(1) = \lim_{n \to \infty} f_n(1) = \lim_{n \to \infty} 0 = 0,$$

and for any fixed $x \in (0, 1)$ there is $n_x \in \mathbb{N}$ such that $n_x > \frac{1}{x}$, which implies $\frac{1}{n} < \frac{1}{n_x} < x$ for $\forall n > n_x$ and, consequently, $f(x) = \lim_{n \to \infty} f_n(x) = 0$. Moreover, this convergence is uniform, because $|f_n(x) - f(x)| \leq \frac{1}{n} \xrightarrow[n \to \infty]{} 0$ simultaneously for all $x \in [0, 1]$. Nevertheless, the limit function $f(x) \equiv 0$ is continuous on $[0, 1]$, while each of the functions $f_n(x)$ has discontinuities at the points $x = 0$ and $x = \frac{1}{n}$:

$$\lim_{x \to 0_+} f_n(x) = \frac{1}{n} \neq 0 = f_n(0), \; \lim_{x \to \frac{1}{n}_-} f_n(x) = \frac{1}{n} \neq 0 = f_n\left(\frac{1}{n}\right), \forall n \in \mathbb{N}.$$

Remark 1. For a series of functions, the respective example is as follows: all the terms of a uniformly convergent on X series $\sum u_n(x)$ are discontinuous at some point $x_0 \in X$, but the series sum is continuous at x_0. One can use the counterexample with the functions $u_0(x) = -\text{sgn } x$ and $u_n(x) = \frac{\text{sgn } x}{n(n+1)}$, $\forall n \in \mathbb{N}$ defined on $X = \mathbb{R}$. All these functions have the discontinuity point $x_0 = 0$, but the sum of this telescoping series is zero function continuous at any point. Indeed, the partial sums can be found as follows:

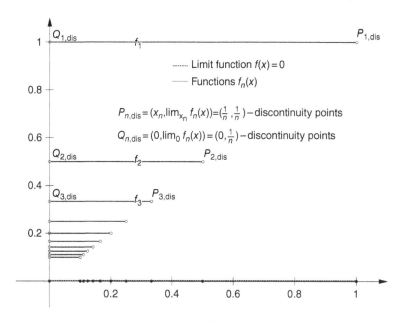

Figure 2.7 Example 11, sequence $f_n(x) = \begin{cases} \frac{1}{n}, 0 < x < \frac{1}{n} \\ 0, x = 0, \frac{1}{n} \le x \le 1 \end{cases}$.

$$f_n(x) = \sum_{k=0}^{n} u_k(x)$$

$$= -\operatorname{sgn} x + \operatorname{sgn} x \left(1 - \frac{1}{2} + \frac{1}{2} - \frac{1}{3} + \cdots + \frac{1}{n} - \frac{1}{n+1}\right) = -\frac{\operatorname{sgn} x}{n+1},$$

and therefore, the sum $f(x) = -\lim_{n \to \infty} \frac{\operatorname{sgn} x}{n+1} = 0, \forall x \in \mathbb{R}$. Note that the series converges uniformly on $\mathbb{R}$ by the Weierstrass test, since for $\forall x \in \mathbb{R}$ the following evaluation holds $|u_n(x)| = \left|\frac{\operatorname{sgn} x}{n(n+1)}\right| \le \frac{1}{n(n+1)} < \frac{1}{n^2}$ and the majorant series $\sum \frac{1}{n^2}$ converges.

Remark 2. A similar example for a function depending on a continuous parameter can be formulated as follows: a function $f(x, y)$ defined on $X \times Y$ has a discontinuity at $x_0 \in X$ for any $y \in Y$ and converges uniformly on X to a limit function $\varphi(x)$, as $y \to y_0$, but $\varphi(x)$ is continuous at x_0. For a counterexample, consider the function $f(x, y) = y \operatorname{sgn} x$ defined on $X \times Y = \mathbb{R} \times (0, +\infty)$ with $y_0 = 0$, which has a discontinuity at $x_0 = 0 \in X$ for $\forall y \in Y$. At the same time, the limit function is zero: $\lim_{y \to 0_+} f(x, y) = \lim_{y \to 0_+} y \operatorname{sgn} x = 0$ and, consequently, has no discontinuity. Checking the character of the convergence $|f(x, y) - 0| = |y \operatorname{sgn} x| \le |y| \underset{y \to 0_+}{\longrightarrow} 0$, we see that the convergence is uniform on $\mathbb{R}$.

Remark 3. The general statement with the opposite conclusion—"if a sequence of discontinuous functions converges uniformly, then the limit function is continuous"—is also false as shown in the next example.

Example 12. A sequence of discontinuous functions converges uniformly on X, and the limit function is also discontinuous on X.

Solution
An elementary example is $f_n(x) = \operatorname{sgn} x$ defined on $X = \mathbb{R}$. Evidently, this sequence converges uniformly to the same function $f(x) = \operatorname{sgn} x$ that has a jump discontinuity at the point $x = 0$.

An example of the functions dependent on n is the sequence $f_n(x) = \operatorname{sgn} x + \frac{1}{n}$ considered on $X = \mathbb{R}$. Each of the functions $f_n(x)$ has a discontinuity (a jump) at the point $x = 0$:

$$\lim_{x \to 0_+} f_n(x) = 1 + \frac{1}{n} \neq \frac{1}{n} = f_n(0), \quad \lim_{x \to 0_-} f_n(x) = -1 + \frac{1}{n} \neq \frac{1}{n} = f_n(0).$$

The limit function $f(x) = \lim_{n \to \infty} f_n(x) = \operatorname{sgn} x$ also has a jump discontinuity at 0. At the same time, the given sequence converges uniformly on $\mathbb{R}$, since $|f_n(x) - f(x)| = \frac{1}{n} \underset{n \to \infty}{\to} 0$ simultaneously for all $x \in \mathbb{R}$.

Remark 1. One can formulate analogous example for a series of functions: a series of discontinuous functions converges uniformly on X, and the sum of the series is also discontinuous on X. A simple counterexample is the series $\sum_{n=1}^{\infty} \frac{\operatorname{sgn} x}{n(n+1)}$ considered on $X = \mathbb{R}$. All the terms of the series have a discontinuity (a jump) at $x = 0$, and the uniform convergence of the series on $\mathbb{R}$ follows from the evaluation shown in Remark 1 to Example 11 for a similar series. Thus, all the statement conditions hold, but the sum of the series is $\operatorname{sgn} x$ that has the jump discontinuity at 0:

$$f_n(x) = \sum_{k=1}^{n} u_k(x)$$

$$= \operatorname{sgn} x \left(1 - \frac{1}{2} + \frac{1}{2} - \frac{1}{3} + \cdots + \frac{1}{n} - \frac{1}{n+1}\right) = \left(1 - \frac{1}{n+1}\right) \operatorname{sgn} x,$$

and therefore, the sum $f(x) = \lim_{n \to \infty} \left(1 - \frac{1}{n+1}\right) \operatorname{sgn} x = \operatorname{sgn} x$.

Remark 2. For a function depending on a continuous parameter, the corresponding example is as follows: a discontinuous in x function $f(x, y)$ converges uniformly on X to a limit function $\varphi(x)$, as $y \to y_0$, which is also discontinuous on X. The function $f(x, y) = \cos y \cdot \operatorname{sgn} x + y^2 \cos x$ defined on $X \times Y = \mathbb{R} \times (0, \frac{\pi}{2})$ with $y_0 = 0$ exemplifies this statement. In fact, $f(x, y)$ has a discontinuity (a jump) at $x_0 = 0$ for each $y \in Y$. The limit function is $\operatorname{sgn} x$ that also has a jump at 0: $\varphi(x) = \lim_{y \to 0_+} f(x,y) = \lim_{y \to 0_+} (\cos y \cdot \operatorname{sgn} x + y^2 \cos x) = \operatorname{sgn} x$. Finally, the convergence is uniform on $\mathbb{R}$ since

$$|f(x,y) - \varphi(x)| = |\cos y \cdot \operatorname{sgn} x + y^2 \cos x - \operatorname{sgn} x|$$
$$= |\operatorname{sgn} x \cdot (\cos y - 1) + y^2 \cos x|$$
$$\leq (1 - \cos y) \cdot |\operatorname{sgn} x| + y^2 \cdot |\cos x|$$
$$\leq 2\sin^2 \frac{y}{2} + y^2 \leq 2\left(\frac{y}{2}\right)^2 + y^2$$
$$= \frac{3}{2} y^2 \underset{y \to 0_+}{\to} 0.$$

Example 13. A sequence of functions $f_n(x)$ converges uniformly on X to a continuous function, but the functions $f_n(x)$ have infinitely many points of discontinuity on X.

Solution

Each of the functions $f_n(x) = \begin{cases} \frac{1}{n}, x \in \mathbb{Q} \\ 0, x \in \mathbb{I} \end{cases}$ is discontinuous at each point of $\mathbb{R}$ (it is just a slight modification of the famous Dirichlet's function $D(x)$, $f_n(x) = \frac{1}{n}D(x)$, which does not have a limit at any point of $\mathbb{R}$). At the same time, the sequence of $f_n(x)$ converges to the continuous on $\mathbb{R}$ function $f(x) \equiv 0$: for $\forall x \in \mathbb{Q}$, it follows $f(x) = \lim_{n \to \infty} f_n(x) = \lim_{n \to \infty} \frac{1}{n} = 0$, and for $\forall x \in \mathbb{I}$, one has $f(x) = \lim_{n \to \infty} f_n(x) = \lim_{n \to \infty} 0 = 0$. And moreover, this convergence is uniform, because $|f_n(x) - f(x)| \leq \frac{1}{n} \underset{n \to \infty}{\to} 0$ simultaneously for all $x \in \mathbb{R}$.

Remark 1. An analogous example for a series of functions is as follows: a series $\sum u_n(x)$ converges uniformly on X to a continuous function, but the terms of the series possess infinitely many discontinuities on X. For a counterexample, consider the series $\sum_{n=0}^{\infty} u_n(x)$, $u_0(x) = -D(x)$, $u_n(x) = \frac{D(x)}{n(n+1)}$, $\forall n \in \mathbb{N}$ defined on $X = \mathbb{R}$. According to the Weierstrass test, this series converges uniformly on $\mathbb{R}$: $|u_n(x)| = \frac{|D(x)|}{n(n+1)} < \frac{1}{n^2}$, $\forall x \in \mathbb{R}$ and the series $\sum \frac{1}{n^2}$ is convergent. To find the sum of the series, note that the series is telescoping:

$$f_n(x) = \sum_{k=0}^{n} u_k(x)$$

$$= -D(x) + D(x)\left(1 - \frac{1}{2} + \frac{1}{2} - \frac{1}{3} + \cdots + \frac{1}{n} - \frac{1}{n+1}\right) = -\frac{D(x)}{n+1},$$

and therefore, $f(x) = \lim_{n \to \infty} -\frac{D(x)}{n+1} = 0$, that is, the sum is a continuous function on $\mathbb{R}$. Nevertheless, all the terms of the series have discontinuity at each point of $\mathbb{R}$.

Remark 2. A similar example can be constructed for functions depending on a continuous parameter: $f(x, y)$ defined on $X \times Y$ converges uniformly on X to a continuous function $\varphi(x)$, as $y \to y_0$, but $f(x, y)$ possesses infinitely many points of discontinuity on X. For a counterexample, consider the function $f(x, y) =$

$$yD(x) = \begin{cases} y, x \in \mathbb{Q} \\ 0, x \in \mathbb{I} \end{cases}$$ on $X \times Y = \mathbb{R} \times (0, 1]$ with $y_0 = 0$. The limit function is

continuous on $\mathbb{R}$: $\varphi(x) = \lim_{y \to 0_+} f(x, y) = \lim_{y \to 0_+} yD(x) = 0$. The convergence is uni-

form on $\mathbb{R}$ since

$$|f(x, y) - \varphi(x)| = |yD(x)| \leq |y| \underset{y \to 0_+}{\to} 0.$$

However, for any fixed $y \in Y$, the function $f(x, y)$ is discontinuous at each real point.

Example 14. A function $f(x, y)$, defined on $X \times Y$, is continuous with respect to x on X for any fixed $y \in Y$, and it has a limit function $\varphi(x)$ on X as y approaches y_0, but $\varphi(x)$ is discontinuous on X.

Solution
The function $f(x, y) = e^{x/y}$ considered on $X \times Y = [-10, 0] \times (0, 1)$ is continu-
ous in x on $[-10, 0]$ for any fixed $y \in (0, 1)$. For the limit point $y_0 = 0$, the limit function is obtained as follows: if $x = 0$, then

$$\varphi(0) = \lim_{y \to 0_+} f(0, y) = \lim_{y \to 0_+} 1 = 1;$$

if $x \in [-10, 0)$, then

$$\varphi(x) = \lim_{y \to 0_+} f(x, y) = \lim_{y \to 0_+} e^{x/y} = 0$$

(since the last power is negative). Therefore, the limit function has a jump discontinuity at 0. Let us show that the convergence is not uniform on $[-10, 0]$. Indeed, choosing for $\forall y \in (0, 1)$ the point $x_y = -y \in (-1, 0) \subset [-10, 0]$, we obtain

$$|f(x_y, y) - \varphi(x_y)| = |e^{-y/y} - 0| = e^{-1} \underset{y \to 0_+}{\not\to} 0.$$

Two other counterexamples are the function $f(x, y) = \arctan \frac{x}{y}$ defined on $X \times Y = [0, +\infty) \times (0, 1]$ with the limit point $y_0 = 0$, and $f(x, y) = \frac{y^2}{x^2 + y^2}$ considered on $X \times Y = [0, 1] \times (0, 1]$ with the limit point $y_0 = 0$.

Remark 1. A similar example can be given for a sequence: a sequence of continuous functions $f_n(x)$ converges on X, but the limit function is discontinuous

on X. It can be exemplified by the sequence $f_n(x) = x^n$ considered on $X = [0, 1]$. In fact, since $\lim_{n\to\infty} f_n(x) = \lim_{n\to\infty} x^n = 0$ for $x \in [0, 1)$ and $\lim_{n\to\infty} f_n(1) = \lim_{n\to\infty} 1 = 1$, the sequence converges on $[0, 1]$ to the function $f(x) = \begin{cases} 0, x \in [0, 1) \\ 1, x = 1 \end{cases}$. Each function $f_n(x) = x^n$ is continuous on $[0, 1]$, but the limit function $f(x) = \begin{cases} 0, x \in [0, 1) \\ 1, x = 1 \end{cases}$ is discontinuous at $x = 1$. Note that the convergence is not uniform on $[0, 1]$ due to the following argument: for any $n \in \mathbb{N}$ there exists $x_n = \frac{1}{\sqrt[n]{2}} \in [0, 1) \subset [0, 1]$ such that

$$|f_n(x_n) - f(x_n)| = \left(\frac{1}{\sqrt[n]{2}} \right)^n - 0 = \frac{1}{2} \underset{n\to\infty}{\nrightarrow} 0.$$

Remark 2. In the case of series, the corresponding example assumes the following form: a series $\sum u_n(x)$ of continuous functions converges on X, but the sum of this series is a discontinuous function. The series of Example 7 $\sum_{n=1}^{\infty}(1 - x)x^n$ considered on $X = [0, 1]$ provides the required counterexample. All the considerations follow those used in Example 7, and we reproduce them here in a brief form just for the sake of completeness. First, the series is convergent, since for $x = 1$ one has $\sum_{n=1}^{\infty} 0 = 0$, and for $x \in [0, 1)$ one obtains the convergent geometric series $\sum_{n=1}^{\infty}(1 - x)x^n = \frac{(1-x)x}{1-x} = x$. Hence, the sum $f(x) = \begin{cases} x, x \in [0, 1) \\ 0, x = 1 \end{cases}$ is discontinuous at $x = 1$, although all the functions $u_n(x) = (1 - x)x^n$ are continuous on $[0, 1]$. Finally, note that the convergence is not uniform on $[0, 1]$, because for $x \in [0, 1)$ the remainder can be written in the form

$$|r_n(x)| = \sum_{k=n+1}^{\infty} (1 - x)x^k = \frac{(1 - x)x^{n+1}}{1 - x} = x^{n+1},$$

and choosing $x_n = \frac{1}{\sqrt[n+1]{2}} \in [0, 1)$ for any given $n \in \mathbb{N}$ we get $|r_n(x_n)| = \frac{1}{2} \underset{n\to\infty}{\nrightarrow} 0$.

Remark 3. In all the above counterexamples, the convergence was not uniform on X, which results in discontinuity of the limit function and the series sum. If the convergence would be uniform on X, then all the corresponding general statements will be true. For instance, in the case of the sequences, the correct general statement goes as follows: if a sequence of continuous functions $f_n(x)$ converges uniformly on X, then the limit function is continuous on X. At the same time, the uniform convergence of continuous functions is just a sufficient condition. The next example shows that the limit function still can be continuous if this condition is not satisfied.

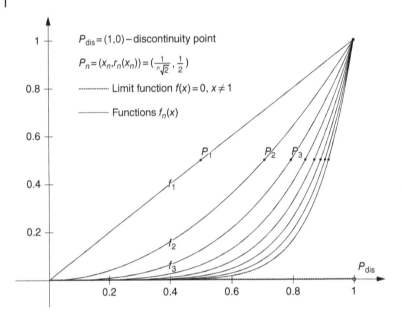

Figure 2.8 Examples 14, 17, 19, 27, and 29, sequence $f_n(x) = x^n$.

Example 15. A function $f(x, y)$, defined on $X \times Y$, is continuous with respect to x on X for any fixed $y \in Y$, and it has a continuous limit function $\varphi(x)$ on X as y approaches y_0, but the convergence is nonuniform on X.

Solution
The function $f(x, y) = \frac{xy}{x^2 + y^2}$ defined on $X \times Y = [0, 1] \times (0, 1]$ is continuous in x on $[0, 1]$ for any fixed $y \in (0, 1]$. The limit function, as y approaches $y_0 = 0$, is

$$\varphi(x) = \lim_{y \to 0_+} f(x, y) = \lim_{y \to 0_+} \frac{xy}{x^2 + y^2} = 0, \forall x \in [0, 1],$$

so $\varphi(x)$ is also continuous on $[0, 1]$. However, the convergence is not uniform on $X = [0, 1]$, since for $\forall y \in (0, 1]$ there exists $x_y = y \in (0, 1]$ such that

$$|f(x_y, y) - \varphi(x_y)| = \frac{y^2}{y^2 + y^2} = \frac{1}{2} \underset{y \to 0_+}{\nrightarrow} 0.$$

Remark 1. A similar example for a sequence is as follows: a sequence of continuous functions $f_n(x)$ converges on X to a continuous function, but the convergence is nonuniform on X. The sequence of the continuous functions $f_n(x) = nxe^{-nx}$, $x \in X = [0, 1]$ provides a counterexample. Indeed, for the limit function we have: $f(0) = \lim_{n \to \infty} f_n(0) = \lim_{n \to \infty} 0 = 0$ if $x = 0$, and

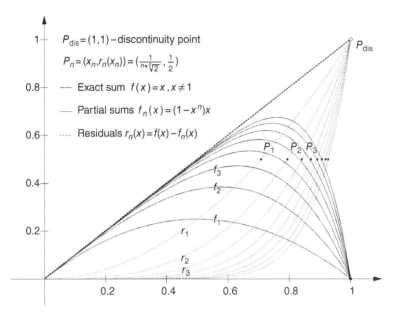

Figure 2.9 Examples 14, 21, 24, 27, and 29, series $\sum_{n=1}^{\infty}(1-x)x^n$.

$$f(x) = \lim_{n\to\infty} f_n(x) = \lim_{n\to\infty} \frac{nx}{e^{nx}} = \lim_{t\to+\infty} \frac{t}{e^t} = \lim_{t\to+\infty} \frac{1}{e^t} = 0$$

if $x \in (0, 1]$ (in the last limit, we apply the change of variable $t = nx$ and l'Hospital's rule). Thus, the sequence converges on $[0, 1]$ to the continuous function $f(x) \equiv 0$, but the convergence is not uniform on $[0, 1]$: for any $n \in \mathbb{N}$ we can choose $x_n = \frac{1}{n} \in [0, 1]$, which results in $|f_n(x_n) - f(x_n)| = e^{-1} \underset{n\to\infty}{\nrightarrow} 0$.

Remark 2. For a series, the example can be formulated as follows: a series $\sum u_n(x)$ of continuous functions converges on X to a continuous function, but the convergence is nonuniform on X. As a counterexample, we can use the series of continuous functions $\sum_{n=1}^{\infty} x^{n-1}$ on $X = (-1, 1)$. The sum of this convergent geometric series $f(x) = \frac{1}{1-x}$ is also continuous on $(-1, 1)$. However, the convergence is not uniform on $(-1, 1)$, because the remainder has the form $r_n(x) = \sum_{k=n+1}^{\infty} u_k(x) = \frac{x^n}{1-x}$ and choosing $x_n = 1 - \frac{1}{n} \in (-1, 1)$ for any given $n \in \mathbb{N}$, we obtain

$$|r_n(x_n)| = \frac{\left(1 - \frac{1}{n}\right)^n}{1 - 1 + \frac{1}{n}} = n\left(1 - \frac{1}{n}\right)^n \underset{n\to\infty}{\to} +\infty.$$

2.3 Conditions of Uniform Convergence. Dini's Theorem

Remark. In this section, we analyze the conditions of Dini's theorem on uniform convergence first for sequences (Examples 16–20) and then for series of functions (Examples 21–25). Example 26 clarifies how many discontinuity points the limit function of a sequence of continuous functions may have.

Example 16. A sequence of functions $f_n(x)$, which are monotone in n for any fixed $x \in X$, converges on a compact set X to a continuous function, but the convergence is nonuniform on X.

Solution

Let us consider the sequence $f_n(x) = \begin{cases} 1, x \in \left(0, \dfrac{1}{n}\right) \\ 0, x = 0, x \in \left[\dfrac{1}{n}, 1\right] \end{cases}$ on the compact set

$X = [0, 1]$. For any fixed $x \in X$, the sequence is monotone in n: $f_{n+1}(x) \le f_n(x)$. It converges on $[0, 1]$ to the continuous function $f(x) \equiv 0$:

$$f(0) = \lim_{n \to \infty} f_n(0) = \lim_{n \to \infty} 0 = 0, \quad f(1) = \lim_{n \to \infty} f_n(1) = \lim_{n \to \infty} 0 = 0,$$

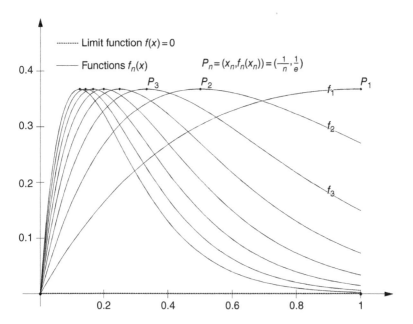

Figure 2.10 Examples 15 and 18, sequence $f_n(x) = nxe^{-nx}$.

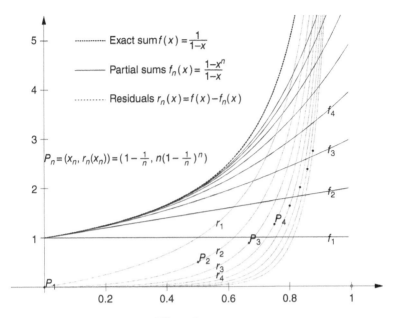

Figure 2.11 Example 15, series $\sum_{n=1}^{\infty} x^{n-1}$.

and for any given $x \in (0,1)$ there exists $N_x \in \mathbb{N}$ such that $N_x > \frac{1}{x}$, and then for $\forall n \geq N_x$ we have $x > \frac{1}{N_x} \geq \frac{1}{n}$, and therefore, $f_n(x) = 0$ for $\forall n \geq N_x$, which implies that $f(x) = \lim_{n\to\infty} f_n(x) = 0$. To see that the convergence is nonuniform on $[0,1]$, we can choose $x_n = \frac{1}{2n} \in \left(0, \frac{1}{n}\right) \subset [0,1]$ for any given $n \in \mathbb{N}$, which results in $|f_n(x_n) - f(x_n)| = 1 \underset{n\to\infty}{\nrightarrow} 0$.

Remark. Note that the functions of the above sequence have a jump discontinuity at the points $x = 0$ and $x = \frac{1}{n}$:

$$\lim_{x\to 0_+} f_n(x) = 1 \neq 0 = f_n(0), \quad \lim_{x\to \frac{1}{n}_-} f_n(x) = 1 \neq 0 = f_n\left(\frac{1}{n}\right), \forall n.$$

Thus, this example shows that a weakened version of Dini's theorem with the omitted condition of the continuity of $f_n(x)$ on X is not true.

Example 17. A sequence of continuous functions $f_n(x)$, which are monotone in n for any fixed $x \in X$, converges on a compact set X, but this convergence is nonuniform on X.

Solution

The sequence $f_n(x) = x^n$ used in Remark 1 to Example 14 consists of the continuous functions on the compact set $X = [0, 1]$. For any fixed $x \in X$, the corresponding numerical sequence is monotone in n: $f_{n+1}(x) = x^{n+1} \leq x^n = f_n(x)$. As shown in Remark 1 to Example 14, this sequence converges on $[0, 1]$ to the function $f(x) = \begin{cases} 0, x \in [0, 1) \\ 1, x = 1 \end{cases}$, but the convergence is not uniform on $[0, 1]$.

Remark. Note that the limit function $f(x)$ in this counterexample has a jump discontinuity at the point $x = 1$. Thus, it is seen that one cannot omit the condition of the continuity of the limit function $f(x)$ on X in the conditions of Dini's theorem.

Example 18. A sequence of continuous functions $f_n(x)$ converges on a compact set X to a continuous function, but this convergence is nonuniform on X.

Solution

The sequence of the continuous functions $f_n(x) = nxe^{-nx}$, defined on the compact set $X = [0, 2]$, converges to the continuous function $f(x) \equiv 0$, but the convergence is not uniform on $[0, 2]$ (see Remark 1 to Example 15 for details, where similar results were derived for $[0, 1]$).

Remark 1. The convergence is not uniform because the condition of monotonicity in Dini's theorem is not satisfied: if $n > 1$, then for $x_n = \frac{1}{n}$ we have $f_{n-1}(x_n) < e^{-1} = f_n(x_n)$, while $f_n(x_n) = e^{-1} > f_{n+1}(x_n)$ (take into account that the function $g(t) = te^{-t}$ is strictly increasing for $t < 1$ and decreasing for $t > 1$).

Remark 2. The general converse statement is true: if a sequence of continuous functions $f_n(x)$ converges uniformly on a set X, then the limit function is continuous on X.

Example 19. A sequence of continuous functions $f_n(x)$, which are monotone in n for any fixed $x \in X$, converges on a set X to a continuous function, but this convergence is nonuniform on X.

Solution

The sequence $f_n(x) = x^n$ consists of the continuous functions on the set $X = [0, 1)$. The sequence is monotone in n for any fixed $x \in X$, and it converges to the continuous function $f(x) \equiv 0$ on $[0, 1)$. However, the convergence is not uniform on $[0, 1)$. (See details in Example 17 and in Remark 1 to Example 14, where the same sequence was considered on $[0, 1]$.)

Remark. The convergence is not uniform because the set $X = [0, 1)$ is not compact (this is one of the conditions of Dini's theorem).

Remark to Examples 16–19. These examples show that if at least one of the four conditions in Dini's theorem is not satisfied, then the convergence may be nonuniform. At the same time, these conditions are sufficient, but not necessary ones for the uniform convergence. It may happen that one, two, or even all the conditions are violated (except for the condition of the convergence on X), and even so the convergence is uniform. For instance, in Example 11, each of

the functions $f_n(x) = \begin{cases} \frac{1}{n}, 0 < x < \frac{1}{n} \\ 0, x = 0, \frac{1}{n} \le x \le 1 \end{cases}$ is discontinuous at the points $x = 0$

and $x = \frac{1}{n}$, but the convergence to $f(x) \equiv 0$ is uniform on $[0, 1]$. In Example 12, three conditions of Dini's theorem are violated: the functions $f_n(x) = \text{sgn } x + \frac{1}{n}$ and the limit function $f(x) = \text{sgn } x$ are discontinuous at 0, and the set $X = \mathbb{R}$ is not compact. However, the convergence is still uniform on $\mathbb{R}$. Slightly modifying the sequence in Example 12, it is possible to construct the example where all four conditions of Dini's theorem do not hold, but the convergence is uniform, as shown in the next example.

Example 20. A sequence of functions that violates all the four conditions of Dini's theorem on a set X, but nevertheless converges uniformly on this set.

Solution
Consider the sequence $g_n(x) = \text{sgn } x + \frac{(-1)^n}{n}$ defined on $X = \mathbb{R}$. The limit function of this sequence is $g(x) = \text{sgn } x$. All the functions $g_n(x)$ and the limit function $g(x)$ are discontinuous at 0, the set $X = \mathbb{R}$ is not compact, and the sequence $g_n(x)$ is not monotone for any fixed x: $g_{2n}(x) = \text{sgn } x + \frac{1}{2n} > \text{sgn } x - \frac{1}{2n+1} = g_{2n+1}(x)$, while $g_{2n+1}(x) = \text{sgn } x - \frac{1}{2n+1} < \text{sgn } x + \frac{1}{2n+2} = g_{2n+2}(x)$, $\forall n \in \mathbb{N}$. Thus, all the four conditions of Dini's theorem are violated. Nevertheless, the convergence is uniform on $\mathbb{R}$, since $|g_n(x) - g(x)| = \frac{1}{n} \underset{n \to \infty}{\to} 0$ simultaneously for all $x \in \mathbb{R}$.

Example 21. A series of continuous nonnegative functions converges on a compact set X, but this convergence is nonuniform on X.

Solution
Consider the series $\sum_{n=1}^{\infty} x^n(1 - x)$ on $X = [0, 1]$ used in Example 7 and Remark 2 to Example 14. Each function $u_n(x) = x^n(1 - x)$ is nonnegative on $[0, 1]$ and continuous on $\mathbb{R}$ (and, in particular, on $[0, 1]$). As shown in Remark 2 to Example 14, this series converges to $f(x) = \begin{cases} x, 0 \le x < 1 \\ 0, x = 1 \end{cases}$. Nevertheless,

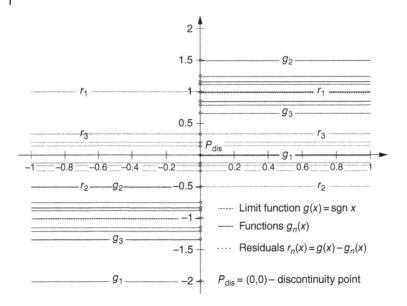

Figure 2.12 Example 20, sequence $g_n(x) = \mathrm{sgn}\, x + \frac{(-1)^n}{n}$.

the convergence on $[0,1]$ is nonuniform, since for $\forall n \in \mathbb{N}$ we can choose $x_n = 1 - \frac{1}{n+1} \in X$ such that

$$|r_n(x_n)| = \left| \sum_{k=n+1}^{\infty} x_n^k (1 - x_n) \right| = \frac{x_n^{n+1}(1 - x_n)}{1 - x_n}$$

$$= \left(1 - \frac{1}{n+1}\right)^{n+1} \xrightarrow[n \to \infty]{} e^{-1} \neq 0.$$

Note that one of the conditions of Dini's theorem is violated—the sum of the series has a discontinuity at $x = 1$—that leads to nonuniformity of the convergence.

Example 22. A series of nonnegative functions converges on a compact set X to a continuous function, but this convergence is nonuniform on X.

Solution
Consider the series $\sum_{n=0}^{\infty} u_n(x)$ with the nonnegative on $X = [0,1]$ terms $u_0(x) =$

$$\begin{cases} 1, x = 0 \\ 0, x \in (0,1] \end{cases}, u_n(x) = \begin{cases} 1, x \in \left(\frac{1}{n+1}, \frac{1}{n}\right] \\ 0, x \in \left[0, \frac{1}{n+1}\right] \cup \left(\frac{1}{n}, 1\right] \end{cases}, \forall n \in \mathbb{N}. \text{ At the point } x = 0,$$

the sum is 1 since $u_0(0) = 1$ and $u_n(0) = 0, \forall n \in \mathbb{N}$. For any fixed $x \in (0,1]$, there exists $n_x \in \mathbb{N}$ such that $x \in \left(\frac{1}{n_x+1}, \frac{1}{n_x}\right]$ (since $\bigcup_{n=1}^{\infty} \left(\frac{1}{n+1}, \frac{1}{n}\right] = (0,1]$). Moreover, such n_x is unique because $\left(\frac{1}{n+1}, \frac{1}{n}\right] \cap \left(\frac{1}{k+1}, \frac{1}{k}\right] = \emptyset$ for $\forall n, k$ such that $n \neq k$. Then $u_{n_x}(x) = 1$ and $u_k(x) = 0$, $\forall k \neq n_x$, that is, exactly one term is equal to 1 and all other terms are 0 in the series $\sum_{n=0}^{\infty} u_n(x)$. Since it is true for any fixed $x \in (0, 1]$, we conclude that $f(x) = \sum_{n=0}^{\infty} u_n(x) = 1$ on $(0, 1]$ and, consequently, on $[0, 1]$. Thus, the sum of the series is continuous (constant) function on X. However, the convergence is nonuniform on $[0, 1]$, which can be seen applying the Cauchy criterion with $p_n = 1$ and $x_n = \frac{1}{n+1} \in X$ for each $n \in \mathbb{N}$:

$$\left| \sum_{k=n+1}^{n+p_n} u_k(x_n) \right| = u_{n+1}\left(\frac{1}{n+1}\right) = 1 \underset{n \to \infty}{\nrightarrow} 0.$$

In this example the nonuniformity of convergence is caused by discontinuity of the series terms: $u_0(x)$ has a jump at 0, $u_1(x)$ at $\frac{1}{2}$ and all other functions $u_n(x)$ have two jump points—$\frac{1}{n+1}$ and $\frac{1}{n}$. Note that the continuity of the series terms is one of the conditions of Dini's theorem.

Example 23. A series of continuous functions converges on a compact set X to a continuous function, but this convergence is nonuniform on X.

Solution
The series $\sum_{n=1}^{\infty} (-1)^{n-1} n (1-x) x^n$ is composed of the functions $u_n(x) = (-1)^{n-1} n (1-x) x^n$ continuous on $\mathbb{R}$ and, in particular, on $X = [0, 1]$. To find the sum of this series, we use the binomial expansion:

$$(1+x)^\alpha = \sum_{n=0}^{\infty} \frac{\alpha(\alpha-1)\cdots(\alpha-n+1)}{n!} x^n, x \in (-1, 1).$$

Setting $\alpha = -2$, we get

$$(1+x)^{-2} = \sum_{n=0}^{\infty} (-1)^n (n+1) x^n = \sum_{k=1}^{\infty} (-1)^{k-1} k x^{k-1}.$$

Comparing this expression with our series, we have

$$\sum_{n=1}^{\infty} (-1)^{n-1} n (1-x) x^n = x(1-x) \sum_{n=1}^{\infty} (-1)^{n-1} n x^{n-1} = \frac{x(1-x)}{(1+x)^2}, \forall x \in (-1, 1).$$

At the point $x = 1$, all the terms are zero and the sum of series is also zero. Thus, for $\forall x \in [0, 1]$, the sum of the series is $f(x) = \sum_{n=1}^{\infty} (-1)^{n-1} n (1-x) x^n = \frac{x(1-x)}{(1+x)^2}$, which is a continuous function on $[0, 1]$.

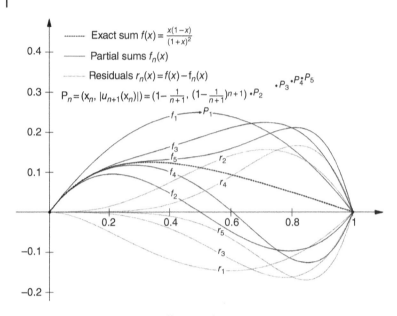

Figure 2.13 Example 23, series $\sum_{n=1}^{\infty} (-1)^{n-1} n(1-x)x^n$.

Now let us check the character of the convergence on $[0,1]$. Applying the Cauchy criterion with $p_n = 1$ and $x_n = 1 - \frac{1}{n+1} \in X$, for each $n \in \mathbb{N}$, we obtain

$$\left| \sum_{k=n+1}^{n+p_n} u_k(x_n) \right| = |(-1)^n (n+1)(1-x_n) x_n^{n+1}|$$

$$= (n+1) \cdot \frac{1}{n+1} \left(1 - \frac{1}{n+1} \right)^{n+1} \xrightarrow[n \to \infty]{} e^{-1} \neq 0,$$

which means that the convergence is nonuniform. Comparing the conditions of the statement with the conditions of Dini's theorem, one can see that the former does not satisfy the condition of the nonnegativity of the series terms.

Example 24. A series of continuous nonnegative functions converges on X to a continuous function, but this convergence is nonuniform on X.

Solution
Consider the series $\sum_{n=1}^{\infty} x^n (1-x)$ on $X = [0,1)$. Each term $u_n(x) = x^n(1-x)$ of this series is a continuous nonnegative function on $[0,1)$. The sum of this series, found in Example 21, is the continuous on $[0,1)$ function $f(x) = \sum_{n=1}^{\infty} x^n (1-x) = x$. However, in the same way as shown in Example 21, one can prove that the convergence to $f(x)$ is nonuniform on $[0,1)$. Comparing to Dini's theorem, one sees that the example is missing the condition of compactness of X.

Remark to Examples 21–24. In each of the last four examples, only one condition of Dini's theorem was omitted, but it causes the nonuniformity of the convergence. Thus, only one condition missing invalidates the result of the Dini's theorem. At the same time, the set of conditions of Dini's theorem is sufficient and not necessary, meaning that a violation of one or a few of these conditions does not imply the nonuniformity of the convergence. This situation is illustrated in Example 25.

Example 25. Some conditions of Dini's theorem are violated, but a series still converges uniformly.

Solution
First, let us consider an example when the condition of the nonnegativity of the terms of series in Dini's theorem is not satisfied. The series $\sum_{n=1}^{\infty} (-1)^{n-1} \frac{x^n}{n}$ is a convergent expansion of $\ln(1 + x)$ on $(-1, 1]$ and, in particular, on $X = [0, 1]$. The terms $(-1)^{n-1} \frac{x^n}{n}$ and the sum $\ln(1 + x)$ of the series are continuous functions on the compact set $[0, 1]$, but infinitely many terms are negative on $(0, 1]$. The type of the convergence can be find out by Leibniz's test for alternating series:

$$|r_n(x)| \le |u_{n+1}(x)| = \left|(-1)^n \frac{x^{n+1}}{n+1}\right| \le \frac{1}{n+1} \xrightarrow[n\to\infty]{} 0,$$

that is, the series converges uniformly, despite the fact that the condition of the nonnegativity of the terms of series in Dini's theorem does not hold.

Another example involves violation of all the conditions of Dini's theorem. All the terms $u_n(x) = \frac{1}{n(n+1)} \tilde{D}(x)$, $\tilde{D}(x) = \begin{cases} 1, x \in \mathbb{Q} \\ -1, x \in \mathbb{I} \end{cases}$ of the series $\sum_{n=1}^{\infty} u_n(x)$ are discontinuous functions at each point $x \in X = \mathbb{R}$. The partial sums of this series are

$$f_n(x) = \sum_{k=1}^{n} u_k(x)$$

$$= \tilde{D}(x)\left(1 - \frac{1}{2} + \frac{1}{2} - \frac{1}{3} + \cdots + \frac{1}{n} - \frac{1}{n+1}\right) = \tilde{D}(x)\left(1 - \frac{1}{n+1}\right),$$

and consequently, the sum is

$$f(x) = \lim_{n\to\infty} f_n(x) = \lim_{n\to\infty} \tilde{D}(x)\left(1 - \frac{1}{n+1}\right) = \tilde{D}(x),$$

that is, the sum is everywhere discontinuous function too. Further, the functions $u_n(x)$ do not keep the sign: $u_n(x) > 0$ for $\forall x \in \mathbb{Q}$, while $u_n(x) < 0$ for $\forall x \in \mathbb{I}$. Finally, the set $X = \mathbb{R}$ is not compact. Thus, all the conditions of Dini's

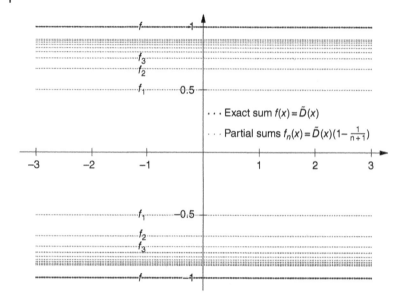

Figure 2.14 Example 25, series $\sum_{n=1}^{\infty} u_n(x)$, $u_n(x) = \frac{1}{n(n+1)}\tilde{D}(x)$, $\tilde{D}(x) = \begin{cases} 1, x \in \mathbb{Q} \\ -1, x \in \mathbb{I} \end{cases}$.

theorem are violated. Nevertheless, the series converges uniformly on $X = \mathbb{R}$ according to the Weierstrass test:

$$|u_n(x)| = \left| \frac{\tilde{D}(x)}{n(n+1)} \right| = \frac{1}{n(n+1)} < \frac{1}{n^2}, \forall x \in \mathbb{R},$$

and the series $\sum \frac{1}{n^2}$ is convergent.

Example 26. A sequence of continuous functions converges on X to a function $f(x)$, which has infinitely many points of discontinuity.

Solution
Suppose that any rational number in $[0, 1]$ is represented in the form $x = \frac{p}{q}$, where $p, q \in \mathbb{N}$ and the fraction $\frac{p}{q}$ is in lowest terms for $x \in (0, 1)$, and additionally, $q = 1$ for $x = 0$ and $x = 1$. In this case, the numbers p and q are uniquely determined in the representation $x = \frac{p}{q}$. For each $n \geq 2$, $n \in \mathbb{N}$, the construction of the function $f_n(x)$ defined on $[0, 1]$ is as follows: for each fixed n, one considers the values $q = 1, \dots, n-1$, each of which forms a set of the special points $\frac{p}{q}, p = 0, \dots, q$, employed to define the nth function of the sequence

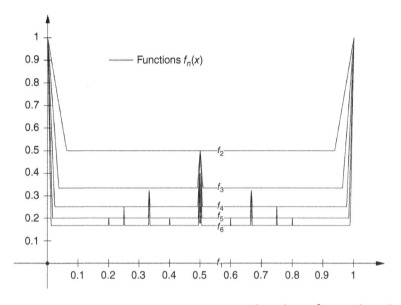

Figure 2.15 Example 26, sequence $f_n(x) = \begin{cases} \frac{1}{q} + 2n^2 \left(x - \frac{p}{q} \right), x \in \left[\frac{p}{q} - \frac{1}{2n^2} \left(\frac{1}{q} - \frac{1}{n} \right), \frac{p}{q} \right] \\ \frac{1}{q} - 2n^2 \left(x - \frac{p}{q} \right), x \in \left[\frac{p}{q}, \frac{p}{q} + \frac{1}{2n^2} \left(\frac{1}{q} - \frac{1}{n} \right) \right] \\ \frac{1}{n}, \text{otherwise} \end{cases}$.

$$f_n(x) = \begin{cases} \frac{1}{q} + 2n^2 \left(x - \frac{p}{q} \right), x \in \left[\frac{p}{q} - \frac{1}{2n^2} \left(\frac{1}{q} - \frac{1}{n} \right), \frac{p}{q} \right] \\ \frac{1}{q} - 2n^2 \left(x - \frac{p}{q} \right), x \in \left[\frac{p}{q}, \frac{p}{q} + \frac{1}{2n^2} \left(\frac{1}{q} - \frac{1}{n} \right) \right] \\ \frac{1}{n}, \text{otherwise} \end{cases}.$$

By construction, at each special point $\frac{p}{q}$, $q < n$, the function $f_n(x)$ has a local maximum equal to $\frac{1}{q} > \frac{1}{n}$, it is increasing in the left-hand and decreasing in the right-hand neighborhood of $\frac{p}{q}$. The values of $f_n(x)$ are larger than $\frac{1}{n}$ only in a neighborhood of each $\frac{p}{q}$, $q < n$, and the diameter $\frac{1}{n^2} \left(\frac{1}{q} - \frac{1}{n} \right)$ of such a neighborhood decreases for larger values of n, approaching 0 as n approaches infinity. Note also that $f_{n+1}(x)$ is smaller than $f_n(x)$ at all points except for the special points $\frac{p}{q}$, $q < n + 1$, at which the two functions coincide.

Since $0 < f_{n+1}(x) \le f_n(x)$ for $\forall n \ge 2$ and $\forall x \in [0, 1]$, the sequence is convergent on $[0, 1]$: $\lim_{n \to \infty} f_n(x) = f(x)$ (because at each point we have the decreasing numerical sequence bounded below). Note that $f_n(0) = f_n(1) = 1$, $\forall n \ge 2$ and, consequently, $f(0) = f(1) = 1$. Let us consider a rational point $x_0 = \frac{p}{q}$ in $(0, 1)$

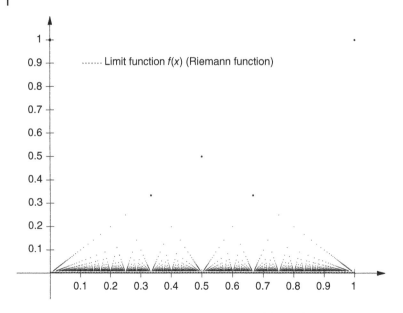

Figure 2.16 Example 26, limit function $f(x) = R(x) = \begin{cases} \frac{1}{q}, x = \frac{p}{q} \in \mathbb{Q} \\ 0, x \in \mathbb{I} \end{cases}$.

with the fraction $\frac{p}{q}$ in lowest terms. For each $n > q$, the point x_0 is used for construction of the function $f_n(x)$, and therefore, $f_n(x_0) = f_n\left(\frac{p}{q}\right) = \frac{1}{q}, \forall n > q$, that is, $\lim_{n \to \infty} f_n\left(\frac{p}{q}\right) = \frac{1}{q} = f\left(\frac{p}{q}\right)$.

Now let us choose an irrational $x_0 \in (0, 1)$ and, by using the method of mathematical induction, show that there exists $n_k \in \mathbb{N}$ such that $f_n(x_0) \leq \frac{1}{k}$ for $\forall n \geq n_k$. For $k = 2$, denote the distance between x_0 and the endpoints 0 and 1 by d_0 and d_1, respectively (both distances are positive because x_0 is irrational and the endpoints are rational). There exists $n_2 \in \mathbb{N}$ such that $\frac{1}{2n_2^2} < \min(d_0, d_1)$, which means that x_0 lies outside of the intervals $\left[0, \frac{1}{2n_2^2}\left(1 - \frac{1}{n_2}\right)\right]$ and $\left[1 - \frac{1}{2n_2^2}\left(1 - \frac{1}{n_2}\right), 1\right]$, implying that $f_n(x_0) \leq \frac{1}{2}$ for $\forall n \geq n_2$. For $k = 3$, besides the endpoints, there appears the point $\frac{1}{2}$ used in the construction of the functions $f_n(x)$, $n \geq 3$. Denote the distance between x_0 and $\frac{1}{2}$ by d_2 and choose $n_3 \in \mathbb{N}$ such that $\frac{1}{2n_3^2} < d_2$ and $n_3 \geq n_2$. Then x_0 lies outside the interval $\left[\frac{1}{2} - \frac{1}{2n_3^2}\left(\frac{1}{2} - \frac{1}{n_3}\right), \frac{1}{2} + \frac{1}{2n_3^2}\left(\frac{1}{2} - \frac{1}{n_3}\right)\right]$, as well as outside the intervals $\left[0, \frac{1}{2n_3^2}\left(1 - \frac{1}{n_3}\right)\right]$ and $\left[1 - \frac{1}{2n_3^2}\left(1 - \frac{1}{n_3}\right), 1\right]$, and consequently,

$f_n(x_0) \leq \frac{1}{3}$ for $\forall n \geq n_3$. Let us suppose now that at $(k-1)$th step $f_n(x_0) \leq \frac{1}{k-1}$ for $\forall n \geq n_{k-1}$ and prove that a similar inequality holds for k.

On kth step, a few (at most $k-2$) new points $\frac{p}{k-1}$, $1 \leq p \leq k-2$, associated with the denominator $q = k-1$, appear in the construction of the functions $f_n(x)$, $n \geq k$. Denote the smallest distance between x_0 and each of these points by d_k and choose $n_k \in \mathbb{N}$ such that $\frac{1}{2n_k^2} < d_k$ and $n_k \geq n_{k-1}$. Therefore, x_0 lies outside of each of the intervals around the points $\frac{p}{q}$, $q < k$, (the intervals used in the first two sentences of the definition of $f_n(x)$) and consequently, $f_n(x_0) \leq \frac{1}{k}$ for $\forall n \geq n_k$. Thus, the last property is proved. Passing to the limit, as k approaches infinity, in the inequality $f_n(x_0) \leq \frac{1}{k}$, we obtain $\lim_{n\to\infty} f_n(x_0) = 0 = f(x_0)$ at all irrational points in $(0,1)$. Thus, the limit function has the following form: $f(x) = \begin{cases} \frac{1}{q}, & x = \frac{p}{q} \in \mathbb{Q} \\ 0, & x \in \mathbb{I} \end{cases}$. This is the Riemann function, which is discontinuous at all the rational points (and continuous at all the irrational points) in $[0,1]$. Therefore, the limit function has infinitely many discontinuity points on $[0,1]$, although each of the functions $f_n(x)$ is continuous on $[0,1]$.

Remark 1. A similar example can be given for a functions defined on $\mathbb{R}$, if we extend the definition of $f_n(x)$ periodically (with the period 1) on $\mathbb{R}$.

Remark 2. Of course, the convergence of the considered sequence is not uniform on $[0,1]$ (or on any subinterval in $[0,1]$), since otherwise the limit function would be continuous (see the corresponding theorem on the continuity of the limit function of a uniformly convergent sequence). The fact that the convergence is nonuniform on $[0,1]$ is also easy to check by choosing the irrational points $x_n \in \left(0, \frac{1}{2n^3}\left(1 - \frac{1}{n}\right)\right)$ $\forall n \in \mathbb{N}$, and finding the limit

$$|f_n(x_n) - f(x_n)| = f_n(x_n) \geq f_n\left(\frac{1}{2n^3}\left(1 - \frac{1}{n}\right)\right)$$
$$= 1 - 2n^2 \frac{1}{2n^3}\left(1 - \frac{1}{n}\right) = 1 - \frac{1}{n}\left(1 - \frac{1}{n}\right) \xrightarrow[n\to\infty]{} 1 \neq 0.$$

2.4 Convergence and Uniform Continuity

Example 27. Each of functions $f_n(x)$ is uniformly continuous on X and the sequence $f_n(x)$ converges on X to $f(x)$, but the limit function is discontinuous on X.

Solution

Each function of the sequence $f_n(x) = x^n$ is continuous on $X = [0, 1]$ and, according to the Cantor theorem, is uniformly continuous on $[0, 1]$. However, the limit function $f(x) = \lim_{n \to \infty} f_n(x) = \lim_{n \to \infty} x^n = \begin{cases} 0, x \in [0, 1) \\ 1, x = 1 \end{cases}$ is discontinuous

at $x = 1$. Note that this sequence converges nonuniformly on $[0, 1]$: choosing $x_n = \left(1 - \frac{1}{n}\right) \in [0, 1]$ for each $n \in \mathbb{N}$, one has

$$|f_n(x_n) - f(x_n)| = x_n^n = \left(1 - \frac{1}{n}\right)^n \underset{n \to \infty}{\to} e^{-1} \neq 0.$$

Remark. For a series, we have an analogous example: a series of uniformly continuous functions converges on X, but the sum of the series is a discontinuous on X function. In fact, consider $\sum_{n=1}^{\infty}(1 - x)x^n$ on $X = [0, 1]$. Each term $u_n(x) = (1 - x)x^n$ is a continuous function on the compact set $X = [0, 1]$ and, consequently, is uniformly continuous on $[0, 1]$. Let us find the sum of this series $f(x)$. At the end points, one has $u_n(0) = u_n(1) = 0, \forall n \in \mathbb{N}$ and, therefore, $f(0) = f(1) = \sum_{n=1}^{\infty} 0 = 0$. For any $x \in (0, 1)$, the series is a convergent geometric series with the sum $f(x) = \sum_{n=1}^{\infty}(1 - x)x^n = \frac{(1-x)x}{1-x} = x$. Thus, $f(x) = \begin{cases} x, x \in [0, 1) \\ 0, x = 1 \end{cases}$, and this function has a discontinuity at $x = 1$. Checking the nature of the convergence, one can evaluate the residual at the points $x_n = 1 - \frac{1}{n+1}, \forall n \in \mathbb{N}$

$$|r_n(x_n)| = \left| \sum_{k=n+1}^{\infty}(1 - x_n)x_n^k \right| = \frac{(1 - x_n)x_n^{n+1}}{1 - x_n} = \left(1 - \frac{1}{n+1}\right)^{n+1} \underset{n \to \infty}{\to} e^{-1} \neq 0$$

and find out that the convergence is not uniform.

Example 28. A sequence of uniformly continuous on X functions $f_n(x)$ converges on X to a continuous function $f(x)$, but the limit function is not uniformly continuous on X.

Solution

Consider the sequence $f_n(x) = \frac{n}{nx+1}$ on $X = (0, 1]$. The functions $f_n(x)$ are continuous on a compact $X = [0, 1]$ for $\forall n \in \mathbb{N}$ and, according to the Cantor theorem, are uniformly continuous on $[0, 1]$ and, consequently, on $(0, 1]$. The limit function

$$f(x) = \lim_{n \to \infty} f_n(x) = \lim_{n \to \infty} \frac{n}{nx + 1} = \frac{1}{x}, \forall x \in (0, 1]$$

is continuous on $(0, 1]$, but this continuity is not uniform. In fact, for the pair of points $x_k = \frac{1}{k}$ and $x_{k+1} = \frac{1}{k+1}$, both in $(0, 1]$, $\forall k \in \mathbb{N}$, one has $|x_k - x_{k+1}| =$

$\left|\frac{1}{k} - \frac{1}{k+1}\right| \underset{k\to\infty}{\to} 0$, but $|f(x_k) - f(x_{k+1})| = |k - (k+1)| = 1 \underset{k\to\infty}{\nrightarrow} 0$. Let us verify the nature of the convergence of $f_n(x)$ on $(0, 1]$. If one picks up $x_n = \frac{1}{n} \in (0, 1]$ for each $n \in \mathbb{N}$, then

$$|f_n(x_n) - f(x_n)| = \left|\frac{n}{nx_n + 1} - \frac{1}{x_n}\right| = \left|\frac{n}{2} - n\right| = \frac{n}{2} \underset{n\to\infty}{\to} \infty,$$

which means that the convergence is not uniform on $(0, 1]$.

Remark. An analogous example for a series is as follows: a series of uniformly continuous functions converges on X to a continuous function $f(x)$, but $f(x)$ is not uniformly continuous on X. For a counterexample, consider $\sum_{n=0}^{\infty} x^n$ on $X = [0, 1)$. Each function $u_n(x) = x^n$ is continuous on the compact set $X = [0, 1]$ and, consequently, is uniformly continuous on $[0, 1]$, which implies the uniform continuity on $[0, 1)$. The series converges (as a geometric series with the ratio in $[0, 1)$) to the function $f(x) = \sum_{n=0}^{\infty} x^n = \frac{1}{1-x}$, which is continuous on $[0, 1)$. However, this continuity is not uniform: for the pair of points $x_k = 1 - \frac{1}{k}$ and $x_{k+1} = 1 - \frac{1}{k+1}$, both in $[0, 1)$, $\forall k \in \mathbb{N}$, one has $|x_k - x_{k+1}| = \left|1 - \frac{1}{k} - 1 + \frac{1}{k+1}\right| = \left|\frac{1}{k} - \frac{1}{k+1}\right| \underset{k\to\infty}{\to} 0$, but $|f(x_k) - f(x_{k+1})| = \left|\frac{1}{1-x_k} - \frac{1}{1-x_{k+1}}\right| = |k + 1 - k| = 1 \underset{k\to\infty}{\nrightarrow} 0$. Note that the series does not converge uniformly on $[0, 1)$, because the general term does not converge uniformly to 0: choosing $x_n = \frac{1}{\sqrt[n]{2}} \in [0, 1), \forall n \in \mathbb{N}$, one gets

$$|u_n(x_n)| = \left(\frac{1}{\sqrt[n]{2}}\right)^n = \frac{1}{2} \underset{n\to\infty}{\nrightarrow} 0.$$

Remark to Examples 27 and 28. It is well known that the limit function/sum of a series is uniformly continuous on a set X if the terms of the sequence/series are uniformly continuous on X and the convergence is uniform on X. In the two above examples, the violation of the condition of the uniform convergence causes an absence of uniform continuity of the limit function/sum of the series (and even discontinuity in Example 27). At the same time, the condition of the uniform convergence is sufficient and not necessary as shown in the next example.

Example 29. A sequence of uniformly continuous on X functions $f_n(x)$ converges on X to an uniformly continuous function $f(x)$, but this convergence is nonuniform on X.

Solution
Consider the sequence $f_n(x) = x^n$ on $X = [0, 1)$. It was shown in Example 27 that each $f_n(x)$ is uniformly continuous on $[0, 1]$ that implies the uniform continuity

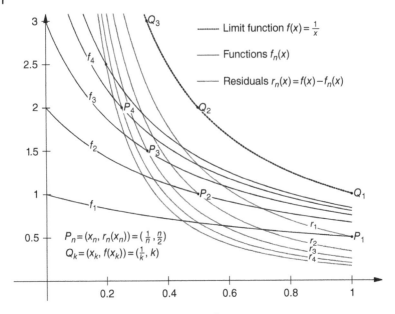

Figure 2.17 Example 28, sequence $f_n(x) = \frac{n}{nx+1}$.

on $[0, 1)$. The limit function $f(x) = \lim_{n\to\infty} f_n(x) = 0, \forall x \in [0, 1)$ is uniformly continuous on $[0, 1)$. However, the convergence is nonuniform on $[0, 1)$ (just apply the same reasoning as in Example 27).

Remark. The corresponding formulation for a series: a series of uniformly continuous functions converges on X to an uniformly continuous function $f(x)$, but this convergence is nonuniform. For a counterexample, use the series $\sum_{n=1}^{\infty}(1 - x)x^n$ on $X = [0, 1)$. It was shown in Remark to Example 27 that the functions $u_n(x) = (1 - x)x^n$ are uniformly continuous on $[0, 1]$, that implies the uniform continuity on $[0, 1)$, and also that the sum of the series is $f(x) = x, \forall x \in [0, 1)$, which is a uniform continuous function on $[0, 1)$. However, using the same reasoning as in the mentioned Remark, one can prove that the convergence is nonuniform on $[0, 1)$.

Example 30. A sequence of nonuniformly continuous on X functions converges on X, but the limit function is uniformly continuous on X.

Solution
Consider the sequence $f_n(x) = x^{\frac{n+1}{n}}$ on $X = [1, +\infty)$. Each function $f_n(x)$ is continuous, but this continuity is nonuniform on $[1, +\infty)$. Indeed, the distance

between the two points $x_k = k^{\frac{n}{n+1}}$ and $x_{k+1} = (k+1)^{\frac{n}{n+1}}$ converges to 0 as k approaches ∞:

$$\lim_{k\to\infty} |x_{k+1} - x_k| = \lim_{t\to+\infty} \left((t+1)^{\frac{n}{n+1}} - t^{\frac{n}{n+1}} \right)$$

$$= \lim_{t\to+\infty} t^{\frac{n}{n+1}} \left(\left(1+\frac{1}{t}\right)^{\frac{n}{n+1}} - 1 \right)$$

$$= \lim_{t\to+\infty} \frac{\left(1+\frac{1}{t}\right)^{\frac{n}{n+1}} - 1}{t^{-\frac{n}{n+1}}} = \lim_{t\to+\infty} \frac{\frac{n}{n+1}\left(1+\frac{1}{t}\right)^{\frac{n}{n+1}-1}\frac{1}{t^2}}{-\frac{n}{n+1}t^{-\frac{n}{n+1}-1}}$$

$$= \lim_{t\to+\infty} \left(1+\frac{1}{t}\right)^{-\frac{1}{n+1}} t^{-\frac{1}{n+1}} = 1\cdot 0 = 0.$$

(Here, we apply the change of the discrete variable k by the continuous variable t, then transform an indeterminate form $\infty - \infty$ to $\infty \cdot 0$ and to $\frac{0}{0}$, consecutively, and finally apply l'Hospital's rule to solve the last indeterminate form.) At the same time, the distance between the respective ordinates does not approaches 0:

$$|f_n(x_{k+1}) - f_n(x_k)| = |k+1-k| = 1 \underset{k\to\infty}{\nrightarrow} 0,$$

which shows that the continuity is nonuniform on $[1, +\infty)$. On the other hand, the limit function

$$f(x) = \lim_{n\to\infty} f_n(x) = \lim_{n\to\infty} x^{\frac{n+1}{n}} = x, \forall x \in [1, +\infty)$$

is uniformly continuous on $[1, +\infty)$.

As for the character of the convergence, choosing $x_n = 2^n \in [1, +\infty)$ for each $n \in \mathbb{N}$, one obtains

$$|f_n(x_n) - f(x_n)| = |2^{n+1} - 2^n| = 2^n \underset{n\to\infty}{\to} \infty,$$

which means that the convergence is not uniform on $[1, +\infty)$.

Remark 1. An analogous example for a series: a series of functions $u_n(x)$, each of which is nonuniformly continuous on X, converges on X, but the sum of series is uniformly continuous on X. For a counterexample, let us construct the series $\sum_{n=1}^{\infty} u_n(x)$ with the partial sums equal on $X = [1, +\infty)$ to $f_n(x)$ in the given counterexample. It is easy to show that the terms of such a series can be chosen as $u_1(x) = x^2$, $u_n(x) = x^{\frac{n+1}{n}} - x^{\frac{n}{n-1}}, \forall n \geq 2$. To prove that the continuity of each function $u_n(x)$, $n \geq 2$ is nonuniform on $[1, +\infty)$, let us use the same points $x_k = k^{\frac{n}{n+1}}$ and $x_{k+1} = (k+1)^{\frac{n}{n+1}}$ as for the sequence. The distance between x_k and

x_{k+1} approaches 0, but the respective distance between $u_n(x_{k+1})$ and $u_n(x_k)$ does not. In fact,

$$\lim_{k\to\infty} |u_n(x_{k+1}) - u_n(x_k)| = \lim_{k\to\infty} \left| k+1 - (k+1)^{\frac{n^2}{n^2-1}} - k + k^{\frac{n^2}{n^2-1}} \right|$$

and, applying the same techniques as in the evaluation of $\lim_{k\to\infty} |x_{k+1} - x_k|$, one can calculate the principal part of the limit as follows:

$$\lim_{k\to\infty} \left((k+1)^{\frac{n^2}{n^2-1}} - k^{\frac{n^2}{n^2-1}} \right) = \lim_{t\to+\infty} \left((t+1)^{\frac{n^2}{n^2-1}} - t^{\frac{n^2}{n^2-1}} \right)$$

$$= \lim_{t\to+\infty} \frac{\left(1+\frac{1}{t}\right)^{\frac{n^2}{n^2-1}} - 1}{t^{-\frac{n^2}{n^2-1}}}$$

$$= \lim_{t\to+\infty} \left(1+\frac{1}{t}\right)^{\frac{1}{n^2-1}} t^{\frac{1}{n^2-1}} = \infty,$$

that is, $\lim_{k\to\infty} |u_n(x_{k+1}) - u_n(x_k)| = \infty$, which means a nonuniform continuity of $u_n(x)$ on $[1, +\infty)$. (Of course, the proof of a nonuniform continuity of $u_1(x)$ is trivial, for example, by using the points $\tilde{x}_k = \sqrt{k}$ and $\tilde{x}_{k+1} = \sqrt{k+1}$.)

On the other hand, it was shown in the above counterexample for the sequence that the series converges nonuniformly on $[1, +\infty)$ to the function $f(x) = x$, which is uniformly continuous on $[1, +\infty)$.

Remark 2. It can also happen that a sequence of nonuniformly continuous on X functions converges on X to a nonuniformly continuous function. For such an example, consider the sequence $f_n(x) = x^{\frac{2n+1}{n}}$ on $X = [1, +\infty)$. Let us show that each function $f_n(x)$ is nonuniformly continuous on $[1, +\infty)$. The distance between the two points $x_k = k^{\frac{n}{2n+1}}$ and $x_{k+1} = (k+1)^{\frac{n}{2n+1}}$ approaches 0 as k approaches ∞:

$$\lim_{k\to\infty} |x_{k+1} - x_k| = \lim_{t\to+\infty} \left((t+1)^{\frac{n}{2n+1}} - t^{\frac{n}{2n+1}} \right)$$

$$= \lim_{t\to+\infty} \frac{\left(1+\frac{1}{t}\right)^{\frac{n}{2n+1}} - 1}{t^{-\frac{n}{2n+1}}}$$

$$= \lim_{t\to+\infty} \frac{\frac{n}{2n+1}\left(1+\frac{1}{t}\right)^{\frac{n}{2n+1}-1}\frac{1}{t^2}}{\frac{n}{2n+1}t^{-\frac{n}{2n+1}-1}}$$

$$= \lim_{t\to+\infty} \left(1+\frac{1}{t}\right)^{-\frac{n+1}{2n+1}} t^{-\frac{n+1}{2n+1}} = 1\cdot 0 = 0.$$

(Here, we apply the same techniques as for the evaluation of a similar limit in the first counterexample.) At the same time,

$$|f_n(x_{k+1}) - f_n(x_k)| = |k + 1 - k| = 1 \underset{k\to\infty}{\nrightarrow} 0,$$

that is, the continuity is nonuniform on $[1, +\infty)$. The limit function

$$f(x) = \lim_{n\to\infty} f_n(x) = \lim_{n\to\infty} x^{\frac{2n+1}{n}} = x^2, \forall x \in [1, +\infty)$$

is also continuous nonuniformly on $[1, +\infty)$ (as for $u_1(x)$, one can use the points $\tilde{x}_k = \sqrt{k}$ and $\tilde{x}_{k+1} = \sqrt{k+1}$). Finally, the convergence is not uniform on $[1, +\infty)$, since for the points $x_n = 2^n \in [1, +\infty), \forall n \in \mathbb{N}$, one has

$$|f_n(x_n) - f(x_n)| = |2^{2n+1} - 2^{2n}| = 2^{2n} \underset{n\to\infty}{\to} \infty.$$

A similar counterexample for a series can be constructed using $\sum_{n=1}^{+\infty} u_n(x)$ on $X = [1, +\infty)$ for which the functions $f_n(x) = x^{\frac{2n+1}{n}}$ represent the partial sums: $u_1(x) = f_1(x) = x^3, u_n(x) = x^{\frac{2n+1}{n}} - x^{\frac{2n-1}{n-1}}, \forall n \geq 2$. As shown above, the sum of this series $f(x) = \lim_{n\to\infty} f_n(x) = x^2$ is nonuniformly continuous on $[1, +\infty)$ function and the convergence to $f(x)$ is nonuniform on $[1, +\infty)$. It remains to see that the terms of the series are nonuniformly continuous on $[1, +\infty)$. To this end, choosing the same points $x_k = k^{\frac{n}{2n+1}}$ and $x_{k+1} = (k+1)^{\frac{n}{2n+1}}, \forall n > 1$, one obtains

$$|u_n(x_{k+1}) - u_n(x_k)| = |k + 1 - (k+1)^{\frac{n(2n-1)}{(n-1)(2n+1)}} - k + k^{\frac{n(2n-1)}{(n-1)(2n+1)}}| \underset{k\to\infty}{\to} +\infty,$$

since

$$\lim_{k\to\infty} \left((k+1)^{\frac{n(2n-1)}{(n-1)(2n+1)}} - k^{\frac{n(2n-1)}{(n-1)(2n+1)}} \right) = \lim_{t\to+\infty} \frac{\left(1 + \frac{1}{t}\right)^{\frac{n(2n-1)}{(n-1)(2n+1)}} - 1}{t^{-\frac{n(2n-1)}{(n-1)(2n+1)}}}$$

$$= \lim_{t\to+\infty} \frac{\frac{2n^2-n}{2n^2-n-1}\left(1 + \frac{1}{t}\right)^{\frac{2n^2-n}{2n^2-n-1}-1} \frac{1}{t^2}}{\frac{2n^2-n}{2n^2-n-1} t^{-\frac{2n^2-n}{2n^2-n-1}-1}}$$

$$= \lim_{t\to+\infty} \left(1 + \frac{1}{t}\right)^{\frac{1}{2n^2-n-1}} t^{\frac{1}{2n^2-n-1}} = +\infty.$$

Example 31. A sequence of nonuniformly continuous on X functions converges uniformly on X, but nevertheless the limit function is uniformly continuous on X.

Solution

Each function $f_n(x) = \frac{1}{n} \sin \frac{1}{x}$, $\forall n \in \mathbb{N}$, is continuous on $X = (0, 1]$, but the continuity is not uniform since the distance between the two points $x_k = \frac{2}{\pi(4k+1)}$ and $x'_k = \frac{2}{\pi(4k-1)}$, both in $(0, 1]$, $\forall k \in \mathbb{N}$, approaches 0 as k approaches ∞, but

$$|f_n(x_k) - f_n(x'_k)| = \frac{1}{n}\left|\sin\frac{1}{x_k} - \sin\frac{1}{x'_k}\right| = \frac{1}{n}|1 - (-1)| = \frac{2}{n} \xrightarrow{k\to\infty} \frac{2}{n} \neq 0.$$

The limit function is zero: $f(x) = \lim_{n\to\infty} \frac{1}{n}\sin\frac{1}{x} = 0$, $\forall x \in (0, 1]$, and the convergence is uniform since for all $x \in (0, 1]$ simultaneously we have $|f_n(x) - f(x)| = \frac{1}{n}\left|\sin\frac{1}{x}\right| \le \frac{1}{n} \xrightarrow{n\to\infty} 0$. However, differently from $f_n(x)$, the limit function is uniformly continuous on $(0, 1]$.

Remark 1. An analogous example for a series: a series of nonuniformly continuous on X functions converges uniformly on X, but nevertheless the sum of the series is uniformly continuous on X. A counterexample can be constructed using the series $\sum_{n=0}^{\infty} u_n(x)$ with the following terms: $u_0(x) = -\sin\frac{1}{x}$, $u_n(x) = \frac{1}{n(n+1)}\sin\frac{1}{x}$, $\forall n \in \mathbb{N}$, defined on $X = (0, 1]$. Each function $u_n(x)$ is continuous on $(0, 1]$, but the continuity is nonuniform (it can be shown in the same way as for $f_n(x)$). The uniform convergence on $(0, 1]$ follows from the Weierstrass test: $|u_n(x)| \le \frac{1}{n(n+1)} < \frac{1}{n^2}$, $\forall x \in (0, 1]$, and the p-series $\sum \frac{1}{n^2}$ is convergent. Therefore, the sum of this series is continuous due to the theorem on uniform convergent series of continuous functions. However, to evaluate if this continuity is uniform, we should find an explicit form of the sum of the series. The partial sums can be easily calculated using the telescoping form of this series:

$$\sum_{k=0}^{n} u_k(x) = -\sin\frac{1}{x} + \sum_{k=1}^{n}\left(\frac{1}{k} - \frac{1}{k+1}\right)\sin\frac{1}{x}$$

$$= \sin\frac{1}{x}\left(-1 + 1 - \frac{1}{2} + \frac{1}{2} - \frac{1}{3} + \cdots + \frac{1}{n} - \frac{1}{n+1}\right)$$

$$= -\frac{1}{n+1}\sin\frac{1}{x},$$

and, consequently, the sum is

$$f(x) = \lim_{n\to\infty} f_n(x) = -\lim_{n\to\infty}\frac{1}{n+1}\sin\frac{1}{x} = 0, \forall x \in (0, 1].$$

Therefore, the sum is uniformly continuous function on $(0, 1]$.

Remark 2. A sequence of nonuniformly continuous on X functions can also converge uniformly on X to a nonuniformly continuous function. For such an example, consider the sequence $f_n(x) = \frac{1}{x} + \frac{1}{n}$ on $X = (0, +\infty)$. Each function $f_n(x)$ is continuous, but nonuniformly on $(0, +\infty)$ since the distance between

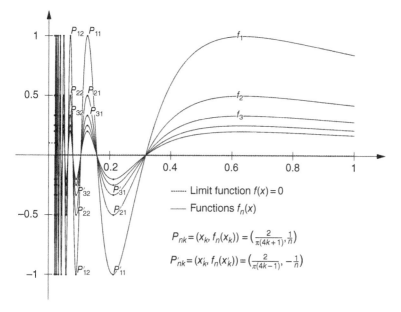

Figure 2.18 Example 31, sequence $f_n(x) = \frac{1}{n} \sin \frac{1}{x}$.

the pair of the points $x_k = \frac{1}{k}$ and $x_{k+1} = \frac{1}{k+1}$ approaches 0 as k approaches ∞, while

$$|f_n(x_{k+1}) - f_n(x_k)| = |k + 1 + \frac{1}{n} - k - \frac{1}{n}| = 1 \underset{k\to\infty}{\nrightarrow} 0.$$

The limit function

$$f(x) = \lim_{n\to\infty} f_n(x) = \lim_{n\to\infty} \left(\frac{1}{x} + \frac{1}{n} \right) = \frac{1}{x}, \forall x \in (0, +\infty)$$

is also continuous nonuniformly on $(0, +\infty)$, which can be proved using the same points x_k and x_{k+1}. At the same time, the convergence is uniform on $(0, +\infty)$, since for all $x \in (0, +\infty)$ simultaneously one has

$$|f_n(x) - f(x)| = \left| \frac{1}{x} + \frac{1}{n} - \frac{1}{x} \right| = \frac{1}{n} \underset{n\to\infty}{\to} 0.$$

As for a series, one can use the series $\sum_{n=1}^{+\infty} u_n(x)$ with $u_n(x) = \frac{1}{n(n+1)} \sin \frac{1}{x}$ defined on $X = (0, 1]$ for a similar counterexample.

Remark to Examples 30 and 31. These two examples show that if the terms of a sequence/series are nonuniformly continuous functions and the limit function/sum of the series is continuous, then the nature of the convergence has no effect on the nature of the continuity of the limit function/sum of the series.

Exercises

1 Show that the sequence of the functions $f_n(x) = \min(n, x)$ on $X = [0, +\infty)$ provides a counterexample to Example 1. Investigate the type of the convergence of $f_n(x)$ on X.

2 Verify that each of the following sequences

a) $f_n(x) = \begin{cases} n \sin nx, x \in [0, \frac{\pi}{n}] \\ 0, x \in (\frac{\pi}{n}, +\infty) \end{cases}$ on $X = [0, +\infty)$

b) $f_n(x) = \begin{cases} n^2, x \in (0, \frac{1}{n}) \\ 0, x = 0, x \in [\frac{1}{n}, 1] \end{cases}$ on $X = [0, 1]$

c) $f_n(x) = \begin{cases} n^3 x \arctan \frac{1}{n^2 x}, x \in (0, \frac{1}{n}) \\ 0, x = 0, x \in [\frac{1}{n}, +\infty) \end{cases}$ on $X = [0, +\infty)$

gives a counterexample to Example 2.

3 Use the sequence $f_n(x) = \begin{cases} \frac{1}{n^2 x^2}, x \neq 0 \\ 0, x = 0 \end{cases}$ on $X = [0, +\infty)$ to illustrate Example 3. Analyze the form of convergence of $f_n(x)$ on X.

4 Show that the sequences $f_n(x) = \begin{cases} \frac{n^2+1}{n^2 x^2}, 0 < x \leq 1 \\ 0, x = 0 \end{cases}$ on $X = [0, 1]$ and $f_n(x) = \frac{2n^2-1}{n^2} x^2$ on $X = \mathbb{R}$ provide counterexamples to Example 4 and its Remark, respectively.

5 Verify that the sequence $f_n(x) = \begin{cases} \ln \frac{n^2+1}{n^2 x^2}, x \neq 0 \\ 0, x = 0 \end{cases}$ on $X = \mathbb{R}$ can be used for Example 5.

6 Show that sequence $f_n(x) = \begin{cases} nx \sin \frac{1}{nx}, x \neq 0 \\ 1, x = 0 \end{cases}$ in Example 6 can be also considered on $X = \mathbb{R}$. Alternatively, use the sequence $f_n(x) = \begin{cases} n^2 x \arctan \frac{1}{n^2 x}, x \neq 0 \\ 1, x = 0 \end{cases}$ on $X = \mathbb{R}$ to construct another counterexample in Example 6.

7 Provide a counterexample to the statement on sequences in Remark 1 of Example 9 such that the left-hand iterated limit $\lim_{x \to x_0} \lim_{n \to \infty} f_n(x)$ does not exist. (Hint: consider $f_n(x) = \begin{cases} x - x^{n+1}, x \in \mathbb{Q} \cap X \\ 0, x \in \mathbb{I} \cap X \end{cases}$ on $X = (0, 1)$ with the

limit point $x_0 = 1$ and prove that $\lim\limits_{n\to\infty} \lim\limits_{x\to 1_-} f_n(x) = \lim\limits_{n\to\infty} 0 = 0$, but $f(x) =$

$\lim\limits_{n\to\infty} f_n(x) = \begin{cases} x, x \in \mathbb{Q} \cap X \\ 0, x \in \mathbb{I} \cap X \end{cases}$ and, consequently, $\lim\limits_{x\to 1_-} f(x)$ does not exist.)

8 Let function $f(x, y)$ be defined on $X \times Y$. Find the limit function $\varphi(x) = \lim\limits_{y\to y_0} f(x, y)$. Verify if there exists $\lim\limits_{x\to x_0} \varphi(x)$ and the possibility of the interchange of the limit order. Investigate the character of the convergence of $f(x, y)$ to $\varphi(x)$ on X. Do this for the following functions:

a) $f(x, y) = \frac{x^4 - y^4}{x^4 + y^4}$, $X \times Y = (0, 1] \times (0, 1]$, $x_0 = 0$, $y_0 = 0$

b) $f(x, y) = \sqrt{\frac{x}{y}} \sin\frac{y}{x}$, $X \times Y = (0, +\infty) \times (0, +\infty)$, $x_0 = 0$, $y_0 = 0$

c) $f(x, y) = \frac{x}{y} \sin\frac{y}{x}$, $X \times Y = (0, +\infty) \times (0, +\infty)$, $x_0 = 0$, $y_0 = 0$

d) $f(x, y) = \begin{cases} \frac{5x^2 - 2y^2}{x^2 + y^2}, x \in \mathbb{Q} \\ -2, x \in \mathbb{I} \end{cases}$, $X \times Y = (0, 1] \times (0, 1]$, $x_0 = 0$, $y_0 = 0$

e) $f(x, y) = \begin{cases} \frac{5x^2 - 2y^2}{x^2 + y^2}, y \in \mathbb{Q} \\ 5, y \in \mathbb{I} \end{cases}$, $X \times Y = (0, 1] \times (0, 1]$, $x_0 = 0$, $y_0 = 0$

f) $f(x, y) = \frac{x^2}{y^2} e^{-\frac{x^2}{y^2}}$, $X \times Y = \mathbb{R} \times (0, +\infty)$, $x_0 = 0$, $y_0 = 0$

g) $f(x, y) = \arctan\frac{x}{y}$, $X \times Y = (0, +\infty) \times (0, +\infty)$, $x_0 = 0$, $y_0 = 0$

h) $f(x, y) = x \sin\frac{y}{x}$, $X \times Y = (0, +\infty) \times (0, +\infty)$, $x_0 = 0$, $y_0 = 0$

i) $f(x, y) = \frac{x}{y} \arctan\frac{y}{x}$, $X \times Y = (0, +\infty) \times (0, +\infty)$, $x_0 = 0$, $y_0 = 0$.

Derive a conclusion on a paper of the uniform convergence of $f(x, y)$ on X for the possibility to interchange the limit order. Compare with Examples 7–10.

9 Find the limit function $f(x) = \lim\limits_{n\to\infty} f_n(x)$ of the sequence $f_n(x)$ defined on X and calculate $\lim\limits_{x\to x_0} f(x)$ if this limit exists. Analyze the nature of the convergence of $f_n(x)$ on X and check if the relation $\lim\limits_{x\to x_0} \lim\limits_{n\to\infty} f_n(x) = \lim\limits_{n\to\infty} \lim\limits_{x\to x_0} f_n(x)$ holds. Do this for the following sequences:

a) $f_n(x) = (-1)^n \frac{\arctan nx}{nx}$, $X = (0, +\infty)$, $x_0 = 0$

b) $f_n(x) = \frac{n^2 x^2}{n^4 x^4 + 1}$, $X = \mathbb{R}$, $x_0 = 0$

c) $f_n(x) = \frac{nx}{1 + n^2 x^2}$, $X = (0, 1]$, $x_0 = 0$

d) $f_n(x) = \begin{cases} \frac{3n^2 x^2 - 4}{n^2 x^2 + 2}, x \in \mathbb{Q} \\ -2, x \in \mathbb{I} \end{cases}$, $X = \mathbb{R}$, $x_0 = 0$

e) $f_n(x) = \frac{2n^2 x^2 - 3}{n^2 x^2 + 1}$, $X = (0, +\infty)$, $x_0 = 0$

f) $f_n(x) = n^2 x e^{-n^2 x}$, $X = (0, +\infty)$, $x_0 = 0$

g) $f_n(x) = \begin{cases} \frac{n^2 x^2 - 1}{n^2 x^2 + 1}, x \in \mathbb{Q} \\ -1, x \in \mathbb{I} \end{cases}$, $X = (0, 1)$, $x_0 = 0$

h) $f_n(x) = \begin{cases} 1 - (1-x)^n, x \in \mathbb{Q} \\ 0, x \in \mathbb{I} \end{cases}$, $X = (0, 1)$, $x_0 = 1$

i) $f_n(x) = n^2 x^2 e^{-n^2 x}$, $X = (0, +\infty)$, $x_0 = 0$

j) $f_n(x) = nx \sin \frac{1}{nx}$, $X = (0, +\infty)$, $x_0 = 0$.

Compare with Examples 7–10.

10 Verify if series $\sum u_n(x)$ converges on X and, if so, find the sum $f(x)$ of this series. Calculate $\lim_{x \to x_0} f(x)$ if this limit exists. Analyze the nature of the convergence of the series and the possibility to calculate the limit by term-by-term rule. Do this for the following series:

a) $\sum_{n=1}^{\infty} (-1)^{n-1} nx^n$, $X = (-1, 1)$, $x_0 = 1$

b) $\sum_{n=0}^{\infty} u_n(x)$, $u_n(x) = \begin{cases} x(1-x)^n, x \in \mathbb{Q} \\ 0, x \in \mathbb{I} \end{cases}$, $X = (0, 1)$, $x_0 = 0$

c) $\sum_{n=1}^{\infty} \left(\frac{x^2}{1+x^2}\right)^n$, $X = \mathbb{R}$, $x_0 = 0$

d) $\sum_{n=1}^{\infty} (-1)^{n-1} n(1-x)x^n$, $X = [0, 1)$, $x_0 = 1$

e) $\sum_{n=0}^{\infty} \frac{x^2}{(1+x^2)^n}$, $X = (0, 1]$, $x_0 = 0$

f) $\sum_{n=0}^{\infty} (-1)^n \frac{x^2}{(1+x^2)^n}$, $X = \mathbb{R}$, $x_0 = 0$

g) $\sum_{n=1}^{\infty} (-1)^{n-1} \left(\frac{x^2}{1+x^2}\right)^n$, $X = \mathbb{R}$, $x_0 = 0$

h) $\sum_{n=0}^{\infty} \frac{x}{(1+x^2)^n}$, $X = (0, +\infty)$, $x_0 = 0$.

Compare with Examples 7–10.

11 Investigate the continuity of a function $f(x, y)$, sequence $f_n(x)$, terms of series $\sum u_n(x)$, and their limit functions and sums of series on a set X in the following cases:

a) $f(x, y) = \begin{cases} y, 0 < x < y \\ 0, x = 0, y \leq x \leq 1 \end{cases}$, $X \times Y = [0, 1] \times (0, 1]$, $y_0 = 0$

b) $f(x, y) = \sin y \cdot D(x)$, $X \times Y = \mathbb{R} \times (0, \frac{\pi}{2}]$, $y_0 = 0$ (recall that $D(x)$ denotes Dirichlet's function)

c) $f(x, y) = \begin{cases} y^2, 0 < x < y \\ 0, x = 0, y \leq x \leq 2 \end{cases}$, $X \times Y = [0, 2] \times (0, 2]$, $y_0 = 0$

d) $f(x, y) = \frac{\sin y}{y} \cdot D(x)$, $X \times Y = \mathbb{R} \times (0, +\infty)$, $y_0 = 0$

e) $f(x, y) = \text{sgn } x + y^2$, $X \times Y = \mathbb{R} \times \mathbb{R}$, $y_0 = 0$

f) $f(x, y) = D(x) + \cos y$, $X \times Y = \mathbb{R} \times \mathbb{R}$, $y_0 = 0$

g) $f(x, y) = yR(x)$, $X \times Y = [0, 1] \times (0, 1]$, $y_0 = 0$ (recall that $R(x)$ denotes the Riemann function)

h) $f_n(x) = D(x) + \frac{1}{n^2}$, $X = \mathbb{R}$

i) $f_n(x) = \frac{1}{n} \text{sgn } x$, $X = \mathbb{R}$

j) $f_n(x) = \begin{cases} \frac{1}{n}, n - \frac{1}{n} < x < n + \frac{1}{n} \\ 0, \text{otherwise} \end{cases}$, $X = [0, +\infty)$

k) $f_n(x) = \frac{1}{n^2}D(x)$, $X = \mathbb{R}$

l) $f_n(x) = \frac{1}{n}R(x)$, $X = [0, 1]$

m) $f_n(x) = \begin{cases} \frac{1}{n^2}, 0 < x < \frac{1}{n} \\ 0, x = 0, \frac{1}{n} \leq x \leq 1 \end{cases}$, $X = [0, 1]$

n) $f_n(x) = \frac{2n+1}{n}$ sgn x, $X = \mathbb{R}$

o) $f_n(x) = $ sgn $x + \frac{\cos x}{n^2}$, $X = \mathbb{R}$

p) $\sum_{n=1}^{\infty} u_n(x)$, $u_1(x) = \frac{3}{4}$ sgn x, $u_n(x) = \frac{\text{sgn } x}{1-n^2}$, $\forall n \geq 2$, $X = \mathbb{R}$

q) $\sum_{n=1}^{\infty} u_n(x)$, $u_1(x) = -\frac{3}{4}D(x)$, $u_n(x) = \frac{D(x)}{n^2-1}$, $\forall n \geq 2$, $X = \mathbb{R}$

r) $\sum_{n=2}^{\infty} \frac{D(x)}{n^2-1}$, $X = \mathbb{R}$

s) $\sum_{n=1}^{\infty} \frac{\text{sgn } x}{n^2}$, $X = \mathbb{R}$.

Analyze the nature of the convergence on X. Formulate false statements for which these functions $f(x, y)$, sequences $f_n(x)$, and series $\sum u_n(x)$ provide counterexamples. Compare with Examples 11–13.

12 Find the limit function for function depending on a parameter $f(x, y)$ and sequence of functions $f_n(x)$, find the sum of series $\sum u_n(x)$. Verify if $f(x, y)$, $f_n(x)$ and $u_n(x)$ and their limit functions and sums of series are continuous on a set X. Do this in the following cases:

a) $f(x, y) = \frac{x^2}{y^2}e^{-\frac{x^2}{y^2}}$, $X \times Y = [0, 1] \times (0, 1]$, $y_0 = 0$

b) $f(x, y) = \arctan \frac{x}{y}$, $X \times Y = [0, +\infty) \times (0, 1]$, $y_0 = 0$

c) $f(x, y) = \frac{y^2}{x^2+y^2}$, $X \times Y = \mathbb{R} \times (0, 1]$, $y_0 = 0$

d) $f(x, y) = \frac{xy^3}{x^4+2y^4}$, $X \times Y = [0, 1] \times (0, 1]$, $y_0 = 0$

e) $f_n(x) = \arctan nx$, $X = [0, +\infty)$

f) $f_n(x) = \frac{n^2x^2}{n^4x^4+1}$, $X = \mathbb{R}$

g) $f_n(x) = e^{-n^2x^2}$, $X = \mathbb{R}$

h) $\sum_{n=1}^{\infty} x(1-x)^n$, $X = [0, 1]$

i) $\sum_{n=0}^{\infty} (-1)^n x^n$, $X = (-1, 1)$

j) $\sum_{n=1}^{\infty} (-1)^{n-1} nx^n$, $X = [0, 1)$

k) $\sum_{n=0}^{\infty} \frac{x}{(1+x^2)^n}$, $X = [-1, 1]$

l) $\sum_{n=1}^{\infty} (-1)^{n-1}\left(\frac{x^2}{1+x^2}\right)^n$, $X = \mathbb{R}$.

Analyze the nature of the convergence on X. Formulate false statements for which these functions $f(x, y)$, sequences $f_n(x)$, and series $\sum u_n(x)$ provide counterexamples. Compare with Examples 14 and 15.

13 Verify what conditions of Dini's theorem are violated for the following sequences on the specified sets:

a) $f_n(x) = \cos^n x$, $X = [0, \frac{\pi}{2}]$

b) $f_n(x) = \frac{n+1}{n}$ sgn x, $X = \mathbb{R}$

c) $f_n(x) = \begin{cases} \frac{1}{n^2}, 0 < x < \frac{1}{n} \\ 0, x = 0, \frac{1}{n} \le x \le 1 \end{cases}, X = [0,1]$

d) $f_n(x) = \sin^n x, X = [0, \frac{\pi}{2})$

e) $f_n(x) = \cos^n x, X = (0, \frac{\pi}{2}]$

f) $f_n(x) = D(x) + \frac{(-1)^n}{n^2}, X = \mathbb{R}$

g) $f_n(x) = \tan^n x, X = [0, \frac{\pi}{4}]$

h) $f_n(x) = \tan^n x, X = [0, \frac{\pi}{4})$

i) $f_n(x) = (1-x)^n, X = (0,1]$

j) $f_n(x) = \frac{nx}{n^2 x^2 + 1}, X = [0,1]$

k) $f_n(x) = \begin{cases} \frac{1}{n}, 0 < x < \frac{1}{n} \\ 1, x = 0, \frac{1}{n} \le x \le 1 \end{cases}, X = [0,1]$

l) $f_n(x) = \begin{cases} \frac{n+1}{n}, 0 < x < \frac{1}{n} \\ 0, x = 0, \frac{1}{n} \le x \le 1 \end{cases}, X = [0,1]$

m) $f_n(x) = \begin{cases} \frac{n+1}{n}, 0 < x < \frac{1}{n} \\ 1, x = 0, \frac{1}{n} \le x \le 1 \end{cases}, X = [0,1]$

n) $f_n(x) = n^2 x^2 e^{-n^2 x^2}, X = [0,1]$.

Analyze the nature of the convergence of $f_n(x)$ on X. Formulate false statements for which these sequences provide counterexamples.

14 Verify what conditions of Dini's theorem are violated for the following series on the specified sets:

a) $\sum_{n=1}^{\infty} x(1-x)^n, X = [0,1]$

b) $\sum_{n=1}^{\infty} x(1-x)^n, X = (0,1]$

c) $\sum_{n=1}^{\infty} (-1)^n x(1-x)^n, X = [0,1]$

d) $\sum_{n=0}^{\infty} \frac{x}{(1+x^2)^n}, X = [0,1]$

e) $\sum_{n=1}^{\infty} \left(\frac{x^2}{1+x^2}\right)^n, X = \mathbb{R}$

f) $\sum_{n=1}^{\infty} (-1)^{n-1} nx(1-x)^n, X = [0,1]$

g) $\sum_{n=0}^{\infty} (-1)^n \frac{x^2}{(1+x^2)^n}, X = \mathbb{R}$

h) $\sum_{n=1}^{\infty} (-1)^{n-1} D(x) \frac{x^n}{n}, X = [0,1)$

i) $\sum_{n=0}^{\infty} u_n(x), u_0(x) = \begin{cases} 1, x = 0 \\ 0, x \in (0, \frac{\pi}{2}] \end{cases}$,

$u_n(x) = \begin{cases} \cos x, x \in (\frac{\pi}{2(n+1)}, \frac{\pi}{2n}] \\ 0, x \in [0, \frac{\pi}{2(n+1)}] \cup (\frac{\pi}{2n}, \frac{\pi}{2}] \end{cases}, \forall n \in \mathbb{N}, X = [0, \frac{\pi}{2}]$.

Analyze the nature of the convergence of the series on X. Formulate false statements for which these series provide counterexamples.

15 Find the limit function of a sequence $f_n(x)$ on a given set X:

a) $f_n(x) = \sin\frac{n\pi}{n\pi x + 1}$, $X = (0, 2]$

b) $f_n(x) = \frac{n+1}{nx}$, $X = (0, 1]$

c) $f_n(x) = \frac{1}{n}\cos\frac{1}{x}$, $X = (0, +\infty)$

d) $f_n(x) = \frac{1}{n}x^{\frac{n+1}{n}}$, $X = (0, +\infty)$

e) $f_n(x) = \frac{n+1}{n}x^{\frac{3n+2}{n}}$, $X = (0, +\infty)$

f) $f_n(x) = (1 - x)^n$, $X = [0, 1]$

g) $f_n(x) = (1 - x)^n$, $X = (0, 1]$

h) $f_n(x) = \frac{1}{n}D(x)$, $X = [0, 1]$ ($D(x)$ is Dirichlet's function).

Analyze the nature of the continuity of $f_n(x)$ and the limit functions on X and also the nature of the convergence on X. Formulate false statements for which these sequences provide counterexamples.

16 Find the sum of a series $\sum u_n(x)$ on a given set X:

a) $\sum_{n=1}^{\infty} x(1 - x)^n$, $X = (0, 1]$

b) $\sum_{n=1}^{\infty} x(1 - x)^n$, $X = [0, 1]$

c) $\sum_{n=1}^{\infty} \frac{2}{4n^2-1}\cos\frac{1}{x}$, $X = (0, +\infty)$

d) $\sum_{n=0}^{\infty} u_n(x)$, $u_0(x) = -\cos\frac{1}{x}$, $u_n(x) = \frac{2}{4n^2-1}\cos\frac{1}{x}$, $\forall n \geq 1$, $X = (0, +\infty)$

e) $\sum_{n=1}^{\infty} \frac{1}{n(n+1)x^2}$, $X = (0, 1]$

f) $\sum_{n=0}^{\infty} u_n(x)$, $u_0(x) = -\frac{1}{x^2}$, $u_n(x) = \frac{1}{n(n+1)x^2}$, $\forall n \geq 1$, $X = (0, 1]$.

Analyze the nature of the continuity of $u_n(x)$ and the sum of the series on X and also the nature of the convergence on X. Formulate false statements for which these series provide counterexamples.

Further Reading

S. Abbott. *Understanding Analysis*, Springer, New York, 2002.

D. Bressoud, *A Radical Approach to Real Analysis*, MAA, Washington, DC, 2007.

T.J.I. Bromwich, *An Introduction to the Theory of Infinite Series*, AMS, Providence, RI, 2005.

B.M. Budak and S.V. Fomin, *Multiple Integrals, Field Theory and Series*, Mir Publisher, Moscow, 1978.

G.M. Fichtengolz, *Differential- und Integralrechnung, Vol. 1–3*, V.E.B. Deutscher Verlag Wiss., Berlin, 1968.

V.A. Ilyin and E.G. Poznyak, *Fundamentals of Mathematical Analysis, Vol.1,2*, Mir Publisher, Moscow, 1982.

K. Knopp, *Theory and Applications of Infinite Series*, Dover Publication, Mineola, NY, 1990.

C.H.C. Little, K.L. Teo and B. Brunt, *Real Analysis via Sequences and Series*, Springer, New York, 2015.

W. Rudin, *Principles of Mathematical Analysis*, McGraw-Hill, New York, 1976.

V.A. Zorich, *Mathematical Analysis I, II*, Springer, Berlin, 2004.

CHAPTER 3

Properties of the Limit Function: Differentiability and Integrability

3.1 Differentiability of the Limit Function

Example 1. A sequence $f_n(x)$ of differentiable on X functions converges on X to a function $f(x)$, but the limit function is not differentiable on X or $f'(x) \neq \lim_{n\to\infty} f_n'(x)$.

Solution
Consider the sequence of functions $f_n(x) = \frac{nx}{1+n^2x^2}$, each of which is differentiable on $X = \mathbb{R}$: $f_n'(x) = \frac{n(1-n^2x^2)}{(1+n^2x^2)^2}$, $\forall n \in \mathbb{N}$. For any fixed $x \in \mathbb{R}$, this sequence converges to 0, that is, $\lim_{n\to\infty} f_n(x) = 0 = f(x)$. Therefore, $f'(x) = 0$, $\forall x \in \mathbb{R}$, but $f_n'(0) = n$ and $\lim_{n\to\infty} f_n'(0) = \infty$.

Note that the convergence of $f_n(x)$ is not uniform on $\mathbb{R}$ (or any interval containing 0), because for any $n \in \mathbb{N}$ there exists $x_n = \frac{1}{n} \in X$ such that

$$|f_n(x_n) - f(x_n)| = \frac{n \cdot \frac{1}{n}}{1 + n^2 \cdot \frac{1}{n^2}} = \frac{1}{2} \underset{n\to\infty}{\not\to} 0.$$

Another interesting example is $f_n(x) = \frac{1}{1+nx}$ considered on $X = [0,1]$. The sequence converges on X to the function $f(x) = \begin{cases} 1, x = 0 \\ 0, x \neq 0 \end{cases}$, which is discontinuous and, consequently, nondifferentiable at 0, although the derivative of each function $f_n(x)$ exists for $\forall x \in X$: $f_n'(x) = \frac{-n}{(1+nx)^2}$. Note that the convergence is not uniform on $[0,1]$: for $\forall n \in \mathbb{N}$, there exists $x_n = \frac{1}{n} \in X$ such that $|f_n(x_n) - f(x_n)| = \frac{1}{2} \underset{n\to\infty}{\not\to} 0$.

Remark 1. Analogous example for a function depending on a parameter is as follows: a function $f(x,y)$ defined on $X \times Y$ is differentiable in $x \in X$ at any fixed

Counterexamples on Uniform Convergence: Sequences, Series, Functions, and Integrals, First Edition. Andrei Bourchtein and Ludmila Bourchtein.
© 2017 John Wiley & Sons, Inc. Published 2017 by John Wiley & Sons, Inc.
Companion website: www.wiley.com/go/bourchtein/counterexamples_on_uniform_convergence

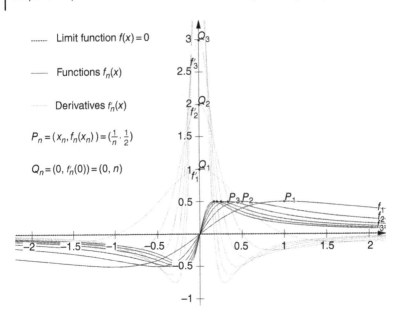

Figure 3.1 Example 1, sequence $f_n(x) = \frac{nx}{1+n^2x^2}$.

$y \in Y$, and there exists $\lim_{y \to y_0} f(x, y) = \varphi(x)$, but $\varphi(x)$ is not differentiable on X or $\varphi'(x) \neq \lim_{y \to y_0} f_x(x, y)$.

The corresponding counterexample is $f(x, y) = \frac{xy}{x^2+y^2}$ with $X \times Y = \mathbb{R} \times (0, 1]$ and $y_0 = 0$. For any fixed $x \in \mathbb{R}$, we have $\lim_{y \to 0_+} f(x, y) = 0 = \varphi(x)$, and therefore, $\varphi'(x) = 0$ for $\forall x \in \mathbb{R}$, including at $x = 0$. However, $f_x(x, y) = \frac{y(y^2-x^2)}{(x^2+y^2)^2}$, $\forall x \in \mathbb{R}, y \in Y$, and the corresponding limit is $\lim_{y \to 0_+} f_x(0, y) = \lim_{y \to 0_+} \frac{1}{y} = +\infty$.

Note that the convergence is not uniform: choosing $x_y = y \in X$ for $\forall y \in Y$, we obtain

$$|f(x_y, y) - \varphi(x_y)| = \frac{y^2}{2y^2} - 0 = \frac{1}{2} \underset{y \to 0}{\not\to} 0.$$

Another interesting counterexample is $f(x, y) = e^{-x^2/y^2}$ considered on $X \times Y = \mathbb{R} \times (0, 1]$ and $y_0 = 0$. The limit function $\varphi(x) = \lim_{y \to 0_+} f(x, y) = \begin{cases} 1, x = 0 \\ 0, x \neq 0 \end{cases}$ is discontinuous and, consequently, nondifferentiable at $x = 0$. However, the partial derivative in x exists on $\mathbb{R} \times (0, 1] : f_x(x, y) = -\frac{2x}{y^2}e^{-x^2/y^2}$. Again the convergence is nonuniform on $\mathbb{R}$: for $\forall y \in (0, 1]$, one can choose $x_y = y \in X$ to obtain $|f(x_y, y) - \varphi(x_y)| = e^{-1} \underset{y \to 0}{\not\to} 0$.

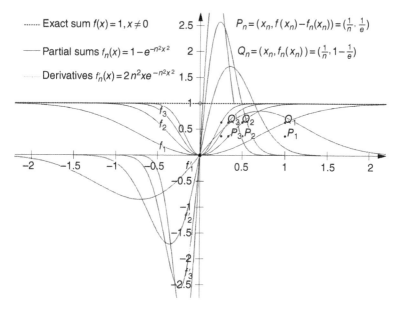

Figure 3.2 Example 1, series $\sum_{n=1}^{\infty}[e^{-(n-1)^2 x^2} - e^{-n^2 x^2}]$.

Remark 2. Analogous example for a series can be formulated in the following form: a series $\sum u_n(x)$ of differentiable functions converges on X, but this series cannot be differentiated term by term on X.

The corresponding counterexample is $\sum_{n=1}^{\infty}[e^{-(n-1)^2 x^2} - e^{-n^2 x^2}]$ with $X = \mathbb{R}$. The partial sums of this telescoping series can be easily evaluated as follows:

$$f_n(x) = \sum_{k=1}^{n}[e^{-(k-1)^2 x^2} - e^{-k^2 x^2}]$$

$$= 1 - e^{-x^2} + e^{-x^2} - e^{-2^2 x^2} + \cdots + e^{-(n-1)^2 x^2} - e^{-n^2 x^2} = 1 - e^{-n^2 x^2}.$$

Therefore, the sum of the series is $f(x) = \lim_{n\to\infty} f_n(x) = \begin{cases} 0, x = 0 \\ 1, x \neq 0 \end{cases}$ for any $x \in \mathbb{R}$. Although the series is convergent on $\mathbb{R}$ and each function $u_n(x) = e^{-(n-1)^2 x^2} - e^{-n^2 x^2}$ is infinitely differentiable on $\mathbb{R}$, but $f(x)$ is discontinuous and, therefore, nondifferentiable at $x = 0$.

Let us show that the convergence is nonuniform on $\mathbb{R}$ (or any interval containing 0). Indeed, for $\forall n \in \mathbb{N}$ there is $x_n = \frac{1}{n}$ such that

$$|r_n(x_n)| = \left| \sum_{k=n+1}^{\infty} u_k(x_n) \right| = |f_n(x_n) - f(x_n)|$$

$$= |1 - e^{-n^2 \cdot 1/n^2} - 1| = e^{-1} \nrightarrow_{n\to\infty} 0.$$

Another telescoping series $\sum_{n=1}^{\infty}\left(\frac{(n-1)x}{1+(n-1)^4x^4} - \frac{nx}{1+n^4x^4}\right)$ provides one more coun-terexample. Its partial sums are easily found:

$$f_n(x) = \sum_{k=1}^{n}\left(\frac{(k-1)x}{1+(k-1)^4x^4} - \frac{kx}{1+k^4x^4}\right)$$

$$= -\frac{x}{1+x^4} + \frac{x}{1+x^4} - \frac{2x}{1+2^4x^4} + \cdots - \frac{nx}{1+n^4x^4} = -\frac{nx}{1+n^4x^4},$$

from which it follows that the series converges to zero for any $x \in \mathbb{R}$: $f(x) = \lim_{n\to\infty} f_n(x) = 0$. Therefore, $f'(x) = 0$ for any $x \in \mathbb{R}$, but the series of derivatives does not converge at the origin: its partial sums can be expressed in the form

$$f_n'(x) = \sum_{k=1}^{n} u_k'(x)$$

$$= -\frac{1-3x^4}{(1+x^4)^2} + \sum_{k=2}^{n}\left[\frac{(k-1)(1-3(k-1)^4x^4)}{(1+(k-1)^4x^4)^2} - \frac{k(1-3k^4x^4)}{(1+k^4x^4)^2}\right]$$

$$= -\frac{n(1-3n^4x^4)}{(1+n^4x^4)^2},$$

and for $x = 0$ we have $f_n'(0) = -n \to -\infty$. Note that the series convergence is nonuniform on $\mathbb{R}$ (and on any interval containing 0): for $\forall n \in \mathbb{N}$, there is $x_n = \frac{1}{n}$ such that

$$|r_n(x_n)| = |f_n(x_n) - f(x_n)| = \frac{1}{2} \nrightarrow 0.$$

Remark 3. In all the above examples, the convergence of a sequence, series and function depending on a parameter is nonuniform. Note that the uniform con-vergence still does not guarantee the differentiability of the limit function or, if the derivative exists, the possibility of the differentiation inside of the limit or term by term.

Example 2. A sequence $f_n(x)$ of differentiable on X functions converges uni-formly on X to a function $f(x)$, but the limit function is not differentiable on X or $f'(x) \neq \lim_{n\to\infty} f_n'(x)$.

Solution
Consider the sequence of functions $f_n(x) = \frac{1}{n}\arctan x^n$, each of which is differentiable on $\mathbb{R}$: $f_n'(x) = \frac{x^{n-1}}{1+x^{2n}}$, $\forall n \in \mathbb{N}$. Since $|\arctan x^n| \leq \frac{\pi}{2}$, $\forall x \in \mathbb{R}$, the sequence is convergent for any fixed x: $\lim_{n\to\infty} f_n(x) = 0 = f(x)$, $\forall x \in \mathbb{R}$, and

consequently, $f'(x) = 0$, $\forall x \in \mathbb{R}$. Also, the convergence is uniform on $\mathbb{R}$, because simultaneously for all $x \in \mathbb{R}$ one has

$$|f_n(x) - f(x)| = \frac{1}{n}|\arctan x^n| \leq \frac{\pi}{2n} \underset{n\to\infty}{\to} 0.$$

However, at the point $x = 1$ the sequence of the derivatives $f_n'(1) = \frac{1}{2}$, $\forall n \in \mathbb{N}$ does not converge to $f'(1) = 0$.

Let us perform an additional analysis on the convergence of sequence of the derivatives $f_n'(x) = \frac{x^{n-1}}{1+x^{2n}}$: if $|x| < 1$, then $\lim\limits_{n\to\infty} x^n = 0$, and therefore, $\lim\limits_{n\to\infty} f_n'(x) = 0$; if $|x| = 1$, then $f_n'(1) = \frac{1}{2}$ and $f_n'(-1) = \frac{(-1)^{n-1}}{2}$, which implies that $\lim\limits_{n\to\infty} f_n'(1) = \frac{1}{2}$ and the limit at $x = -1$ does not exist; if $|x| > 1$, then $\lim\limits_{n\to\infty} \frac{1}{x^n} = 0$, and therefore, $\lim\limits_{n\to\infty} f_n'(x) = \lim\limits_{n\to\infty} \frac{(1/x)^{n+1}}{(1/x)^{2n}+1} = 0$. Thus, the sequence of the derivatives converges on $\mathbb{R} \setminus \{-1\}$ to the limit function $g(x) = \lim\limits_{n\to\infty} f_n'(x) = \begin{cases} 1/2, x = 1 \\ 0, x \neq \pm 1 \end{cases}$. However, the convergence is not uniform on any interval containing the point $x = 1$, since for $\forall n \in \mathbb{N}$ one can choose $x_n = \frac{1}{\sqrt[n]{2}}$, which gives

$$|f_n'(x_n) - g(x_n)| = \frac{(1/2)^{\frac{n-1}{n}}}{1+1/4} \underset{n\to\infty}{\to} \frac{2}{5} \neq 0.$$

Another sequence with similar properties is $f_n(x) = \frac{x}{1+n^2x^2}$, $x \in \mathbb{R}$. It converges for any fixed x to $f(x) = 0$, and the convergence is uniform, because simultaneously for all $x \in \mathbb{R}$ it holds that

$$|f_n(x) - f(x)| = \left|\frac{x}{1+n^2x^2}\right| = \frac{1}{2n}\frac{2n|x|}{1+n^2x^2} \leq \frac{1}{2n} \underset{n\to\infty}{\to} 0.$$

At the same time, the derivatives $f_n'(x) = \frac{1-n^2x^2}{(1+n^2x^2)^2}$ at the point $x = 0$ give the numerical sequence $f_n'(0) = 1$, $\forall n \in \mathbb{N}$, which does not converge to $f'(0) = 0$. Note, that again the sequence of the derivatives converges on $\mathbb{R}$: $g(x) = \lim\limits_{n\to\infty} f_n'(x) = \begin{cases} 1, x = 0 \\ 0, x \neq 0 \end{cases}$, but this convergence is not uniform on any interval containing the point $x = 0$: for $\forall n \in \mathbb{N}$, one can choose $x_n = \frac{1}{2n}$ to obtain

$$|f_n'(x_n) - g(x_n)| = \frac{1 - n^2\frac{1}{4n^2}}{\left(1 + n^2\frac{1}{4n^2}\right)^2} = \frac{12}{25} \underset{n\to\infty}{\nrightarrow} 0.$$

Remark 1. For a function depending on a parameter, the analogous example is as follows: a function $f(x, y)$ defined on $X \times Y$ is differentiable in $x \in X$ at any fixed $y \in Y$, and $f(x, y)$ converges uniformly on X to a function $\varphi(x)$ as y approaches y_0, but $\varphi'(x) \neq \lim\limits_{y \to y_0} f_x(x, y)$, $\forall x \in X$.

The function $f(x, y) = xe^{-x^2/y^2}$ considered on $X \times Y = \mathbb{R} \times (0, 1]$ with $y_0 = 0$ provides the required counterexample. The partial derivative of $f(x, y)$ with respect to x exists for any $y \in (0, 1]$: $f_x(x, y) = \left(1 - \frac{2x^2}{y^2}\right) e^{-x^2/y^2}$. Further, for any fixed $x \in \mathbb{R}$, this function converges to 0, as y approaches 0, so $\varphi(x) \equiv 0$. Moreover, the convergence is uniform on $\mathbb{R}$ as the following evaluation shows. For an arbitrary fixed $y \in Y$, the critical points of $f(x, y) - \varphi(x) = xe^{-x^2/y^2}$ are found from the equation

$$(f(x, y) - \varphi(x))_x = \left(1 - \frac{2x^2}{y^2}\right) e^{-x^2/y^2} = 0,$$

which implies $x^2 = \frac{y^2}{2}$, or $x_y = \pm\frac{y}{\sqrt{2}}$. Since $1 - \frac{2x^2}{y^2} > 0$ for $|x| < \frac{|y|}{\sqrt{2}}$ and $1 - \frac{2x^2}{y^2} < 0$ for $|x| > \frac{|y|}{\sqrt{2}}$, it follows that $x_{y+} = \frac{y}{\sqrt{2}}$ is the maximum point and $x_{y-} = -\frac{y}{\sqrt{2}}$ is the minimum point (recall that $y > 0$). Since $|f(x, y) - \varphi(x)|$ assumes the same value at x_{y+} and x_{y-}, we obtain

$$\sup_{x \in \mathbb{R}} |f(x, y) - \varphi(x)| = \max_{x \in \mathbb{R}} |f(x, y) - \varphi(x)|$$

$$= |x_{y+} e^{-x_{y+}^2/y^2}| = \frac{y}{\sqrt{2}} e^{-y^2/2y^2} = \frac{y}{\sqrt{2}} e^{-1/2} \underset{y \to 0}{\to} 0,$$

which proves the uniform convergence on $\mathbb{R}$.

Let us turn to the convergence of the partial derivatives. Since

$$\lim_{\substack{y \to 0 \\ x \neq 0}} \frac{x^2}{y^2} e^{-x^2/y^2} = \lim_{t \to +\infty} \frac{t}{e^t} = \lim_{t \to +\infty} \frac{1}{e^t} = 0$$

(apply the substitution $t = \frac{x^2}{y^2}$ and l'Hospital's rule), we conclude that $g(x) = \lim\limits_{y \to 0} f_x(x, y) = 0$ for $\forall x \neq 0$. Additionally, at the point 0 we have $g(0) = \lim\limits_{y \to 0} f_x(0, y) = 1 \cdot e^0 = 1$. At the same time, $(\lim\limits_{y \to 0} f(x, y))_x(x = 0) = \varphi'(0) = 0 \neq 1 = \lim\limits_{y \to 0} f_x(0, y)$. Note, that the convergence of the derivatives is nonuniform on any interval with the point 0 included, because for $\forall y \in (0, 1]$ one can choose $x_y = y \in X$ to obtain $|f_x(x_y, y) - g(x_y)| = e^{-1} \underset{y \to 0}{\not\to} 0$.

Remark 2. For a series of functions, the example can be formulated as follows: a series $\sum u_n(x)$ of differentiable functions converges uniformly on X, but this series cannot be differentiated term by term on X.

It is substantiated by the counterexample with $\sum_{n=1}^{\infty}\frac{(-1)^{n-1}x^n}{n}$ considered on $X = [0,1]$. This series converges uniformly on $[0,1]$ according to the Abel test: the numerical alternating series $\sum_{n=1}^{\infty}\frac{(-1)^{n-1}}{n}$ converges by Leibniz's test and the sequence x^n is (nonstrictly) decreasing and uniformly bounded on $[0,1]$. Recalling the expansion of the logarithmic function in the Taylor series, we immediately conclude that $\sum_{n=1}^{\infty}\frac{(-1)^{n-1}x^n}{n} = \ln(1+x) = f(x)$ on $[0,1]$. All the functions $u_n(x) = \frac{(-1)^{n-1}x^n}{n}$ and $f(x) = \ln(1+x)$ are differentiable on $[0,1]$: $u_n'(x) = (-1)^{n-1}x^{n-1}$ and $f'(x) = \frac{1}{1+x}$, but the series of derivatives and $f'(x)$ give different results at $x = 1$. Indeed, $\sum_{n=1}^{\infty}u_n'(1) = \sum_{n=1}^{\infty}(-1)^{n-1}$ is divergent, since the general term $(-1)^{n-1}$ does not approach 0; at the same time, $f'(1) = \frac{1}{2}$. Note that the series $\sum_{n=1}^{\infty}u_n'(x)$ converges on $[0,1)$, but nonuniformly: for $\forall n \in \mathbb{N}$, there is $x_n = \frac{1}{\sqrt[n]{2}} \in [0,1)$ such that

$$|r_n(x_n)| = \left|\sum_{k=n+1}^{\infty}u_k'(x_n)\right| = \left|\sum_{k=n+1}^{\infty}(-1)^{k-1}x_n^{k-1}\right|$$
$$= \frac{x_n^n}{1+x_n} = \frac{1/2}{1+1/\sqrt[n]{2}} > \frac{1}{4} \nrightarrow_{n\to\infty} 0.$$

Remark 3. In all the above examples, the convergence of the original sequence (function with a parameter or series) is uniform on the considered domain, but the convergence of the corresponding sequence (function with a parameter or series) of the derivatives is not uniform on the same domain. This is the cause of the impossibility of differentiation under the sign of the limit (or differentiating the series term by term). Recall that in Example 1 we have considered such sequences (functions with a parameter or series) that do not converge uniformly on a set X. This implies that the convergence of derivatives (even when it takes place) cannot be uniform on X. The condition of the uniform convergence of a sequence (function with a parameter or series) of derivatives guarantees the validity of differentiation under the sign of the limit or infinite sum (an additional minor condition of the convergence of original sequence, function, or series in some point should also be fulfilled). At the same time, this condition, albeit very important, is sufficient but not necessary as shown in Example 3.

Example 3. A sequence $f_n(x)$ of differentiable on X functions converges on X to a function $f(x)$ and $f'(x) = \lim_{n\to\infty}f_n'(x)$, $\forall x \in X$, however, the convergence of $f_n'(x)$ to $f'(x)$ is nonuniform on X.

Solution
Consider the sequence $f_n(x) = \frac{1}{2n}\ln(1+n^2x^2)$, $X = [0,1]$. If $x = 0$, then $f_n(0) = 0$, $\forall n \in \mathbb{N}$ and $\lim_{n\to\infty}f_n(0) = 0$. If $x \neq 0$, then

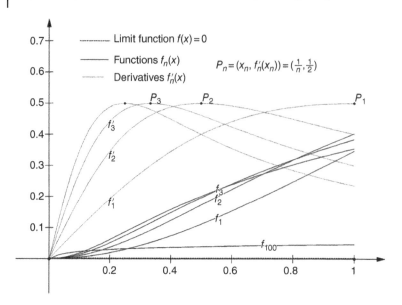

Figure 3.3 Example 3, sequence $f_n(x) = \frac{1}{2n}\ln(1 + n^2x^2)$.

$$\lim_{n\to\infty} f_n(x) = \lim_{n\to\infty} \frac{\ln(1 + n^2x^2)}{2n} = \lim_{t\to+\infty} \frac{\ln(1 + t^2x^2)}{2t}$$
$$= \lim_{t\to+\infty} \frac{2tx^2}{2(1 + t^2x^2)} = 0$$

(the discrete variable n was substituted by the continuous variable t and then l'Hospital's rule was applied). Thus, $\lim_{n\to\infty} f_n(x) = 0 = f(x)$, $\forall x \in [0, 1]$. For the derivatives, we have: $f_n'(x) = \frac{nx}{1+n^2x^2}$, $f'(x) = 0$, and $\lim_{n\to\infty} f_n'(x) = \lim_{n\to\infty} \frac{nx}{1+n^2x^2} = 0 = f'(x)$, $\forall x \in [0, 1]$. Therefore, the conditions of the statement are satisfied, however, the convergence of the derivatives is nonuniform: for any $n \in \mathbb{N}$, there exists $x_n = \frac{1}{n} \in [0, 1]$ such that

$$|f_n'(x_n) - f'(x_n)| = \frac{n \cdot \frac{1}{n}}{1 + n^2 \cdot \frac{1}{n^2}} = \frac{1}{2} \not\to 0.$$

It can be shown that in this counterexample, the convergence of $f_n(x)$ to $f(x)$ is uniform on X: for all $x \in [0, 1]$, it is verified that

$$|f_n(x) - f(x)| = \frac{1}{2n}\ln(1 + n^2x^2) \leq \frac{1}{2n}\ln(1 + n^2) \underset{n\to\infty}{\to} 0.$$

However, it is easy to construct another counterexample where the convergence of the original sequence is nonuniform. For instance, if one considers the same sequence of functions on $X = \mathbb{R}$, then all the above properties are true, in particular, $\lim\limits_{n\to\infty} f_n'(x) = \lim\limits_{n\to\infty} \frac{nx}{1+n^2x^2} = 0 = f'(x)$, $\forall x \in \mathbb{R}$, but the original sequence converges nonuniformly on X: choosing $x_n = e^n$ for each $n \in \mathbb{N}$, we obtain

$$|f_n(x_n) - f(x_n)| = \frac{1}{2n}\ln(1 + n^2 e^n) > \frac{1}{2n}\ln e^n = \frac{1}{2} \underset{n\to\infty}{\nrightarrow} 0.$$

Remark 1. The corresponding example for functions depending on a parameter goes as follows: a function $f(x, y)$ defined on $X \times Y$ is differentiable in $x \in X$ at any fixed $y \in Y$, $f(x, y)$ converges on X to a function $\varphi(x)$, as y approaches y_0, and $\varphi'(x) = \lim\limits_{y\to y_0} f_x(x, y)$, $\forall x \in X$, however, the convergence of $f_x(x, y)$ is not uniform.

It can be illustrated by the function $f(x, y) = y\, e^{-x^2/y^2}$ considered on $X \times Y = \mathbb{R} \times (0, 1]$ with $y_0 = 0$. Indeed, $\lim\limits_{y\to 0} y\, e^{-x^2/y^2} = 0 = \varphi(x)$ for any fixed $x \in \mathbb{R}$. For the partial derivative $f_x(x, y) = -\frac{2x}{y} e^{-x^2/y^2}$, we have $\lim\limits_{y\to 0} f_x(0, y) = \lim\limits_{y\to 0} 0 = 0$ and

$$\lim_{\substack{y\to 0_+ \\ x\neq 0}} f_x(x, y) = \lim_{\substack{y\to 0_+ \\ x\neq 0}} \left(-\frac{2x}{y}\right) e^{-x^2/y^2} = \lim_{t\to\infty} \frac{-2t}{e^{t^2}} = \lim_{t\to\infty} \frac{-2}{2te^{t^2}} = 0$$

(apply the change of variable $t = \frac{x}{y}$ and l'Hospital's rule). Thus,

$$\lim_{y\to 0_+} f_x(x, y) = 0 = g(x), \forall x \in \mathbb{R},$$

and, therefore

$$\varphi'(x) = \left(\lim_{y\to 0_+} f(x, y)\right)_x = 0 = \lim_{y\to 0_+} f_x(x, y) = g(x), \forall x \in \mathbb{R}.$$

Let us check the type of convergence of $f(x, y)$ and $f_x(x, y)$ on $X = \mathbb{R}$. Since the evaluation $|f(x, y) - \varphi(x)| = |y|e^{-x^2/y^2} \leq |y|$ is true simultaneously for all $x \in \mathbb{R}$ and $|y| \underset{y\to 0}{\to} 0$, it follows that the convergence of $f(x, y)$ is uniform on $\mathbb{R}$. On the other hand, for the derivatives one can choose $x_y = y \in X$ for $\forall y \in (0, 1]$ to obtain

$$|f_x(x_y, y) - g(x_y)| = \frac{2y}{y} e^{-y^2/y^2} = 2e^{-1} \underset{y\to 0}{\nrightarrow} 0,$$

which means that the convergence of $f_x(x, y)$ is not uniform on $\mathbb{R}$ (and on any interval containing 0).

Another interesting counterexample for the same statement is the function $f(x, y) = y\ln\left(1 + \frac{x^2}{y^2}\right)$, $X \times Y = [-1, 1] \times (0, 1]$, $y_0 = 0$. The limit function is 0:

$$\lim_{y\to 0^-} f(x, y) = \lim_{y\to 0} y\ln\left(1 + \frac{x^2}{y^2}\right) = 0 = \varphi(x), \forall x \in \mathbb{R}$$

(for $x = 0$ the result is evident and for $x \neq 0$ one can apply l'Hospital's rule for the ratio $\frac{\ln(1+x^2/y^2)}{1/y}$). For the partial derivative $f_x(x, y) = \frac{2xy}{x^2+y^2}$ the limit function is also 0:

$$\lim_{y \to 0} f_x(x, y) = \lim_{y \to 0} \frac{2xy}{x^2 + y^2} = 0 = g(x)$$

(for $x = 0$ the result is evident and for $x \neq 0$ just apply the arithmetic rules of the limits). Therefore, $(\lim_{y \to 0_+} f(x, y))_x = 0 = \lim_{y \to 0_+} f_x(x, y)$ for $\forall x \in [-1, 1]$. Thus, all the conditions of the statement hold. Additionally, the convergence of $f(x, y)$ is uniform on $X = [-1, 1]$. In fact, for any fixed y, the function $f(x, y)$ is strictly decreasing on $X = [-1, 0]$ and strictly increasing on $[0, 1]$, which ensures the following evaluation simultaneously for all $x \in [-1, 1]$:

$$|f(x, y) - \varphi(x)| = y \ln\left(1 + \frac{x^2}{y^2}\right) \leq y \ln\left(1 + \frac{1}{y^2}\right).$$

Since the right-hand side of the last inequality approaches 0:

$$\lim_{y \to 0_+} y \ln\left(1 + \frac{1}{y^2}\right) = \lim_{t \to +\infty} \frac{\ln(1 + t^2)}{t} = \lim_{t \to +\infty} \frac{2t}{1 + t^2} = 0$$

(apply the substitution $t = \frac{1}{y}$ and l'Hospital's rule), it guarantees that the convergence of $f(x, y)$ is uniform on $[-1, 1]$. Nevertheless, the convergence of the derivatives is not uniform on $[-1, 1]$: for $\forall y \in (0, 1]$, there exists $x_y = y \in X$ such that

$$|f_x(x_y, y) - g(x_y)| = \frac{2y \cdot y}{y^2 + y^2} = 1 \underset{y \to 0}{\nrightarrow} 0.$$

If we consider a slightly modified counterexample—$f(x, y) = y \ln\left(1 + \frac{x^2}{y^2}\right)$, $X \times Y = \mathbb{R} \times (0, 1]$, $y_0 = 0$ (only the set X was changed)—then all the above properties hold (in particular, $(\lim_{y \to 0_+} f(x, y))_x = 0 = \lim_{y \to 0_+} f_x(x, y)$ for $\forall x \in X$), but the convergence of $f(x, y)$ to $\varphi(x) \equiv 0$ is not uniform on $X = \mathbb{R}$. Indeed, choosing $x_y = y e^{\frac{1}{2y}}$ for $\forall y \in (0, 1]$, we obtain

$$|f(x_y, y) - \varphi(x_y)| = y \ln\left(1 + e^{\frac{1}{y}}\right) > 1 \underset{y \to 0}{\nrightarrow} 0.$$

Remark 2. For a series of functions, a similar example has the following formulation: a series $\sum u_n(x)$ of differentiable functions converges on X and can be differentiated term by term on X, however, the series $\sum u'_n(x)$ converges nonuniformly on X.

Let us consider the series $\sum_{n=1}^{\infty} \left(\frac{1}{n}x^n - \frac{1}{n+1}x^{n+1} \right)$ on $X = [0, 1)$. For the partial sums of this telescoping series, we have

$$f_n(x) = \sum_{k=1}^{n} \left(\frac{1}{k}x^k - \frac{1}{k+1}x^{k+1} \right)$$

$$= x - \frac{1}{2}x^2 + \frac{1}{2}x^2 - \frac{1}{3}x^3 + \cdots + \frac{1}{n}x^n - \frac{1}{n+1}x^{n+1} = x - \frac{1}{n+1}x^{n+1}.$$

Then, $f(0) = \lim_{n\to\infty} f_n(0) = 0$ and

$$f(x) = \lim_{n\to\infty} f_n(x) = \lim_{n\to\infty} \left(x - \frac{1}{n+1}x^{n+1} \right) = x$$

for $x \in (0, 1)$. Thus, the sum of the series is $f(x) = x$ for $x \in [0, 1)$. Note that simultaneously for all $x \in [0, 1)$, the following evaluation for the series remainder holds:

$$|r_n(x)| = |f_n(x) - f(x)| = \frac{1}{n+1}x^{n+1} < \frac{1}{n+1} \xrightarrow[n\to\infty]{} 0,$$

which means the uniform convergence of the series.

The series of the derivatives $\sum_{n=1}^{\infty} u_n'(x) = \sum_{n=1}^{\infty} (x^{n-1} - x^n)$ is also telescoping with the partial sums

$$f_n'(x) = \sum_{k=1}^{n} u_k'(x) = 1 - x + x - x^2 + \cdots + x^{n-1} - x^n = 1 - x^n.$$

For $x \in [0, 1)$, it follows that $\lim_{n\to\infty} f_n'(x) = \lim_{n\to\infty} (1 - x^n) = 1$. Thus,

$$\left(\sum_{n=1}^{\infty} u_n(x) \right)' = \left(\sum_{n=1}^{\infty} \left(\frac{1}{n}x^n - \frac{1}{n+1}x^{n+1} \right) \right)' = 1$$

$$= \sum_{n=1}^{\infty} (x^{n-1} - x^n) = \sum_{n=1}^{\infty} u_n'(x), \forall x \in [0, 1).$$

At the same time, the series of derivatives does not converge uniformly on $[0, 1)$, since for $\forall n \in \mathbb{N}$ there exists $x_n = \frac{1}{\sqrt[n]{2}} \in [0, 1)$ such that

$$|f_n'(x_n) - f'(x_n)| = |1 - x_n^n - 1| = \left(\frac{1}{\sqrt[n]{2}} \right)^n = \frac{1}{2} \xrightarrow[n\to\infty]{} 0.$$

Remark 3. In the given examples, it is shown that the condition of the uniform convergence of a sequence (function with a parameter or series) of derivatives is sufficient, but not necessary for the differentiation under the sign of the limit (or differentiating the series term by term). It happens that the condition of the uniform convergence of the original sequence (function with a parameter or series) is not also necessary as shown in Example 4.

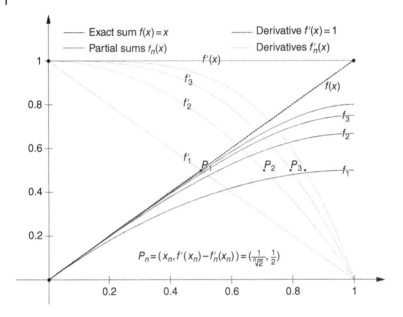

Figure 3.4 Example 3, series $\sum_{n=1}^{\infty} \left(\frac{1}{n}x^n - \frac{1}{n+1}x^{n+1} \right)$.

Example 4. A sequence $f_n(x)$ of differentiable functions converges nonuniformly on X to a function $f(x)$, but nevertheless $f'(x) = \lim_{n\to\infty} f_n'(x)$ on X.

Solution

Consider the sequence $f_n(x) = \frac{nx^2}{1+n^2x^4}$ on $X = \mathbb{R}$. If $x = 0$, then $f_n(0) = 0, \forall n \in \mathbb{N}$ and $\lim_{n\to\infty} f_n(0) = 0$. If $x \neq 0$, then

$$\lim_{n\to\infty} f_n(x) = \lim_{n\to\infty} \frac{nx^2}{1 + n^2x^4} = \lim_{n\to\infty} \frac{\frac{x^2}{n}}{\frac{1}{n^2} + x^4} = 0.$$

Thus, $\lim_{n\to\infty} f_n(x) = 0 = f(x)$, $\forall x \in \mathbb{R}$, and therefore, $f'(x) = 0$, $\forall x \in \mathbb{R}$. For the derivatives, we have: $f_n'(x) = \frac{2nx(1-n^2x^4)}{(1+n^2x^4)^2}$, and again $\lim_{n\to\infty} f_n'(0) = \lim_{n\to\infty} 0 = 0$, and for $x \neq 0$ we obtain

$$\lim_{n\to\infty} f_n'(x) = \lim_{n\to\infty} \frac{2nx(1 - n^2x^4)}{(1 + n^2x^4)^2} = \lim_{n\to\infty} \frac{2\frac{x}{n}\left(\frac{1}{n^2} - x^4\right)}{\left(\frac{1}{n^2} + x^4\right)^2} = 0.$$

Thus, $\lim_{n\to\infty} f_n'(x) = 0 = f'(x)$, $\forall x \in \mathbb{R}$. Let us show that both sequences $f_n(x)$ and $f_n'(x)$ converge nonuniformly on $X = \mathbb{R}$ (and on any interval containing 0).

Indeed, for the first sequence, by choosing $x_n = \frac{1}{\sqrt{n}}$ for any $n \in \mathbb{N}$, we obtain

$$|f_n(x_n) - f(x_n)| = \frac{n \cdot \frac{1}{n}}{1 + n^2 \frac{1}{n^2}} = \frac{1}{2} \underset{n \to \infty}{\nrightarrow} 0,$$ and for the second we can choose $\tilde{x}_n = \frac{1}{2\sqrt{n}}$ for any $n \in \mathbb{N}$ to get

$$|f_n'(\tilde{x}_n) - f'(\tilde{x}_n)| = \frac{2n \cdot \frac{1}{2\sqrt{n}}\left(1 - n^2 \frac{1}{4n^2}\right)}{\left(1 + n^2 \cdot \frac{1}{4n^2}\right)^2} = \frac{12}{25}\sqrt{n} \underset{n \to \infty}{\nrightarrow} 0.$$

Remark 1. For a function depending on a parameter, the example can be formulated as follows: a function $f(x, y)$ defined on $X \times Y$ is differentiable in $x \in X$ at any fixed parameter $y \in Y$, $f(x, y)$ converges nonuniformly on X to a function $\varphi(x)$ as y approaches y_0, but nevertheless $\varphi'(x) = \lim_{y \to y_0} f_x(x, y)$ on X.

It can be illustrated by the function $f(x, y) = \frac{x^2 y^2}{x^4 + y^4}$ considered on $X \times Y = \mathbb{R} \times (0, 1]$ with $y_0 = 0$. Indeed, $\lim_{y \to 0} \frac{x^2 y^2}{x^4 + y^4} = 0 = \varphi(x)$ for any fixed $x \in \mathbb{R}$. For the partial derivative, we have $\lim_{y \to 0} f_x(x, y) = \lim_{y \to 0} \frac{2xy^2(y^4 - x^4)}{(x^4 + y^4)^2} = 0, \forall x \in \mathbb{R}$. Thus, $\varphi'(x) = (\lim_{y \to 0} f(x, y))_x = 0 = \lim_{y \to 0} f_x(x, y)$ for $\forall x \in \mathbb{R}$. However, the convergence of $f(x, y)$ and $f_x(x, y)$ is not uniform on $X = \mathbb{R}$ (and on any interval containing 0). Indeed, for the first function, one can use $x_y = y \in X$ for $\forall y \in (0, 1]$ to obtain $|f(x_y, y) - \varphi(x_y)| = \frac{y^4}{2y^4} = \frac{1}{2} \underset{y \to 0}{\nrightarrow} 0$, and, for the derivatives, one can choose $\tilde{x}_y = \frac{y}{2} \in X$ for $\forall y \in (0, 1]$ to get

$$|f_x(\tilde{x}_y, y) - \varphi'(\tilde{x}_y)| = \frac{2\frac{y}{2}y^2\left(y^4 - \frac{y^4}{16}\right)}{\left(\frac{y^4}{16} + y^4\right)^2} = \frac{240}{289}\frac{1}{y} \underset{y \to 0}{\nrightarrow} 0.$$

Remark 2. For a series of functions, the corresponding formulation is as follows: a series $\sum u_n(x)$ of differentiable functions converges nonuniformly on X, but nevertheless this series can be differentiated term by term on X.

Let us consider the telescoping series

$$\sum_{n=1}^{\infty} u_n(x) = \sum_{n=1}^{\infty} (nx^2 e^{-n^2 x^4} - (n-1)x^2 e^{-(n-1)^2 x^4})$$

on $X = \mathbb{R}$. The partial sums are easily found as follows:

$$f_n(x) = \sum_{k=1}^{n}(kx^2 e^{-k^2 x^4} - (k-1)x^2 e^{-(k-1)^2 x^4})$$
$$= x^2 e^{-x^4} + 2x^2 e^{-2^2 x^4} - x^2 e^{-x^4} + \cdots + nx^2 e^{-n^2 x^4} - (n-1)x^2 e^{-(n-1)^2 x^4}$$
$$= nx^2 e^{-n^2 x^4}.$$

Then, $f(0) = \lim_{n \to \infty} f_n(0) = 0$ and

$$f(x) = \lim_{n \to \infty} f_n(x) = \lim_{n \to \infty} \frac{nx^2}{e^{n^2 x^4}} = \lim_{t \to +\infty} \frac{t}{e^{t^2}} = \lim_{t \to +\infty} \frac{1}{2te^{t^2}} = 0$$

for $x \neq 0$ (the change of the variable $t = nx^2$ and l'Hospital's rule have been applied). Thus, the sum of the series is $f(x) = 0, \forall x \in \mathbb{R}$.

Let us consider the series of the derivatives:

$$\sum_{n=1}^{\infty} u'_n(x) = \sum_{n=1}^{\infty} (2nxe^{-n^2 x^4} - 4n^3 x^5 e^{-n^2 x^4} - 2(n-1)xe^{-(n-1)^2 x^4}$$
$$+ 4(n-1)^3 x^5 e^{-(n-1)^2 x^4}).$$

This is again a telescoping series, and its partial sums are

$$f'_n(x) = \sum_{k=1}^{n} u'_k(x)$$
$$= 2xe^{-x^4} - 4x^5 e^{-x^4} + 2 \cdot 2xe^{-2^2 x^4} - 4 \cdot 2^3 x^5 e^{-2^2 x^4}$$
$$- 2xe^{-x^4} + 4x^5 e^{-x^4} \cdots + 2nxe^{-n^2 x^4}$$
$$- 4n^3 x^5 e^{-n^2 x^4} - 2(n-1)xe^{-(n-1)^2 x^4} + 4(n-1)^3 x^5 e^{-(n-1)^2 x^4}$$
$$= 2nxe^{-n^2 x^4} - 4n^3 x^5 e^{-n^2 x^4}.$$

Further, $\lim_{n \to \infty} f'_n(0) = \lim_{n \to \infty} 0 = 0$, and for $x \neq 0$ we have

$$\lim_{n \to \infty} f'_n(x) = \lim_{n \to \infty} \frac{2nx(1 - 2n^2 x^4)}{e^{n^2 x^4}} = \lim_{t \to +\infty} \frac{2}{x} \frac{t - 2t^3}{e^{t^2}}$$
$$= \frac{2}{x} \lim_{t \to +\infty} \frac{1 - 6t^2}{2te^{t^2}} = \frac{1}{x} \left(\lim_{t \to +\infty} \frac{1}{te^{t^2}} - \lim_{t \to +\infty} \frac{6t}{e^{t^2}} \right)$$
$$= \frac{1}{x} \left(\lim_{t \to +\infty} \frac{1}{te^{t^2}} - \lim_{t \to +\infty} \frac{6}{2te^{t^2}} \right) = 0.$$

Therefore,

$$\sum_{n=1}^{\infty} u'_n(x) = 0 = f'(x) = \left(\sum_{n=1}^{\infty} u_n(x) \right)', \forall x \in \mathbb{R}.$$

Let us check the type of convergence of the series $\sum_{n=1}^{\infty} u_n(x)$ and $\sum_{n=1}^{\infty} u'_n(x)$. For the first series, by choosing $x_n = \frac{1}{\sqrt{n}}$ for $\forall n \in \mathbb{N}$, we obtain

$$|r_n(x_n)| = |f_n(x_n) - f(x_n)| = nx_n^2 e^{-n^2 x_n^4} = e^{-1} \not\to_{n \to \infty} 0,$$

that is, this series converges nonuniformly on $\mathbb{R}$ (or on any interval containing 0). Similarly, using the same $x_n = \frac{1}{\sqrt{n}}$ for the second series, we get

$$\left| \sum_{k=n+1}^{\infty} u_k'(x_k) \right| = |f_n'(x_n) - f'(x_n)|$$

$$= |2nx_n(1 - 2n^2 x_n^4)e^{-n^2 x_n^4}| = 2\sqrt{n}e^{-1} \underset{n\to\infty}{\nrightarrow} 0,$$

which shows that the convergence for the series of derivatives is also nonuniform.

Remark 3. For a series of functions, the example can be strengthened to the following form: a series $\sum u_n(x)$ of differentiable functions converges nonuniformly on X and both the partial sums and the sum of the series cannot be expressed through elementary functions, but nevertheless this series can be differentiated term by term on X.

For a counterexample in this case, let us consider the series $\sum_{n=1}^{\infty} u_n(x) = \sum_{n=1}^{\infty} \frac{1}{n^x}$ on $X = (1, +\infty)$, which represents the real-valued Riemann zeta function (a restriction of the famous complex-valued Riemann zeta function to the real axis). Note that this series converges on X, since this is a p-series with $p = x > 1$. However, the convergence is not uniform on X, as is seen from the Cauchy criterion: substituting $p_n = n$ and $x_n = 1 + \frac{1}{n}$ for each n in $\sum_{k=n+1}^{n+p} u_k(x)$, we obtain

$$\left| \sum_{k=n+1}^{n+p_n} u_k(x_n) \right| = \frac{1}{(n+1)^{x_n}} + \cdots + \frac{1}{(n+p_n)^{x_n}} > \frac{p_n}{(n+p_n)^{x_n}}$$

$$= \frac{n}{(2n)^{1+1/n}} = \frac{1}{2} \cdot 2^{-1/n} n^{-1/n} \underset{n\to\infty}{\to} \frac{1}{2} \neq 0$$

(notice that $\lim_{n\to\infty} 2^{-1/n} = 1$ and $\lim_{n\to\infty} n^{-1/n} = \lim_{t\to+\infty} t^{-1/t} = \lim_{t\to+\infty} e^{-\ln t/t} = \exp\left(-\lim_{t\to+\infty} \frac{\ln t}{t}\right) = \exp\left(-\lim_{t\to+\infty} \frac{1}{t}\right) = e^0 = 1$).

On the other hand, for any $a > 1$ the same series converges uniformly on $X_a = [a, +\infty)$ by the Weierstrass test: $|u_n(x)| = \frac{1}{n^x} \leq \frac{1}{n^a}$, $\forall x \in X_a$ and $\forall n \in \mathbb{N}$, and the numerical series $\sum_{n=1}^{\infty} \frac{1}{n^a}$ is convergent (p-series with $p = a > 1$).

Note that the functions $u_n(x) = \frac{1}{n^x}$ are infinitely differentiable and the mth-order derivative has the form $u_n^{(m)}(x) = (-1)^m \frac{\ln^m n}{n^x}$. Let us consider the series of the mth derivatives $\sum_{n=1}^{\infty} u_n^{(m)}(x) = (-1)^m \sum_{n=1}^{\infty} \frac{\ln^m n}{n^x}$ and prove that for any $m \geq 1$ this series converges nonuniformly on $X = (1, +\infty)$, and uniformly on $X_a = [a, +\infty)$, $\forall a > 1$. Let us start with convergence of the series of the derivatives on $X = (1, +\infty)$. Consider an arbitrary fixed $x \in (1, +\infty)$. Due to the density of the set of real numbers, there exists c such that $1 < c < x$. The series of the derivatives can be rewritten as follows:

$$\sum_{n=1}^{\infty} u_n^{(m)}(x) = (-1)^m \sum_{n=1}^{\infty} \frac{1}{n^c} \cdot \frac{\ln^m n}{n^{x-c}}.$$

Note that for $\forall \alpha = x - c > 0$ and $\forall m \in \mathbb{N}$, the limit $\lim_{t \to +\infty} \frac{\ln^m t}{t^\alpha}$ can be calculated just applying m times l'Hospital's rule:

$$\lim_{t \to +\infty} \frac{\ln^m t}{t^\alpha} = \lim_{t \to +\infty} \frac{m \ln^{m-1} t}{\alpha t^\alpha}$$

$$= \lim_{t \to +\infty} \frac{m(m-1) \ln^{m-2} t}{\alpha^2 t^\alpha} = \cdots = \lim_{t \to +\infty} \frac{m!}{\alpha^m t^\alpha} = 0.$$

Therefore, there exists N such that $\frac{\ln^m n}{n^{x-c}} < 1, \forall n > N$, that is, $\frac{1}{n^c} \cdot \frac{\ln^m n}{n^{x-c}} < \frac{1}{n^c}, \forall n > N$. Since the series $\sum_{n=1}^{\infty} \frac{1}{n^c}$ converges (p-series with $p = c > 1$), by the comparison test of positive series it follows that the series $\sum_{n=1}^{\infty} \frac{\ln^m n}{n^x}$ also converges for any fixed $x \in (1, +\infty)$, which implies the convergence of $\sum_{n=1}^{\infty} u_n^{(m)}(x)$ on $X = (1, +\infty)$.

To show that this convergence is not uniform on X, we can use the Cauchy criterion with the same parameters $p_n = n$ and $x_n = 1 + \frac{1}{n}$ for each n as for the original series. Then $\ln^m n > 1$ for $\forall n \geq 3$, and we obtain

$$\left| \sum_{k=n+1}^{n+p_n} u_k^{(m)}(x_n) \right| = \frac{\ln^m(n+1)}{(n+1)^{x_n}} + \cdots + \frac{\ln^m(n+p_n)}{(n+p_n)^{x_n}}$$

$$> \frac{1}{(n+1)^{x_n}} + \cdots + \frac{1}{(n+p_n)^{x_n}}$$

$$> \frac{p_n}{(n+p_n)^{x_n}} = \frac{n}{(2n)^{1+1/n}}$$

$$= \frac{1}{2} \cdot 2^{-1/n} n^{-1/n} \xrightarrow[n \to \infty]{} \frac{1}{2} \neq 0.$$

This means that the convergence is not uniform on X. At the same time, this series converges uniformly on $X_a = [a, +\infty)$ for any $a > 1$. Indeed, choosing d between 1 and a ($1 < d < a$), evaluating the derivatives for sufficiently large n in the form

$$|u_n^{(m)}(x)| = \frac{\ln^m n}{n^x} \leq \frac{\ln^m n}{n^a} = \frac{1}{n^d} \cdot \frac{\ln^m n}{n^{a-d}} \leq \frac{1}{n^d}$$

(again use the fact that $\lim_{n \to \infty} \frac{\ln^m n}{n^\alpha} = 0, \forall \alpha > 0, \forall m \in \mathbb{N}$), and noting that the series $\sum_{n=1}^{\infty} \frac{1}{n^d}$ is convergent (p-series with $p = d > 1$), we can conclude that by the Weierstrass test the series $\sum_{n=1}^{\infty} u_n^{(m)}(x)$ converges uniformly on $X_a = [a, +\infty)$, $\forall a > 1$.

Finally, using the fact that the differentiability is a local property, we can show that the original series can be differentiated term by term on $X = (1, +\infty)$ as many times as we want. For any $x \in X$, there is a such that $1 < a < x$. It was already proved that the original series and the series of mth derivatives, $\forall m \in \mathbb{N}$, converge uniformly on $[a, +\infty)$. Therefore, the original series is infinitely differentiable on $[a, +\infty)$ and its mth derivative can be found by differentiating term by term m times. Since $x \in [a, +\infty)$, the same is true at the point x. Recalling that x is an arbitrary point in X, it follows that the original series can be differentiated term by term m times on $X = (1, +\infty)$.

Example 5. A sequence $f_n(x)$ of infinitely differentiable functions converges uniformly on X to a function $f(x)$, but the sequence $f_n'(x)$ diverges at each point of X.

Solution
Each of the functions $f_n(x) = \frac{\sin nx}{\sqrt{n}}$ is infinitely differentiable on $X = \mathbb{R}$ and the limit function of this sequence is zero: $\lim_{n \to \infty} f_n(x) = 0 = f(x)$, $\forall x \in \mathbb{R}$. Moreover, the convergence is uniform on $\mathbb{R}$, because simultaneously for all $x \in \mathbb{R}$ it holds that $|f_n(x) - f(x)| = \frac{|\sin nx|}{\sqrt{n}} \leq \frac{1}{\sqrt{n}} \underset{n \to \infty}{\to} 0$. Let us consider now the sequence of the derivatives $f_n'(x) = \sqrt{n} \cos nx$ and show that it is divergent at each x. If $x = 0$, then $f_n'(0) = \sqrt{n} \underset{n \to \infty}{\to} \infty$, that is, the sequence diverges at 0. If $x \neq 0$, then the sequence of the derivatives is unbounded: if $n \in \mathbb{N}$ is such that $|\cos nx| \geq \frac{1}{2}$, then $|f_n'(x)| = |\sqrt{n} \cos nx| \geq \frac{\sqrt{n}}{2}$; otherwise (if $n \in \mathbb{N}$ is such that $|\cos nx| < \frac{1}{2}$), we have $|\cos 2nx| = |2\cos^2 nx - 1| = 1 - 2\cos^2 nx > \frac{1}{2}$, that is, $|f_{2n}'(x)| = |\sqrt{2n} \cos 2nx| > \frac{\sqrt{2n}}{2}$. Since the sequence is unbounded, it is divergent.

Example 6. A sequence $f_n(x)$ of infinitely differentiable functions converges uniformly on X, but the sequence of the derivatives $f_n'(x)$ is convergent and divergent at infinitely many points of X.

Solution
The sequence of infinitely differentiable functions $f_n(x) = \frac{\cos nx}{n}$ converges uniformly on $\mathbb{R}$ to zero:

$$\lim_{n \to \infty} f_n(x) = 0; \quad |f_n(x) - f(x)| = \frac{|\cos nx|}{n} \leq \frac{1}{n} \underset{n \to \infty}{\to} 0, \forall x \in \mathbb{R}.$$

However, there are two infinite sets of points, the first of which contains the convergence points of the sequence $f_n'(x) = -\sin nx$, and the second one—the divergence points. Evidently, the first set can be chosen as $x_k = k\pi$, $\forall k \in \mathbb{Z}$, since $f_n'(x_k) = 0$, $\forall n \in \mathbb{N}$, and therefore $\lim_{n \to \infty} f_n'(x_k) = 0$, $\forall x_k = k\pi$, $\forall k \in \mathbb{Z}$.

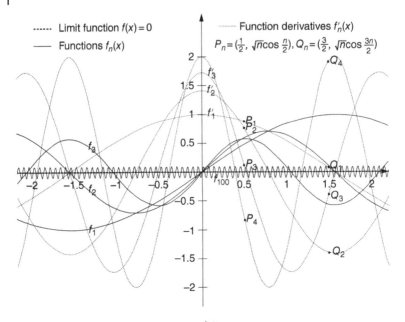

Figure 3.5 Example 5, sequence $f_n(x) = \frac{\sin nx}{\sqrt{n}}$.

To construct the second set, let us prove that for any $x \in \mathbb{R}$ such that $\sin x \neq 0$, that is, $x \neq k\pi$, $\forall k \in \mathbb{Z}$, the limit $\lim\limits_{n\to\infty} \sin nx$ does not exist. Suppose, by absurd, that $\exists x \in \mathbb{R}$, $\sin x \neq 0$ such that $\lim\limits_{n\to\infty} \sin nx$ exists and denote this limit by a. Then the following two limits also exist: $\lim\limits_{n\to\infty} \cos nx = \lim\limits_{n\to\infty} \pm\sqrt{1 - \sin^2 nx} = \pm\sqrt{1 - a^2}$ and $\lim\limits_{n\to\infty} \sin 2nx = \lim\limits_{n\to\infty} 2 \sin nx \cdot \cos nx = 2a \cdot (\pm\sqrt{1 - a^2})$ (note that if $a \neq 0$, then the square root should be chosen with the plus sign). On the other hand, $\lim\limits_{n\to\infty} \sin 2nx$ is a partial limit of $\lim\limits_{n\to\infty} \sin nx$ and, therefore, it equals a. From the equality $a = 2a \cdot (\pm\sqrt{1 - a^2})$, one gets $a_1 = 0$ or $1 - a^2 = \frac{1}{4}$, that is, $a_{2,3} = \pm\frac{\sqrt{3}}{2}$. In the first case, let us consider the additional trigonometric equality $\sin(n + 1)x = \sin nx \cdot \cos x + \cos nx \cdot \sin x$ and pass to limit in both sides as $n \to \infty$. Since $\lim\limits_{n\to\infty} \sin nx = 0$ and $\lim\limits_{n\to\infty} \cos nx = \pm\sqrt{1 - a_1^2} = \pm 1$, one obtains $0 = 0 \cdot \cos x \pm 1 \cdot \sin x$. It follows then that $\sin x = 0$ that contradicts the choice of the point x. In the case $a_{2,3} = \pm\frac{\sqrt{3}}{2}$, let us consider the limit of the additional equality $\cos 2nx = \cos^2 nx - \sin^2 nx$, which gives $\frac{1}{2} = \left(\frac{1}{2}\right)^2 - \left(\pm\frac{\sqrt{3}}{2}\right)^2 = -\frac{1}{2}$ (since $\lim\limits_{n\to\infty} \cos nx = \sqrt{1 - a_{2,3}^2} = \frac{1}{2}$). Therefore, again one arrives to a contradiction. Thus, the supposition that

$\lim\limits_{n\to\infty} \sin nx$ exists is wrong, that is, the sequence $\sin nx$ diverges for $\forall x \neq k\pi$, $\forall k \in \mathbb{Z}$.

Example 7. A series $\sum u_n(x)$ of continuous functions converges uniformly on X, but the sum of this series is not differentiable at an infinite number of points in X.

Solution
Consider all the rational points of the interval $[0, 1]$ and order them in the form of a numerical sequence r_n, $n = 1, 2, \ldots$ (this can be done, since the set of all the rational numbers of any interval is countable). Define the function $f(x)$ as follows: $f(x) = \sum_{n=1}^{\infty} u_n(x) = \sum_{n=1}^{\infty} \frac{|x-r_n|}{3^n}$, $x \in [0, 1]$. Since $x, r_n \in [0, 1]$, $\forall n \in \mathbb{N}$, we have $|x - r_n| \leq 1$, that is, $|u_n(x)| = \frac{|x-r_n|}{3^n} \leq \frac{1}{3^n}$. The numerical series $\sum_{n=1}^{\infty} \frac{1}{3^n}$ converges, and consequently, the series for $f(x)$ converges uniformly on $[0, 1]$ by the Weierstrass test. Additionally, the continuity of the functions $\frac{|x-r_n|}{3^n}$ on $[0, 1]$ for $\forall n \in \mathbb{N}$ ensures the continuity of $f(x)$ on $[0, 1]$.

Each of $u_n(x) = \frac{|x-r_n|}{3^n}$ is differentiable at any $x \neq r_n$, $u'_n(x) = \frac{1}{3^n} \begin{cases} -1, x < r_n \\ 1, x > r_n \end{cases}$, and has one-sided derivatives at the point $x = r_n$: $u'_n(r_{n_-}) = -\frac{1}{3^n}$, $u'_n(r_{n_+}) = \frac{1}{3^n}$. Since $|u'_n(x_-)| = |u'_n(x_+)| = \frac{1}{3^n}$ for any x and the series $\sum_{n=1}^{\infty} \frac{1}{3^n}$ is convergent, the series $\sum_{n=1}^{\infty} u'_n(x_-)$ and $\sum_{n=1}^{\infty} u'_n(x_+)$ converge uniformly (and absolutely) on $\mathbb{R}$ (and, in particular, on $[0, 1]$).

Let us prove that $f(x)$ has one-sided derivatives at each point x_0 in $[0, 1]$. For an arbitrary fixed $x \in [0, 1)$, we choose $h > 0$ such that $x + h \in [0, 1]$ and consider the series

$$\frac{f(x+h) - f(x)}{h} = \sum_{n=1}^{\infty} \frac{u_n(x+h) - u_n(x)}{h}$$

$$= \sum_{n=1}^{\infty} \frac{|x+h-r_n| - |x-r_n|}{h \cdot 3^n} = \sum_{n=1}^{\infty} v_n(h).$$

For any $h \in (0, 1-x)$, the following evaluation holds:

$$|v_n(h)| = \frac{||x+h-r_n| - |x-r_n||}{h \cdot 3^n} \leq \frac{|x+h-r_n - (x-r_n)|}{h \cdot 3^n} = \frac{1}{3^n},$$

and since the series $\sum_{n=1}^{\infty} \frac{1}{3^n}$ converges, by the Weierstrass test it implies the uniform convergence of the series $\sum_{n=1}^{\infty} v_n(h)$ on $(0, 1-x)$. Also, for any fixed n we have $\lim\limits_{h\to 0_+} v_n(h) = \lim\limits_{h\to 0_+} \frac{u_n(x+h)-u_n(x)}{h} = u'_n(x_+)$, and the series $\sum_{n=1}^{\infty} u'_n(x_+)$

converges. Therefore, the limit of the series exists and can be calculated term by term:

$$f'(x_+) = \lim_{h \to 0_+} \frac{f(x+h) - f(x)}{h} = \lim_{h \to 0_+} \sum_{n=1}^{\infty} v_n(h)$$

$$= \sum_{n=1}^{\infty} \lim_{h \to 0_+} v_n(h) = \sum_{n=1}^{\infty} u_n'(x_+).$$

Similarly, $f'(x_-) = \sum_{n=1}^{\infty} u_n'(x_-)$ for any $x \in (0, 1]$.

Let us show now that $f(x)$ is differentiable at each irrational point x in $[0, 1]$: since $x \neq r_n$, every $u_n(x) = \frac{|x - r_n|}{3^n}$ is differentiable at x, which means that $u_n'(x_+) = u_n'(x_-)$, and consequently, $f'(x_+) = \sum_{n=1}^{\infty} u_n'(x_+) = \sum_{n=1}^{\infty} u_n'(x_-) = f'(x_-)$, that is, $f(x)$ is differentiable.

Finally, let us show that $f(x)$ is not differentiable at any rational point x in $[0, 1]$. For a rational point x_0, there exists k such that $x_0 = r_k$. The function $u_k(x) = \frac{|x - r_k|}{3^k}$ is not differentiable at $x_0 = r_k$, because the one-sided derivatives are different: $u_k'(r_{k_-}) = -\frac{1}{3^k} \neq \frac{1}{3^k} = u_k'(r_{k_+})$. Thus, at any rational point, one of the terms of the series is nondifferentiable, and all others are differentiable:

$$f(x) = \sum_{n=1}^{\infty} u_n(x) = \sum_{n=1, n \neq k}^{\infty} u_n(x) + u_k(x) \equiv g_k(x) + u_k(x).$$

Using the same reasoning as above for an irrational point, we can deduce that the function $g_k(x)$ is differentiable at $x_0 = r_k$. Therefore, $f(x)$ is not differentiable at $x_0 = r_k$ as the sum of differentiable and nondifferentiable functions.

Remark. This example can be also formulated in the form: a function $f(x)$ is continuous on an interval, but it can be nondifferentiable at infinitely many points of this interval.

Example 8. A series $\sum u_n(x)$ of continuous functions converges uniformly on X, but nevertheless the sum of this series is not differentiable at any point of X.

Solution

Let $d(x)$ be the distance from x to the nearest integer, which means that $d(x)$ is a continuous 1-periodic function with the range $\left[0, \frac{1}{2}\right]$ defined on the fundamental interval $[0, 1]$ in the following form: $d(x) = \begin{cases} x, x \in \left[0, \frac{1}{2}\right] \\ 1 - x, x \in \left[\frac{1}{2}, 1\right] \end{cases}$.

Accordingly, $u_n(x) = \frac{1}{10^n} d(10^n x)$ is a continuous $\frac{1}{10^n}$-periodic function with the range $\left[0, \frac{1}{2 \cdot 10^n}\right]$ defined on the fundamental interval $\left[0, \frac{1}{10^n}\right]$ in the form:

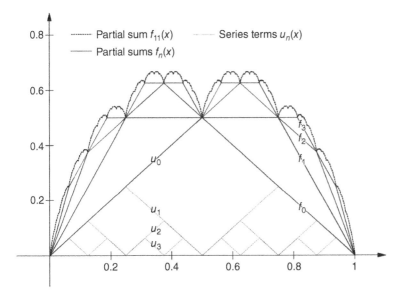

Figure 3.6 Example 8, partial sums of series $\sum_{n=0}^{\infty} \tilde{u}_n(x)$ with $\tilde{u}_n(x) = \frac{1}{2^n} d(2^n x)$ instead of $u_n(x) = \frac{1}{10^n} d(10^n x)$ for better visualization.

$$u_n(x) = \begin{cases} x, x \in \left[0, \frac{1}{2\cdot 10^n}\right] \\ \frac{1}{10^n} - x, x \in \left[\frac{1}{2\cdot 10^n}, \frac{1}{10^n}\right] \end{cases}.$$ Let us consider the series $\sum_{n=0}^{\infty} u_n(x)$ on

$X = [0, 1]$. Since $|u_n(x)| \leq \frac{1}{10^n}$ for all $x \in [0, 1]$ and the numerical series $\sum_{n=0}^{\infty} \frac{1}{10^n}$ converges, by the Weierstrass test the series $\sum_{n=0}^{\infty} u_n(x)$ converges uniformly on $[0, 1]$ to its sum $f(x)$, and the continuity of $u_n(x)$ implies the continuity of $f(x)$.

Now let us show that $f(x)$ is differentiable nowhere in $[0, 1]$. Let us consider any fixed a in $[0, 1)$ with the decimal expansion $a = 0. a_1 a_2 \cdots a_n a_{n+1} \cdots$, and choose a particular sequence h_m approaching 0, as m approaches infinity, in the following form: $h_m = \begin{cases} -10^{-m}, a_m = 4 \text{ or } 9 \\ 10^{-m}, \text{otherwise} \end{cases}$. Then

$$\frac{f(a + h_m) - f(a)}{h_m} = \sum_{n=0}^{\infty} \frac{1}{10^n} \cdot \frac{d(10^n(a + h_m)) - d(10^n a)}{\pm 10^{-m}}$$

$$= \sum_{n=0}^{\infty} \pm 10^{m-n} \cdot [d(10^n(a + h_m)) - d(10^n a)].$$

Since $10^n h_m$ is an integer for $n \geq m$, we get $d(10^n(a + h_m)) - d(10^n a) = 0$. Therefore, the last series is actually a finite sum with m terms. For these terms

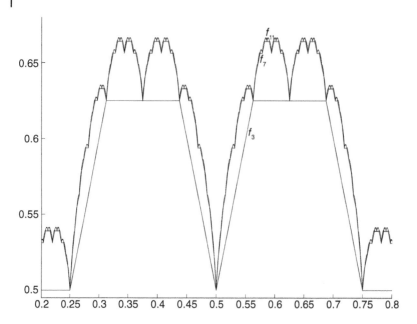

Figure 3.7 Example 8, amplified view of partial sums of series $\sum_{n=0}^{\infty} \tilde{u}_n(x)$.

(when $n < m$) we can represent $10^n a = a_1 \cdots a_n + 0. \, a_{n+1} a_{n+2} \cdots a_m \cdots$ and $10^n(a + h_m) = a_1 \cdots a_n + 0. \, a_{n+1} a_{n+2} \cdots (a_m \pm 1) \cdots$, where $a_1 \cdots a_n$ is the integer part of both numbers. (Note that except for mth decimal digit, other digits are not changed, because we have chosen $h_m = -10^{-m}$ if $a_m = 9$). If it happens that $0. \, a_{n+1} a_{n+2} \cdots a_m \cdots < \frac{1}{2}$, then also $0. \, a_{n+1} a_{n+2} \cdots (a_m \pm 1) \cdots < \frac{1}{2}$ (in the special case $n = m - 1$, the last inequality is true because we have chosen $h_m = -10^{-m}$ when $a_m = 4$). Therefore,

$$d(10^n(a + h_m)) - d(10^n a)$$
$$= d(0.a_{n+1} \cdots (a_m \pm 1) \cdots) - d(0.a_{n+1} \cdots a_m \cdots)$$
$$= 0.a_{n+1} \cdots (a_m \pm 1) \cdots - 0.a_{n+1} \cdots a_m \cdots = \pm 10^{n-m}.$$

Similarly, if $0. \, a_{n+1} a_{n+2} \cdots a_m \cdots \geq \frac{1}{2}$, then $0. \, a_{n+1} a_{n+2} \cdots (a_m \pm 1) \cdots \geq \frac{1}{2}$ (here, between two equivalent representations of the rational numbers $0.a_{n+1} \cdots a_k 00 \cdots = 0.a_{n+1} \cdots (a_k - 1)99 \cdots$ we use the former), and consequently

$$d(10^n(a + h_m)) - d(10^n a)$$
$$= d(0.a_{n+1} \cdots (a_m \pm 1) \cdots) - d(0.a_{n+1} \cdots a_m \cdots)$$
$$= (1 - 0.a_{n+1} \cdots (a_m \pm 1) \cdots) - (1 - 0.a_{n+1} \cdots a_m \cdots) = \mp 10^{n-m}.$$

Thus, for $n < m$ we have

$$10^{m-n} \cdot [d(10^n(a + h_m)) - d(10^n a)] = \pm 1,$$

and, therefore

$$\frac{f(a + h_m) - f(a)}{h_m} = \sum_{n=0}^{m-1} \pm 10^{m-n} \cdot [d(10^n(a + h_m)) - d(10^n a)] = \sum_{n=0}^{m-1} \pm 1.$$

The last sum is an even integer when m is even, and an odd integer when m is odd. Therefore, the sequence $\frac{f(a+h_m)-f(a)}{h_m} = \sum_{n=0}^{m-1} \pm 1$ does not converge as m approaches infinity, because this is a sequence of integers that are alternately odd and even. Hence, we have constructed a sequence $h_m = \pm 10^{-m}$ such that the limit of $\frac{f(a+h_m)-f(a)}{h_m}$ does not exists, which means that $f(x)$ is not differentiable at a.

Remark 1. This is a strengthened version of Example 7.

Remark 2. Of course, all these considerations and conclusions are also true on $\mathbb{R}$, because all the functions $u_n(x) = \frac{1}{10^n} d(10^n x)$, $n = 0, 1, 2, \ldots$ have the common period 1, and so does $f(x)$. Therefore, their properties on $[0, 1]$ are automatically extended to $\mathbb{R}$.

Remark 3. This example can also be formulated in the following form: there exist continuous on an interval functions $f(x)$ that are nondifferentiable at any point of this interval. The provided counterexample was first presented by Takagi in 1903 and later (in 1930) rediscovered by van der Waerden. The first (and more intricate) example of an everywhere continuous and nowhere differentiable function was constructed by Weierstrass as early as 1861.

3.2 Integrability of the Limit Function

Example 9. A sequence $f_n(x)$ of Riemann integrable on $[a, b]$ functions converges on $[a, b]$ to a function $f(x)$, but the limit function is not Riemann integrable on $[a, b]$.

Solution
Consider the set of the rational points in $[0, 1]$ and order it to form a sequence r_n, $n = 1, 2, \ldots$. Define the functions $f_n(x)$ on $[0, 1]$ as follows: $f_n(x) = \begin{cases} 1, x = r_1, r_2, \ldots, r_n \\ 0, \text{otherwise} \end{cases}$. Each of $f_n(x)$ is bounded on $[0, 1]$ and has the finite number of the discontinuities (at the points $r_1, \ldots, r_n$), which implies that $f_n(x)$ is Riemann integrable on $[0, 1]$. Since $f_{n+1}(x) \geq f_n(x)$, $\forall x \in [0, 1]$

$(f_{n+1}(r_{n+1}) = 1 > 0 = f_n(r_{n+1})$ and $f_{n+1}(x) = f_n(x)$, $\forall x \neq r_{n+1})$, the sequence $f_n(x)$ is increasing and bounded with respect to n, and therefore it is convergent for any fixed $x \in [0, 1]$. From the form of $f_n(x)$, it follows that the limit function is Dirichlet's function considered on $[0, 1]$: $D(x) = \begin{cases} 1, x \in \mathbb{Q} \\ 0, x \in \mathbb{I} \end{cases}$, which is not Riemann integrable on any interval. Note, that the convergence of $f_n(x)$ to $D(x)$ is not uniform on $[0, 1]$: for any $n \in \mathbb{N}$, there exists $x_n = r_{n+1} \in \mathbb{Q} \cap [0, 1]$ such that $|f_n(r_{n+1}) - D(r_{n+1})| = |0 - 1| = 1 > \frac{1}{2} = \varepsilon_0$.

Example 10. A sequence $f_n(x)$ converges on $[a, b]$ to a function $f(x)$, and each of the functions $f_n(x)$ is not Riemann integrable on $[a, b]$, but $f(x)$ is Riemann integrable on $[a, b]$.

Solution
The functions $f_n(x) = \begin{cases} 1/n, x \in \mathbb{Q} \\ 0, x \in \mathbb{I} \end{cases}$ considered on $[0, 1]$ are scaled Dirichlet's functions: $f_n(x) = \frac{1}{n}D(x)$ and, therefore, are not Riemann integrable on $[0, 1]$. At the same time, the sequence $f_n(x)$ converges to $f(x) \equiv 0$, $\forall x \in [0, 1]$, which is a Riemann integrable function. Moreover, this convergence is uniform due to the following evaluation held simultaneously for all $\forall x \in [0, 1]$: $|f_n(x) - f(x)| \leq \frac{1}{n} \underset{n \to \infty}{\to} 0$.

Example 11. A sequence of continuous functions $f_n(x)$ converges on $[a, b]$ to a function $f(x)$, but $\lim_{n \to \infty} \int_a^b f_n(x)dx \neq \int_a^b f(x)dx$.

Solution
Let us consider the sequence of continuous functions $f_n(x) = nxe^{-nx^2}$ on $[0, 1]$. This sequence converges to $f(x) \equiv 0$, $\forall x \in [0, 1]$, because for $x = 0$ we have $f_n(0) = 0$, $\forall n$, and for $x \neq 0$ we can apply l'Hospital's rule to obtain

$$f(x) = \lim_{n \to \infty} f_n(x) = \lim_{n \to \infty} \frac{nx}{e^{nx^2}} = \lim_{t \to +\infty} \frac{tx}{e^{tx^2}} = \lim_{t \to +\infty} \frac{x}{x^2 e^{tx^2}} = 0.$$

Therefore, $\int_0^1 f(x)dx = 0$. Each of $f_n(x)$ is also integrable (since it is continuous) and $\int_0^1 f_n(x)dx = -\frac{1}{2}e^{-nx^2}\Big|_0^1 = \frac{1}{2} - \frac{1}{2}e^{-n}$. However,

$$\lim_{n \to \infty} \int_0^1 f_n(x)dx = \lim_{n \to \infty} \frac{1}{2}(1 - e^{-n}) = \frac{1}{2} \neq 0 = \int_0^1 f(x)dx.$$

Let us check the type of convergence. Choosing $x_n = \frac{1}{\sqrt{n}} \in [0, 1]$ for any $n \in \mathbb{N}$, we get

$$|f_n(x_n) - f(x_n)| = n \cdot \frac{1}{\sqrt{n}} e^{-n \cdot 1/n} = \sqrt{n} \cdot e^{-1} \underset{n \to \infty}{\to} +\infty,$$

which means that the convergence is nonuniform on $[0, 1]$.

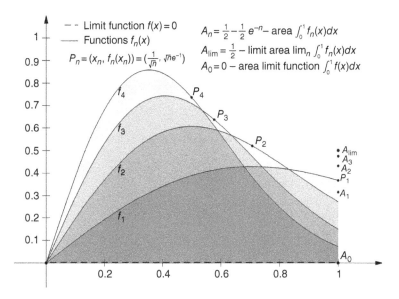

Figure 3.8 Example 11, sequence $f_n(x) = nxe^{-nx^2}$.

Another counterexample can be given using the sequence $f_n(x) = \frac{2nx}{1+n^2x^4}$ on $[0, 1]$. These functions are continuous and the sequence converges to $f(x) \equiv 0$, $\forall x \in [0, 1]$, which implies that $\int_0^1 f(x)dx = 0$. The integral of each $f_n(x)$ is easy to calculate $\int_0^1 f_n(x)dx = \arctan nx^2 |_0^1 = \arctan n$, but the corresponding limit is different from 0:

$$\lim_{n\to\infty} \int_0^1 f_n(x)dx = \lim_{n\to\infty} \arctan n = \frac{\pi}{2} \neq 0 = \int_0^1 f(x)dx.$$

Note that the convergence of $f_n(x)$ is not uniform on $[0, 1]$: for any $n \in \mathbb{N}$, one can choose $x_n = \frac{1}{\sqrt{n}} \in [0, 1]$ to obtain

$$|f_n(x_n) - f(x_n)| = \frac{2n \cdot \frac{1}{\sqrt{n}}}{1 + n^2 \cdot \frac{1}{n^2}} = \sqrt{n} \underset{n\to\infty}{\to} +\infty.$$

Remark 1. Analogous example can be formulated for a function depending on a parameter: a function $f(x, y)$ defined on $X \times Y = [a, b] \times Y$ is continuous in $x \in X$ at any fixed $y \in Y$, and there exists $\lim_{y\to y_0} f(x, y) = \varphi(x)$, but $\int_a^b \varphi(x)dx \neq \lim_{y\to y_0} \int_a^b f(x, y)dx.$

The corresponding counterexample is $f(x, y) = \frac{x}{y^2}e^{-x^2/y^2}$ on $X \times Y = [0, 1] \times (0, 1]$ and $y_0 = 0$. Since $f(0, y) = 0$ and for any fixed $x \in (0, 1]$ the application of l'Hospital's rule gives $\lim\limits_{y\to 0_+} f(x,y) = \lim\limits_{y\to 0_+} \frac{x}{y^2} e^{-x^2/y^2} = \lim\limits_{t\to +\infty} \frac{t}{xe^t} = 0$, one concludes

that $\lim\limits_{y\to 0_+} f(x,y) = 0 = \varphi(x)$, $\forall x \in [0,1]$ and, consequently, $\int_0^1 \varphi(x)dx = 0$. Further, $f(x,y)$ is continuous in $x \in X$, and therefore, Riemann integrable in x, for any fixed $y \in Y$:

$$\int_0^1 f(x,y)dx = -\frac{1}{2}e^{-x^2/y^2}\Big|_0^1 = \frac{1}{2} - \frac{1}{2}e^{-1/y^2}.$$

However,

$$\lim_{y\to 0_+} \int_0^1 f(x,y)dx = \lim_{y\to 0_+} \frac{1}{2}(1 - e^{-1/y^2}) = \frac{1}{2} \ne 0 = \int_0^1 \lim_{y\to 0_+} f(x,y)dx.$$

Note that the convergence is not uniform on $[0,1]$: choosing $x_y = y \in [0,1]$ for $\forall y \in (0,1]$, we obtain

$$|f(x_y,y) - \varphi(x_y)| = \frac{y}{y^2}e^{-y^2/y^2} = \frac{1}{y}e^{-1} \xrightarrow[y\to 0_+]{} +\infty.$$

Remark 2. The corresponding example for a series is as follows: a series of continuous functions converges on $[a,b]$, but it cannot be integrated term by term on $[a,b]$. For a counterexample, one can use the series

$$\sum_{n=1}^\infty u_n(x) = \sum_{n=1}^\infty 2x[n^2 e^{-n^2 x^2} - (n-1)^2 e^{-(n-1)^2 x^2}]$$

on $[0,1]$. The partial sums of this telescoping series can be easily found as follows:

$$f_n(x) = 2x\sum_{k=1}^n [k^2 e^{-k^2 x^2} - (k-1)^2 e^{-(k-1)^2 x^2}]$$

$$= 2x[e^{-x^2} + 2^2 e^{-2^2 x^2} - e^{-x^2} + \cdots + n^2 e^{-n^2 x^2} - (n-1)^2 e^{-(n-1)^2 x^2}]$$

$$= 2xn^2 e^{-n^2 x^2}.$$

Therefore, the sum of the series is $f(x) = \lim\limits_{n\to\infty} f_n(x) = 0$ for any $x \in [0,1]$, and consequently, $\int_0^1 f(x)dx = 0$.

At the same time,

$$\int_0^1 u_n(x)dx = (-e^{-n^2 x^2} + e^{-(n-1)^2 x^2})|_0^1 = -e^{-n^2} + e^{-(n-1)^2}$$

and the numerical series

$$\sum_{n=1}^\infty \int_0^1 u_n(x)dx = \sum_{n=1}^\infty (-e^{-n^2} + e^{-(n-1)^2})$$

is telescoping with the partial sums

$$S_n = \sum_{k=1}^{n} (-e^{-k^2} + e^{-(k-1)^2})$$
$$= -e^{-1} + 1 - e^{-2^2} + e^{-1} + \cdots - e^{-n^2} + e^{-(n-1)^2} = 1 - e^{-n^2}.$$

Therefore,

$$\sum_{n=1}^{\infty} \int_{0}^{1} u_n(x)dx = \sum_{n=1}^{\infty} (-e^{-n^2} + e^{-(n-1)^2})$$
$$= \lim_{n\to\infty} s_n = 1 \neq 0 = \int_{0}^{1} \sum_{n=1}^{\infty} u_n(x)dx.$$

Note that the convergence of the series is nonuniform on $[0, 1]$: for $\forall n \in \mathbb{N}$, there is $x_n = \frac{1}{n}$ such that

$$|r_n(x_n)| = \left| \sum_{k=n+1}^{\infty} u_k(x_n) \right| = |f_n(x_n) - f(x_n)|$$
$$= 2 \cdot \frac{1}{n} \cdot n^2 e^{-n^2 \cdot 1/n^2} = 2ne^{-1} \underset{n\to\infty}{\to} +\infty.$$

Another interesting counterexample is the telescoping series

$$\sum_{n=1}^{\infty} u_n(x) = \sum_{n=1}^{\infty} [nx(1-x^2)^n - (n-1)x(1-x^2)^{n-1}]$$

considered on $[0, 1]$. Its partial sums are

$$f_n(x) = x(1-x^2) + 2x(1-x^2)^2 - x(1-x^2) \cdots + nx(1-x^2)^n$$
$$- (n-1)x(1-x^2)^{n-1} = nx(1-x^2)^n.$$

For $x = 0$ and 1, we have $f_n(x) = 0$, and for $x \in (0, 1)$ we can solve the indeterminate form in the following way:

$$\lim_{n\to\infty} f_n(x) = \lim_{t\to+\infty} x\frac{t}{(1-x^2)^{-t}} = x \lim_{t\to+\infty} \frac{1}{-(1-x^2)^{-t} \ln(1-x^2)} = 0$$

(here the discrete variable n was replaced by the continuous variable t and then l'Hospital's rule was applied). Therefore, the sum of the series is $f(x) = \lim_{n\to\infty} f_n(x) = 0$ for any $x \in [0, 1]$ and, consequently, $\int_0^1 f(x)dx = 0$.

At the same time,

$$\int_{0}^{1} u_n(x)dx = \left(-\frac{1}{2}\frac{n}{n+1}(1-x^2)^{n+1} + \frac{1}{2}\frac{n-1}{n}(1-x^2)^n \right)\Big|_0^1$$
$$= \frac{1}{2}\frac{n}{n+1} - \frac{1}{2}\frac{n-1}{n}$$

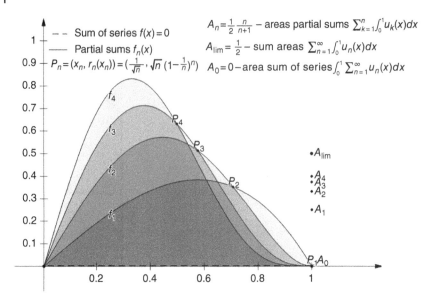

Figure 3.9 Example 11, series $\sum_{n=1}^{\infty}[nx(1-x^2)^n - (n-1)x(1-x^2)^{n-1}]$.

and the resulting numerical series

$$\sum_{n=1}^{\infty}\int_0^1 u_n(x)dx = \sum_{n=1}^{\infty}\frac{1}{2}\left(\frac{n}{n+1}-\frac{n-1}{n}\right)$$

is telescoping. Its partial sums are

$$s_n = \sum_{k=1}^{n}\frac{1}{2}\left(\frac{k}{k+1}-\frac{k-1}{k}\right) = \frac{1}{2}\frac{n}{n+1},$$

and, therefore,

$$\sum_{n=1}^{\infty}\int_0^1 u_n(x)dx = \sum_{n=1}^{\infty}\frac{1}{2}\left(\frac{n}{n+1}-\frac{n-1}{n}\right) = \lim_{n\to\infty}s_n$$

$$= \frac{1}{2}\neq 0 = \int_0^1\sum_{n=1}^{\infty}u_n(x)dx.$$

Note that the convergence of the series is nonuniform on $[0,1]$: for $\forall n \in \mathbb{N}$, there is $x_n = \frac{1}{\sqrt{n}}$ such that

$$|r_n(x_n)| = |f_n(x_n)-f(x_n)| = \sqrt{n}\cdot\left(1-\frac{1}{n}\right)^n \xrightarrow[n\to\infty]{} +\infty.$$

Remark 3. In all the counterexamples, the impossibility to interchange the operations of the integration and the limit or infinite sum is caused by nonuniform convergence of the sequence, function, or series. Although we see that the condition of the uniform convergence is important, it is not necessary for the integration under the sign of limit or infinite sum as shown in the next example.

Example 12. A sequence of continuous functions $f_n(x)$ converges on $[a, b]$ to a function $f(x)$ and $\lim\limits_{n\to\infty} \int_a^b f_n(x)dx = \int_a^b f(x)dx$, but the convergence is nonuniform on $[a, b]$.

Solution
Let us consider the sequence of continuous functions $f_n(x) = nx(1-x)^n$ on $[0, 1]$. This sequence converges to $f(x) \equiv 0$, $\forall x \in [0, 1]$. Indeed, for $x = 0$ and $x = 1$ we have $f_n(0) = f_n(1) = 0$, $\forall n$, and for $x \in (0, 1)$ we can apply l'Hospital's rule to obtain

$$f(x) = \lim_{n\to\infty} f_n(x) = \lim_{t\to+\infty} x\frac{t}{(1-x)^{-t}} = x\lim_{t\to+\infty} \frac{1}{-(1-x)^{-t}\ln(1-x)} = 0.$$

Therefore, $\int_0^1 f(x)dx = 0$. On the other hand, each $f_n(x)$ is also integrable (since it is continuous) and

$$\int_0^1 f_n(x)dx = \left(-\frac{n}{n+1}(1-x)^{n+1} + \frac{n}{n+2}(1-x)^{n+2}\right)\Big|_0^1$$

$$= \frac{n}{n+1} - \frac{n}{n+2}.$$

Therefore,

$$\lim_{n\to\infty} \int_0^1 f_n(x)dx = \lim_{n\to\infty} \frac{n}{(n+1)(n+2)} = 0$$

$$= \int_0^1 \lim_{n\to\infty} f_n(x)dx = \int_0^1 f(x)dx.$$

Thus, all the statement conditions are satisfied. Nevertheless, the convergence of $f_n(x)$ is not uniform on $[0, 1]$: choosing $x_n = \frac{1}{n} \in [0, 1]$ for any $n \in \mathbb{N}$, we get

$$|f_n(x_n) - f(x_n)| = n \cdot \frac{1}{n}\left(1 - \frac{1}{n}\right)^n \xrightarrow[n\to\infty]{} e^{-1} \neq 0.$$

Remark 1. Analogous example can be formulated for a function depending on a parameter: a function $f(x, y)$ defined on $X \times Y = [a, b] \times Y$ is continuous in $x \in X$ at any fixed $y \in Y$, there exists $\lim\limits_{y\to y_0} f(x, y) = \varphi(x)$, and $\int_a^b \varphi(x)dx = \lim\limits_{y\to y_0} \int_a^b f(x, y)dx$, but $f(x, y)$ converges nonuniformly on X to $\varphi(x)$. The corresponding counterexample is $f(x, y) = \frac{xy}{x^2+y^2}$ on $X \times Y = [0, 1] \times (0, 1]$

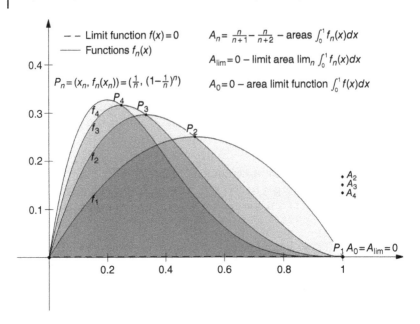

Figure 3.10 Example 12, sequence $f_n(x) = nx(1-x)^n$.

and $y_0 = 0$. For any fixed $x \in [0, 1]$, we have $\lim\limits_{y \to 0_+} f(x, y) = 0 = \varphi(x)$ and, therefore, $\int_0^1 \varphi(x)dx = 0$. Further, $f(x, y)$ is continuous in $x \in X$, and consequently Riemann integrable in x, for any fixed $y \in Y$:

$$\int_0^1 f(x, y)dx = \frac{1}{2}y\ln(x^2 + y^2)\Big|_0^1 = \frac{1}{2}y(\ln(1 + y^2) - \ln y^2).$$

Consequently,

$$\lim_{y \to 0_+} \int_0^1 f(x, y)dx = \lim_{y \to 0_+} \frac{1}{2}y\ln\left(1 + \frac{1}{y^2}\right)$$

$$= \lim_{t \to +\infty} \frac{\ln(1 + t^2)}{2t} = 0$$

$$= \int_0^1 \lim_{y \to 0_+} f(x, y)dx = \int_0^1 \varphi(x)dx.$$

However, the convergence is not uniform on $[0, 1]$: choosing $x_y = y \in [0, 1]$ for $\forall y \in (0, 1]$, we obtain $|f(x_y, y) - \varphi(x_y)| = \frac{y^2}{2y^2} = \frac{1}{2} \nrightarrow 0$.

Remark 2. The corresponding example for a series goes as follows: a series of continuous functions converges on $[a, b]$, and it can be integrated term by term, but this convergence is nonuniform on $[a, b]$.

For a counterexample, consider the series $\sum_{n=1}^{\infty} u_n(x)$, $u_n = x^{\frac{1}{2n+1}} - x^{\frac{1}{2n-1}}$ on $[0, 1]$. This is a telescoping series with the partial sums $f_n(x) = \sum_{k=1}^{n} \left(x^{\frac{1}{2k+1}} - x^{\frac{1}{2k-1}} \right) = x^{\frac{1}{2n+1}} - x$. For $x = 0$, all these sums are 0, and consequently, $f(0) = \lim_{n \to \infty} f_n(0) = 0$; for $x \in (0, 1]$, the total sum of the series is found in the form $f(x) = \lim_{n \to \infty} x^{\frac{1}{2n+1}} - x = 1 - x$. Thus, the series sum is $f(x) = \begin{cases} 0, x = 0 \\ 1 - x, x \in (0, 1] \end{cases}$, and therefore, $\int_0^1 f(x)dx = \int_0^1 (1 - x)dx = \left(x - \frac{x^2}{2} \right)\Big|_0^1 = \frac{1}{2}$ (the value of a function in a single point does not affect the value of the definite integral).

On the other hand,

$$\int_0^1 u_n(x)dx = \left(\frac{2n+1}{2n+2} x^{\frac{2n+2}{2n+1}} - \frac{2n-1}{2n} x^{\frac{2n}{2n-1}} \right)\Big|_0^1 = \frac{2n+1}{2n+2} - \frac{2n-1}{2n}$$

and the numerical series $\sum_{n=1}^{\infty} \int_0^1 u_n(x)dx = \sum_{n=1}^{\infty} \left(\frac{2n+1}{2n+2} - \frac{2n-1}{2n} \right)$ is telescoping with the partial sums $s_n = \frac{2n+1}{2n+2} - \frac{1}{2}$. Therefore,

$$\sum_{n=1}^{\infty} \int_0^1 u_n(x)dx = \lim_{n \to \infty} s_n = \lim_{n \to \infty} \left(\frac{2n+1}{2n+2} - \frac{1}{2} \right) = \frac{1}{2} = \int_0^1 \sum_{n=1}^{\infty} u_n(x)dx,$$

that is, the original series can be integrated term by term. Nevertheless, the convergence of the series is nonuniform on $[0, 1]$: for $\forall n \in \mathbb{N}$, there exists $x_n = 2^{-(2n+1)} \in [0, 1]$ such that

$$|r_n(x_n)| = |f_n(x_n) - f(x_n)| = \left| x_n^{\frac{1}{2n+1}} - x_n - 1 + x_n \right|$$

$$= 1 - (2^{-(2n+1)})^{\frac{1}{2n+1}} = \frac{1}{2} \nrightarrow 0.$$

Another counterexample is the telescoping series $\sum_{n=1}^{\infty} u_n(x) = \sum_{n=1}^{\infty} \left(\frac{1}{1+nx} - \frac{1}{1+(n+1)x} \right)$ considered on $[0, 1]$. Its partial sums are $f_n(x) = \frac{1}{1+x} - \frac{1}{1+(n+1)x}$. For $x = 0$, we have $f_n(0) = 0$, $\forall n$, and for $x \in (0, 1]$ we get $\lim_{n \to \infty} f_n(x) = \frac{1}{1+x}$. Therefore, the sum of the series is $f(x) = \begin{cases} 0, x = 0 \\ \frac{1}{1+x}, x \in (0, 1] \end{cases}$, and consequently,

$$\int_0^1 f(x)dx = \int_0^1 \frac{1}{1+x}dx = \ln(1 + x)\Big|_0^1 = \ln 2$$

(again we use the property that the value of the integral does not depend on the value of a function in a single point).

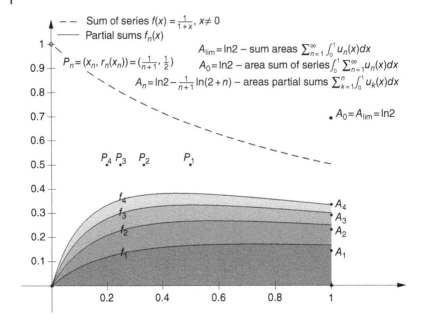

Figure 3.11 Example 12, series $\sum_{n=1}^{\infty} \left(\frac{1}{1+nx} - \frac{1}{1+(n+1)x} \right)$.

On the other hand,

$$\int_0^1 u_n(x)dx = \left(\frac{1}{n} \ln(1+nx) - \frac{1}{n+1} \ln(1+(n+1)x) \right) \Big|_0^1$$

$$= \frac{1}{n} \ln(1+n) - \frac{1}{n+1} \ln(1+(n+1)).$$

The resulting telescoping series

$$\sum_{n=1}^{\infty} \int_0^1 u_n(x)dx = \sum_{n=1}^{\infty} \left(\frac{1}{n} \ln(1+n) - \frac{1}{n+1} \ln(2+n) \right)$$

has the partial sums

$$s_n = \sum_{k=1}^{n} \left(\frac{1}{k} \ln(1+k) - \frac{1}{k+1} \ln(2+k) \right) = \ln 2 - \frac{1}{n+1} \ln(2+n),$$

and therefore,

$$\sum_{n=1}^{\infty} \int_0^1 u_n(x)dx = \lim_{n \to \infty} s_n = \lim_{n \to \infty} \left(\ln 2 - \frac{1}{n+1} \ln(2+n) \right)$$

$$= \ln 2 = \int_0^1 \sum_{n=1}^{\infty} u_n(x)dx = \int_0^1 f(x)dx.$$

Thus, the original series can be integrated term by term, but it does not converge uniformly on $[0, 1]$: for $\forall n \in \mathbb{N}$, there is $x_n = \frac{1}{n+1}$ such that

$$|r_n(x_n)| = |f_n(x_n) - f(x_n)|$$

$$= \left| \frac{1}{1 + x_n} - \frac{1}{1 + x_n} + \frac{1}{1 + (n+1)x_n} \right|$$

$$= \frac{1}{2} \underset{n \to \infty}{\nrightarrow} 0.$$

Example 13. The elements of a convergent on X sequence and its limit function have infinitely many discontinuity points on X, but the formula of term-by-term integration holds.

Solution
A simple example is $f_n(x) = R(x)$ on $X = [0, 1]$, where $R(x)$ is the Riemann function. Evidently, this sequence converges uniformly on $[0, 1]$ to the limit function $R(x)$. It is known that $R(x)$ is discontinuous at all rational points and continuous at all irrational points. It follows from this fact that $R(x)$ is Riemann integrable on $[a, b]$ according to the Lebesgue criterion that asserts that a bounded function is Riemann integrable if the set of the points of its discontinuity is of the Lebesgue measure zero, which is the case of the Riemann function that has countably many discontinuities on any interval $[a, b]$. Further, we can find the value of the integral by noting that for an arbitrary partition $a = x_0 < x_1 < \cdots < x_n = b$ of an interval $[a, b]$ the lower Riemann sums are equal to 0:

$$\sum_{i=1}^{n} \inf_{x \in [x_{i-1}, x_i]} R(x) \cdot (x_i - x_{i-1}) = \sum_{i=1}^{n} 0 \cdot (x_i - x_{i-1}) = 0$$

(since any subinterval $[x_{i-1}, x_i]$ contains irrational points). Therefore, since the Riemann integral exists, its value is equal to the lower Riemann integral, that is, $\int_a^b R(x)dx = 0$. (Of course, anyone familiar with integrability properties of the Riemann function can appeal directly to the last formula.)

Since each function $f_n(x)$ is Riemann integrable on $[0, 1]$ and the convergence is uniform, the Theorem on term-by-term integration of sequences guarantees that $\int_0^1 R(x)dx = \int_0^1 \lim_{n \to \infty} f_n(x)dx = \lim_{n \to \infty} \int_0^1 f_n(x)dx = \lim_{n \to \infty} \int_0^1 R(x)dx = 0$.

Based on the given counterexample, it is also easy to construct an example of a sequence with the terms dependent on n, which maintains the same properties: $f_n(x) = \frac{n}{n+1} R(x)$ on $X = [0, 1]$.

Remark 1. The corresponding example for series is as follows: the terms of a convergent on X series and its sum have infinitely many discontinuity points on X, but the formula of term-by-term integration holds. A simple counterexamples involves again the Riemann function: $\sum_{n=1}^{\infty} u_n(x) = \sum_{n=1}^{\infty} \frac{1}{n^2} R(x)$ considered on $[0, 1]$. This series is convergent uniformly on $[0, 1]$ by the Weierstrass test: $\left|\frac{1}{n^2}R(x)\right| \leq \frac{1}{n^2}$ (since $|R(x)| \leq 1$) and the series $\sum \frac{1}{n^2}$ converges. The sum of this series is $f(x) = \frac{\pi^2}{6}R(x)$. Therefore,

$$\int_0^1 \sum_{n=1}^{\infty} u_n(x)dx = \int_0^1 \sum_{n=1}^{\infty} \frac{1}{n^2}R(x)dx = \frac{\pi^2}{6}\int_0^1 R(x)dx = 0$$

$$= \sum_{n=1}^{\infty} 0 = \sum_{n=1}^{\infty} \int_0^1 \frac{1}{n^2}R(x)dx = \sum_{n=1}^{\infty} \int_0^1 u_n(x)dx,$$

that is also guaranteed by the Theorem on term-by-term integration of series.

Remark 2. Although these counterexamples are in full agreement with the theorems on term-by-term integration of sequences and series, they show that these results are also applied to some "irregular" functions that can be a bit surprising.

Exercises

1 Let $f_n(x)$ be a sequence of differentiable on X functions convergent on X to $f(x)$. Verify whether $f(x)$ is differentiable on X and, if so, whether the relation $f'(x) = \lim_{n \to \infty} f_n'(x)$ holds. Analyze the character of the convergence of $f_n(x)$ and $f_n'(x)$ on X. Do so for the following sequences:
a) $f_n(x) = \frac{n^2x}{1+n^4x^2}, X = \mathbb{R}$
b) $f_n(x) = \frac{n^2x}{1+n^4x^4}, X = \mathbb{R}$
c) $f_n(x) = \frac{n^2x^2}{1+n^4x^4}, X = \mathbb{R}$
d) $f_n(x) = \frac{1}{1+nx^2}, X = \mathbb{R}$
e) $f_n(x) = (-1)^n\frac{x}{1+n^4x^4}, X = \mathbb{R}$
f) $f_n(x) = \frac{1}{n}e^{-n^4x^4}, X = \mathbb{R}$
g) $f_n(x) = nxe^{-n^2x^2}, X = \mathbb{R}$
h) $f_n(x) = x\arctan nx, X = \mathbb{R}$
i) $f_n(x) = n\sin\frac{x}{n}, X = \mathbb{R}$.
Formulate the false statements for which these sequences represent counterexamples. Compare with Examples 1–4 for sequences.

2 Let $f(x, y)$ be defined on $X \times Y$ and differentiable in x on X for each fixed $y \in Y$. Find the limit $\lim_{y \to y_0} f(x, y) = \varphi(x)$, where y_0 is a limit point of Y. Verify if $\varphi(x)$ is differentiable on X and, if so, whether the relation $\varphi'(x) = \lim_{y \to y_0} f_x'(x, y)$ holds. Analyze the character of the convergence of $f(x, y)$ and $f_x(x, y)$ on X as y approaches y_0. Do so for the following functions:
a) $f(x, y) = \frac{y}{y+x^2}, X = \mathbb{R}, Y = (0, +\infty), y_0 = 0$

b) $f(x, y) = x \arctan xy$, $X = \mathbb{R}$, $Y = \mathbb{R}$, $y_0 = 0$

c) $f(x, y) = y \arctan \frac{x}{y}$, $X = \mathbb{R}$, $Y = (0, 1]$, $y_0 = 0$

d) $f(x, y) = \frac{x^3 y^2}{x^4 + y^4}$, $X = \mathbb{R}$, $Y = (0, 1]$, $y_0 = 0$

e) $f(x, y) = y \cos \frac{y}{x}$, $X = (0, +\infty)$, $Y = (0, +\infty)$, $y_0 = 0$

f) $f(x, y) = y \sin \frac{x}{y}$, $X = \mathbb{R}$, $Y = (0, 1]$, $y_0 = 0$

g) $f(x, y) = \frac{1}{y} \sin xy$, $X = \mathbb{R}$, $Y = (0, +\infty)$, $y_0 = 0$

h) $f(x, y) = ye^{-\frac{x}{y}}$, $X = [0, +\infty)$, $Y = (0, 1]$, $y_0 = 0$

i) $f(x, y) = e^{-\frac{x}{y}}$, $X = [0, +\infty)$, $Y = (0, 1]$, $y_0 = 0$

j) $f(x, y) = xe^{-\frac{x}{y}}$, $X = [0, +\infty)$, $Y = (0, 1]$, $y_0 = 0$

k) $f(x, y) = \arctan \frac{x}{y}$, $X = (0, +\infty)$, $Y = (0, 1]$, $y_0 = 0$.

Formulate the false statements for which these functions represent counterexamples. Compare with Examples 1–4 for functions depending on a parameter.

3 Let functions $u_n(x)$ be differentiable on X for $\forall n$. Investigate if the series $\sum u_n(x)$ and $\sum u_n'(x)$ are convergent, the character of the convergence on X and the possibility to differentiate the series $\sum u_n(x)$ term by term on X. Do so for the following series:

a) $\sum_{n=0}^{\infty} (-1)^n xe^{-nx^2}$, $X = \mathbb{R}$

b) $\sum_{n=1}^{\infty} \frac{x^n}{n}$, $X = [0, 1)$

c) $\sum_{n=1}^{\infty} \left(\frac{1}{n} \cos \frac{1}{nx} - \frac{1}{n+1} \cos \frac{1}{(n+1)x} \right)$, $X = (0, +\infty)$

d) $\sum_{n=1}^{\infty} (\arctan nx - \arctan(n-1)x)$, $X = (0, +\infty)$

e) $\sum_{n=1}^{\infty} (\arctan nx - \arctan(n-1)x)$, $X = [0, +\infty)$.

Formulate the false statements for which these series represent counterexamples. Compare with Examples 1–4 for series.

4 Show that the sequence $f_n(x) = \frac{\sin \sqrt{nx}}{\sqrt[4]{n}}$, $X = \mathbb{R}$ gives one more counterexample for Example 5.

5 Verify whether the statement "if a series of infinitely differentiable functions converges uniformly on X, then the series of the derivatives converges at least in one point of X" is false or true. (Hint: compare this statement with that in Example 5. Use $\sum_{n=1}^{\infty} \frac{\sin nx}{n}$, $X = [a, 2\pi - a]$, $\forall a \in (0, \pi)$.)

6 Prove that the statement in Example 6 can be exemplified by the sequence $f_n(x) = \frac{\cos nx}{\sqrt{n}}$ considered on $\mathbb{R}$.

7 Analyze if the following statement is true or false: "if a series of infinitely differentiable functions converges uniformly on X, then the series of the derivatives cannot converge at infinitely many points in X and, at the same time, diverge at infinitely many points in X." (Hint: compare with the statement of Example 6. Use the series $\sum_{n=1}^{\infty} \frac{\sin nx}{n^2}$ on $X = \mathbb{R}$.)

8 Show that the statement "if all the terms $u_n(x)$ of a convergent on $[a, b]$ series $\sum u_n(x)$ are Riemann integrable on $[a, b]$, then the sum of this series is also Riemann integrable on $[a, b]$" is false by constructing a counterexample. (Hint: construct the series whose partial sums are equal to the functions $f_n(x)$ in Example 9.)

9 Show that the statement "if all the terms $u_n(x)$ of a convergent on $[a, b]$ series $\sum u_n(x)$ are not Riemann integrable on $[a, b]$, then the sum of this series is not Riemann integrable on $[a, b]$ also" is false by constructing a counterexample. (Hint: use the series $\sum_{n=0}^{\infty} u_n(x)$, where $u_0(x) = -D(x)$, $u_n(x) = \frac{1}{n(n+1)} D(x)$, $n \geq 1$ and $D(x)$ is Dirichlet's function on $[a, b] = [0, 1]$. Compare with Example 10.)

10 Find the limit function of a given sequence $f_n(x)$ and analyze the character of the convergence on $X = [a, b]$. Verify the possibility to interchange the order of the integration on $[a, b]$ and the passage to the limit. Do so for the following sequences:
a) $f_n(x) = \arctan nx$, $X = [0, 1]$
b) $f_n(x) = 3n^{2/3}x^2(1 - x^3)^n$, $X = [0, 1]$
c) $f_n(x) = 3nx^2(1 - x^3)^n$, $X = [0, 1]$
d) $f_n(x) = n^{3/2}x^3 e^{-n^2 x^4}$, $X = [0, 1]$
e) $f_n(x) = 4n^2 x^3 e^{-n^2 x^4}$, $X = [0, 1]$
f) $f_n(x) = \frac{4n^2 x}{1 + n^4 x^4}$, $X = [0, 1]$.
Formulate the false statements for which these sequences represent counterexamples. Compare with Examples 11 and 12 for sequences.

11 Find the limit function of a function $f(x, y)$ depending on a parameter $y \in Y$ as y approaches y_0. Investigate the character of the convergence on $X = [a, b]$. Verify the possibility to interchange the order of the integration on $[a, b]$ and the passage to the limit. Do so for the following functions:
a) $f(x, y) = \frac{xy^2}{x^4 + y^4}$, $X = [0, 1]$, $Y = (0, 1]$, $y_0 = 0$
b) $f(x, y) = \frac{2xy^3}{x^4 + y^4}$, $X = [0, 1]$, $Y = (0, 1]$, $y_0 = 0$
c) $f(x, y) = \frac{x^3 y}{x^4 + y^4}$, $X = [0, 1]$, $Y = (0, 1]$, $y_0 = 0$.

Formulate the false statements for which these functions represent counterexamples. Compare with Examples 11 and 12 for functions depending on a parameter.

12 Find the sum of a given series and analyze the character of its convergence on $X = [a, b]$. Verify the possibility of term-by-term integration of the series on $[a, b]$. Do so for the following series:

a) $\sum_{n=1}^{\infty} x(1-x)^n$, $X = [0, 1]$

b) $\sum_{n=1}^{\infty} (-1)^{n-1} 2nxe^{-nx^2}$, $X = [0, 1]$

c) $\sum_{n=1}^{\infty} (\arctan nx - \arctan(n-1)x)$, $X = [0, 1]$

d) $\sum_{n=0}^{\infty} \frac{x}{(1+x)^n}$, $X = [0, 1]$

e) $\sum_{n=1}^{\infty} (2nxe^{-nx^2} - 2(n-1)xe^{-(n-1)x^2})$, $X = [0, 1]$.

Formulate the false statements for which these series represent counterexamples. Compare with Examples 11 and 12 for series.

Further Reading

S. Abbott. *Understanding Analysis*, Springer, New York, 2002.

D. Bressoud, *A Radical Approach to Real Analysis*, MAA, Washington, DC, 2007.

T.J.I. Bromwich, *An Introduction to the Theory of Infinite Series*, AMS, Providence, RI, 2005.

B.M. Budak and S.V. Fomin, *Multiple Integrals, Field Theory and Series*, Mir Publisher, Moscow, 1978.

G.M. Fichtengolz, *Differential- und Integralrechnung, Vol.1–3*, V.E.B. Deutscher Verlag Wiss., Berlin, 1968.

V.A. Ilyin and E.G. Poznyak, *Fundamentals of Mathematical Analysis, Vol.1,2*, Mir Publisher, Moscow, 1982.

K. Knopp, *Theory and Applications of Infinite Series*, Dover Publication, Mineola, NY, 1990.

C.H.C. Little, K.L. Teo and B. Brunt, *Real Analysis via Sequences and Series*, Springer, New York, 2015.

W. Rudin, *Principles of Mathematical Analysis*, McGraw-Hill, New York, 1976.

V.A. Zorich, *Mathematical Analysis I, II*, Springer, Berlin, 2004.

CHAPTER 4

Integrals Depending on a Parameter

4.1 Existence of the Limit and Continuity

Example 1. A function $f(x,y)$ defined on $X \times Y = [a,b] \times Y$ is continuous in $x \in X$ at any fixed $y \in Y$, and there exists $\lim\limits_{y \to y_0} f(x,y) = \varphi(x)$, but $\varphi(x)$ is not integrable on X or $\int_a^b \varphi(x)dx \neq \lim\limits_{y \to y_0} \int_a^b f(x,y)dx$.

(See analogous example in Remark 1 to Example 11 in Chapter 3, but with another counterexample.)

Solution
The corresponding counterexample is $f(x,y) = \frac{x}{y}(1-x^2)^{1/y}$ on $X \times Y = [0,1] \times (0,1]$ and $y_0 = 0$. For any fixed $x \in [0,1]$, we have $\lim\limits_{y \to 0_+} f(x,y) = 0 = \varphi(x)$: $f(0,y) = f(1,y) = 0$ and for $x \in (0,1)$ the application of l'Hospital's rule gives

$$\lim_{y \to 0_+} f(x,y) = \lim_{y \to 0_+} \frac{x}{y}(1-x^2)^{1/y} = \lim_{t \to +\infty} x \frac{(1-x^2)^t}{1/t}$$

$$= x \lim_{t \to +\infty} \frac{t}{(1-x^2)^{-t}} = x \lim_{t \to +\infty} \frac{1}{-(1-x^2)^{-t}\ln(1-x^2)} = 0.$$

Therefore, $\int_0^1 \varphi(x)dx = 0$. Further, $f(x,y)$ is continuous in $x \in X$ and, consequently, Riemann integrable in x for any fixed $y \in Y$:

$$\int_0^1 f(x,y)dx = -\frac{1}{2y} \cdot \frac{y}{y+1}(1-x^2)^{1+1/y}\Big|_0^1 = \frac{1}{2(y+1)}.$$

However,

$$\lim_{y \to 0} \int_0^1 f(x,y)dx = \lim_{y \to 0} \frac{1}{2(y+1)} = \frac{1}{2} \neq 0 = \int_0^1 \lim_{y \to 0} f(x,y)dx.$$

Counterexamples on Uniform Convergence: Sequences, Series, Functions, and Integrals, First Edition.
Andrei Bourchtein and Ludmila Bourchtein.
© 2017 John Wiley & Sons, Inc. Published 2017 by John Wiley & Sons, Inc.
Companion website: www.wiley.com/go/bourchtein/counterexamples_on_uniform_convergence

Note that the convergence is not uniform on $[0, 1]$: choosing $x_y = \sqrt{y} \in [0, 1]$ for $\forall y \in (0, 1]$, we obtain

$$|f(x_y, y) - \varphi(x_y)| = \frac{\sqrt{y}}{y}(1 - y)^{1/y} \underset{y \to 0}{\to} +\infty.$$

The second counterexample illustrates the situation when $\varphi(x)$ is not integrable on X. Consider the function $f(x, y) = \frac{x^2 - 2xy}{(x+y)^4}$ continuous on the domain $X \times Y = [0, 1] \times (0, +\infty)$. Choosing the limit point $y_0 = 0$, we can easily find the limit function: $\varphi(x) = \lim_{y \to 0} f(x, y) = \lim_{y \to 0} \frac{x^2 - 2xy}{(x+y)^4} = \frac{1}{x^2}$ for $\forall x \in (0, 1]$ and, additionally, $\varphi(0) = \lim_{y \to 0} f(0, y) = 0$. Therefore, the limit function $\varphi(x)$ is not integrable on $[0, 1]$ (it is unbounded in a neighborhood of 0 and the corresponding improper integral diverges).

At the same time, the integral $F(y) = \int_0^1 f(x, y)dx$ exists for each fixed $y \in (0, +\infty)$:

$$F(y) = \int_0^1 \frac{x^2 - 2xy}{(x+y)^4}dx$$

$$= \int_0^1 \frac{x^2 + 2xy + y^2}{(x+y)^4}dx - 4\int_0^1 \frac{xy + y^2}{(x+y)^4}dx + 3\int_0^1 \frac{y^2}{(x+y)^4}dx$$

$$= \left(-\frac{1}{x+y} + \frac{2y}{(x+y)^2} - \frac{y^2}{(x+y)^3}\right)\bigg|_0^1 = -\frac{1}{(1+y)^3},$$

and, therefore

$$\lim_{y \to 0} F(y) = -\lim_{y \to 0} \frac{1}{(1+y)^3} = -1.$$

Note that the convergence is not uniform on $[0, 1]$: choosing $x_y = y \in (0, 1]$ for $\forall y \in (0, 1]$, we obtain

$$|f(x_y, y) - \varphi(x_y)| = \left|\frac{y^2 - 2y^2}{(2y)^4} - \frac{1}{y^2}\right| = \frac{17}{16y^2} \underset{y \to 0}{\to} +\infty.$$

Example 2. A function $f(x, y)$ defined on $[a, b] \times Y$ is continuous in x for any fixed y, and the function $\varphi(x) = \lim_{y \to y_0} f(x, y)$ is continuous on $[a, b]$, but

$$\lim_{y \to y_0} \int_a^b f(x, y)dx \neq \int_a^b \lim_{y \to y_0} f(x, y)dx.$$

Solution

Let us consider the function $f(x, y) = \begin{cases} \frac{x}{y^2}e^{-x^2/y^2}, y \neq 0 \\ 0, y = 0 \end{cases}$ on $[0, 1] \times \mathbb{R}$ and choose $y_0 = 0$. For any fixed $y \neq 0$, the continuity of $f(x, y) = \frac{x}{y^2}e^{-x^2/y^2}$ follows from the arithmetic and composition rules, and for $y = 0$, the continuity of $f(x, 0) = 0$ is evident. Calculating $\varphi(x) = \lim_{y \to 0} f(x, y)$, we obtain for $x \neq 0$:

$$\varphi(x) = \lim_{y \to 0} \frac{x}{y^2} e^{-x^2/y^2} = \frac{1}{x} \lim_{t \to +\infty} \frac{t}{e^t} = \frac{1}{x} \lim_{t \to +\infty} \frac{1}{e^t} = 0$$

(the change of the variable $t = \frac{x^2}{y^2}$ was used with the subsequent application of l'Hospital's rule); and for $x = 0$: $\varphi(0) = \lim_{y \to 0} f(0, y) = \lim_{y \to 0} 0 = 0$. Therefore, $\varphi(x) \equiv 0$ is continuous on $\mathbb{R}$ and, in particular, on $[0, 1]$, and consequently $\int_0^1 \varphi(x) dx = 0$. However, for any $y \neq 0$ we have

$$\int_0^1 f(x, y) dx = \int_0^1 \frac{x}{y^2} e^{-x^2/y^2} dx = -\frac{1}{2} e^{-x^2/y^2} \Big|_0^1 = -\frac{1}{2} e^{-1/y^2} + \frac{1}{2},$$

and, consequently

$$\lim_{y \to 0} \int_0^1 f(x, y) dx = \lim_{y \to 0} \left(\frac{1}{2} - \frac{1}{2} e^{-1/y^2} \right) = \frac{1}{2} \neq 0 = \int_0^1 \lim_{y \to 0} f(x, y) dx.$$

Let us show that $f(x, y)$ does not satisfy all the conditions of the theorem about the passing to the limit under the sign of integral or the conditions of the corollary to this theorem. In fact, choosing $x_y = y \in [0, 1]$ for $\forall y \in (0, 1]$, we obtain

$$|f(x_y, y) - \varphi(x_y)| = \frac{y}{y^2} e^{-y^2/y^2} = \frac{1}{y} e^{-1} \underset{y \to 0_+}{\to} +\infty,$$

that is, in the mentioned theorem the condition of the uniform convergence on $[0, 1]$ is not satisfied. On the other hand,

$$f_y(x, y) = \left(-\frac{2x}{y^3} + \frac{2x^3}{y^5} \right) e^{-x^2/y^2} = \frac{2x}{y^5} (x^2 - y^2) e^{-x^2/y^2},$$

which implies that for any fixed x the derivative $f_y(x, y)$ changes its sign in the points $y = 0$ and $y = \pm x$. In particular, for $\forall x \in (0, 1)$ it follows that $f_y(x, y) > 0$ if $0 < y < x$ and $f_y(x, y) < 0$ if $x < y < 1$, which means that $f(x, y)$ is not monotone in y, that is, the monotonicity condition in the corollary to the theorem is not satisfied.

Remark to Examples 1 and 2. The condition of the uniform convergence or the condition of monotonicity is important to guarantee passing to the limit under the sign of integral, but these conditions are sufficient and not necessary, as shown in Example 3.

Example 3. A function $f(x, y)$ defined on $X \times Y = [a, b] \times Y$ is continuous in $x \in X$ at any fixed $y \in Y$, there exists $\lim_{y \to y_0} f(x, y) = \varphi(x)$ and

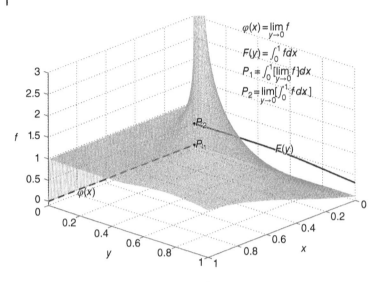

Figure 4.1 Example 2, function $f(x, y) = \begin{cases} \frac{x}{y^2} e^{-x^2/y^2}, y \neq 0 \\ 0, y = 0 \end{cases}$.

$\int_a^b \varphi(x)dx = \lim\limits_{y \to y_0} \int_a^b f(x, y)dx$, but $f(x, y)$ converges nonuniformly on X to $\varphi(x)$.

(See analogous example in Remark 1 to Example 12 in Chapter 3, but with another counterexample.)

Solution

Consider the function $f(x, y) = \frac{x}{y} e^{-x^2/y^2}$ on $X \times Y = [0, 1] \times (0, 1]$ and $y_0 = 0$. If $x = 0$, then $\varphi(0) = \lim\limits_{y \to 0} f(0, y) = \lim\limits_{y \to 0} 0 = 0$, and if $x \neq 0$, then by using the change of the variable $t = \frac{x}{y}$ and l'Hospital's rule, one gets

$$\varphi(x) = \lim\limits_{y \to 0} \frac{x}{y} e^{-x^2/y^2} = \lim\limits_{t \to +\infty} \frac{t}{e^{t^2}} = \lim\limits_{t \to +\infty} \frac{1}{2te^{t^2}} = 0.$$

Thus, $\lim\limits_{y \to 0_+} f(x, y) = 0 = \varphi(x)$ for any fixed $x \in [0, 1]$, and therefore, $\int_0^1 \varphi(x)dx = 0$.

On the other hand, $f(x, y)$ is continuous in $x \in X$ and, consequently, Riemann integrable in x, for any fixed $y \in Y$:

$$\int_0^1 f(x, y)dx = -\frac{1}{2}y\, e^{-x^2/y^2}\Big|_0^1 = \frac{1}{2}y(1 - e^{-1/y^2}).$$

Therefore, we can pass to the limit under the sign of integral:

$$\lim_{y \to 0_+} \int_0^1 f(x, y)dx = \lim_{y \to 0_+} \frac{1}{2}y(1 - e^{-1/y^2}) = 0$$

$$= \int_0^1 \lim_{y \to 0_+} f(x, y)dx = \int_0^1 \varphi(x)dx.$$

Nevertheless, just like in Example 2, the convergence is not uniform on $[0, 1]$, neither the function is monotone in y. In fact, choosing $x_y = y \in [0, 1]$ for $\forall y \in (0, 1]$, we obtain $|f(x_y, y) - \varphi(x_y)| = \frac{y}{y}e^{-y^2/y^2} = e^{-1} \nrightarrow 0$, which shows

that the convergence is nonuniform. Also,

$$f_y(x, y) = \left(-\frac{x}{y^2} + \frac{2x^3}{y^4}\right)e^{-x^2/y^2} = \frac{x}{y^4}(2x^2 - y^2)e^{-x^2/y^2},$$

from which it follows that for any fixed $x \in \left(0, \frac{1}{\sqrt{2}}\right)$ the derivative $f_y(x, y) > 0$ when $0 < y < \sqrt{2}x$ and $f_y(x, y) < 0$ when $\sqrt{2}x < y < 1$, which means that $f(x, y)$ is not monotone in y.

Example 4. A function $f(x, y)$ is continuous on $\mathbb{R}^2$ except at only one point, but $F(y) = \int_a^b f(x, y)dx$ is not continuous.

Solution
The function $f(x, y) = \frac{y(x^2+1)}{x^2+y^2}$ is continuous on $\mathbb{R}^2 \setminus (0, 0)$ due to the arithmetic rules of continuous functions, and it is easy to show that $f(x, y)$ is unbounded and does not have a limit at the origin: along the x-axis we have $\lim_{x \to 0^-} f(x, 0) = \lim_{x \to 0} 0 = 0$, while along the y-axis we get $\lim_{y \to 0} f(0, y) = \lim_{y \to 0} \frac{y}{y^2} = \infty$. The function $F(y) = \int_0^1 f(x, y)dx$ is found as follows: for $y = 0$, we have $F(0) = \int_0^1 0 dx = 0$, and for $y \neq 0$ we obtain

$$F(y) = \int_0^1 \frac{y(x^2 + 1)}{x^2 + y^2}dx = y\left(\int_0^1 \frac{x^2 + y^2}{x^2 + y^2}dx + (1 - y^2)\int_0^1 \frac{1}{x^2 + y^2}dx\right)$$

$$= yx|_0^1 + y(1 - y^2)\frac{1}{y}\arctan\frac{x}{y}\Big|_0^1 = y + (1 - y^2)\arctan\frac{1}{y}.$$

Since

$$\lim_{y \to 0_+} F(y) = \lim_{y \to 0_+} \left(y + (1 - y^2)\arctan\frac{1}{y}\right) = \frac{\pi}{2} \neq 0 = F(0),$$

$F(y)$ is not continuous at zero.

Example 5. A function $f(x, y)$ is continuous in y for any fixed x and also in x for any fixed y, but the function $F(y) = \int_a^b f(x, y)dx$ is not continuous.

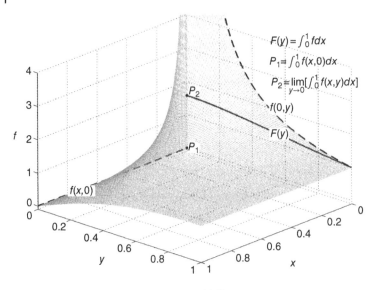

$F(y) = \int_0^1 f dx$

$P_1 = \int_0^1 f(x,0) dx$

$P_2 = \lim_{y \to 0} [\int_0^1 f(x,y) dx]$

$f(0,y)$

$F(y)$

P_2

P_1

$f(x,0)$

Figure 4.2 Example 4, function $f(x,y) = \frac{y(x^2+1)}{x^2+y^2}$.

Solution

Let us consider the function $f(x,y) = \begin{cases} \frac{y^2}{x^3} e^{-y^2/x^2}, & x \neq 0 \\ 0, & x = 0 \end{cases}$ on $\mathbb{R}^2$. For any fixed $x_0 \neq 0$, this function is continuous in y due to the arithmetic and composition rules of continuous functions. For $x_0 = 0$, the definition gives $f(0,y) = 0$ for any y, so again the function is continuous. Analogously, for $y = 0$, the function $f(x,0) = 0$ is continuous. Finally for $y_0 \neq 0$, the function $f(x,y_0)$ is continuous at each $x \neq 0$ according to the arithmetic and composition rules, and at $x = 0$ one gets

$$\lim_{x \to 0} f(x,y_0) = \lim_{x \to 0} \frac{y_0^2}{x^3} e^{-y_0^2/x^2} = \lim_{t \to \infty} \frac{y_0^2 t^3}{e^{y_0^2 t^2}} = \lim_{t \to \infty} \frac{3y_0^2 t^2}{2ty_0^2 e^{y_0^2 t^2}}$$

$$= \frac{3}{2} \lim_{t \to \infty} \frac{t}{e^{y_0^2 t^2}} = \frac{3}{2} \lim_{t \to \infty} \frac{1}{2ty_0^2 e^{y_0^2 t^2}} = 0 = f(0,y_0).$$

(In order to calculate this limit, make the change of the variable $x = \frac{1}{t}$ and then apply l'Hospital's rule.) Hence, the conditions of the statement hold.

Note, however, that $f(x,y)$ is discontinuous at the origin as a function of two variables: approaching zero along the path $y = x$ we have $\lim_{x \to 0} f(x,x) = \lim_{x \to 0} \frac{x^2}{x^3} e^{-x^2/x^2} = \infty$. This causes a discontinuity of $F(y) = \int_0^1 f(x,y) dx$ at zero. Indeed, for $y = 0$ we have $F(0) = \int_0^1 f(x,0) dx = \int_0^1 0\, dx = 0$, while for $y \neq 0$, we obtain

$$F(y) = \int_0^1 \frac{y^2}{x^3} e^{-y^2/x^2} dx = \frac{1}{2} e^{-y^2/x^2} \Big|_0^1 = \frac{1}{2} e^{-y^2},$$

and the limit at zero shows discontinuity:

$$\lim_{y\to 0} F(y) = \lim_{y\to 0} \frac{1}{2} e^{-y^2} = \frac{1}{2} \neq 0 = F(0).$$

Remark to Examples 4 and 5. Both these examples show that the condition of continuity of $f(x, y)$ as a function of two variables is important to guarantee continuity of the function $F(y) = \int_a^b f(x, y)dx$. At the same time, this condition is only sufficient and $F(y)$ still can be continuous when $f(x, y)$ is discontinuous, as it is shown in the next example.

Example 6. A function $f(x, y)$ is discontinuous at infinitely many points of $[a, b] \times Y$, but $F(y) = \int_a^b f(x, y)dx$ is continuous on Y.

Solution

Let us consider the function $f(x, y) = \operatorname{sgn}(x - y) = \begin{cases} -1, x - y < 0 \\ 0, x = y \\ 1, x - y > 0 \end{cases}$ on $S =$

$[0, 1] \times \mathbb{R}$. This function is discontinuous at each point of the line $y = x$,

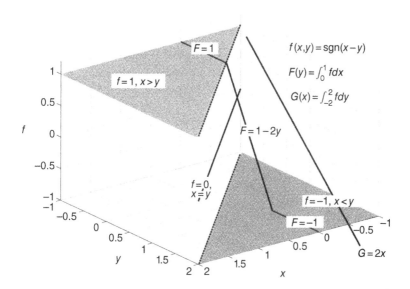

Figure 4.3 Examples 6 and 16, function $f(x, y) = \operatorname{sgn}(x - y)$.

but $F(y) = \int_0^1 f(x,y)dx$ is continuous on $\mathbb{R}$. In fact, for $y < 0$ it follows that $F(y) = \int_0^1 1dx = 1$, for $y > 1$ one gets $F(y) = \int_0^1 -1dx = -1$, and for $0 \le y \le 1$ the integral can be separated in two parts:

$$F(y) = \int_0^y fdx + \int_y^1 fdx = \int_0^y -1dx + \int_y^1 1dx$$
$$= -y + (1 - y) = 1 - 2y.$$

Summarizing, $F(y) = \begin{cases} 1, y < 0 \\ 1 - 2y, 0 \le y \le 1, \\ -1, y > 1 \end{cases}$ which is a continuous function on $\mathbb{R}$.

Remark 1. In the given counterexample, the function $f(x,y)$ is discontinuous and bounded on S. One can also construct a counterexample with unbounded

function. For instance, $f(x,y) = \begin{cases} \frac{xy^{7/3}}{x^4+y^4}, x^2 + y^2 \ne 0 \\ 0, x^2 + y^2 = 0 \end{cases}$ is discontinuous and

unbounded on $S = [0,1] \times \mathbb{R}$, since

$$\lim_{x \to 0^-} f(x,x) = \lim_{x \to 0} \frac{x^{10/3}}{2x^4} = \lim_{x \to 0} \frac{1}{2x^{2/3}} = +\infty$$

(albeit $f(x,y)$ is continuous in $\mathbb{R}^2 \setminus (0,0)$). At the same time, $F(0) = \int_0^1 0dx = 0$ and for $y \ne 0$ one gets

$$F(y) = \int_0^1 \frac{xy^{7/3}}{x^4 + y^4}dx = \frac{y^{7/3}}{2y^2} \arctan \frac{x^2}{y^2}\Big|_0^1 = \frac{y^{1/3}}{2} \arctan \frac{1}{y^2}.$$

Evidently, $F(y)$ is continuous at every $y \ne 0$, and additionally

$$\lim_{y \to 0} F(y) = \lim_{y \to 0} \frac{y^{1/3}}{2} \arctan \frac{1}{y^2} = 0 = F(0).$$

Therefore, $F(y)$ is continuous on $\mathbb{R}$.

Remark 2. The original statement can be strengthened to the following form: a function $f(x,y)$ is discontinuous and unbounded at infinitely many points of $[a,b] \times Y$, but still $F(y) = \int_a^b f(x,y)dx$ is continuous on Y. For a counterex-

ample, one can use the function $f(x,y) = \begin{cases} \frac{y}{\sqrt{x}}, y < x^2, x \ne 0 \\ 2xy^2, y \ge x^2 \\ 0, x = 0 \end{cases}$ considered on

$[0,1] \times \mathbb{R}$. This function is continuous on the sets $D_1 = \{x \in (0,1], y < x^2\}$ and $D_2 = \{x \in [0,1], y > x^2\}$. Let us show that $f(x,y)$ is discontinuous at all the

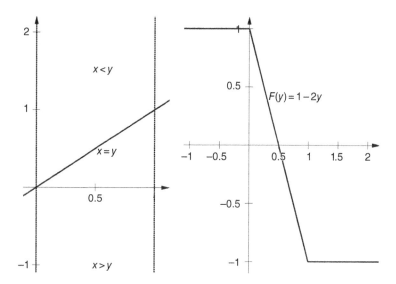

Figure 4.4 Examples 6 and 16, domain of function $f(x,y)$ and function $F(y) = \int_0^1 f(x,y)dx$.

points of the parabola $P = \{y = x^2, x \in (0,1]\}$ except for $x_1 = 2^{-2/7}$. Indeed, approaching the points of the parabola P from within D_1, one gets

$$\lim_{\substack{(x,y)\to(x,x^2) \\ (x,y)\in D_1}} f(x,y) = \lim_{(x,y)\to(x,x^2)} \frac{y}{\sqrt{x}} = x^{3/2},$$

while approaching P from within D_2, one has

$$\lim_{\substack{(x,y)\to(x,x^2) \\ (x,y)\in D_2}} f(x,y) = \lim_{(x,y)\to(x,x^2)} 2xy^2 = 2x^5 = f(x,x^2).$$

Then the continuity condition $x^{3/2} = 2x^5$ is satisfied for the only point $x_1 = 2^{-2/7}$ of P. Besides, $f(x,y)$ is discontinuous and unbounded at all the points of the negative part of the y-axis:

$$\lim_{\substack{(x,y)\to(0,y) \\ y<0}} f(x,y) = \lim_{\substack{(x,y)\to(0,y) \\ y<0}} \frac{y}{\sqrt{x}} = -\infty.$$

Note that the origin is also the point of discontinuity and unboundedness—for example, along the path $y = -\sqrt[3]{x}$ one obtains:

$$\lim_{\substack{(x,y)\to(0,0) \\ y=-\sqrt[3]{x}}} f(x,y) = \lim_{x\to 0} \frac{-\sqrt[3]{x}}{\sqrt{x}} = -\infty.$$

Therefore, $f(x,y)$ is discontinuous and unbounded at infinitely many points of $[0,1] \times \mathbb{R}$.

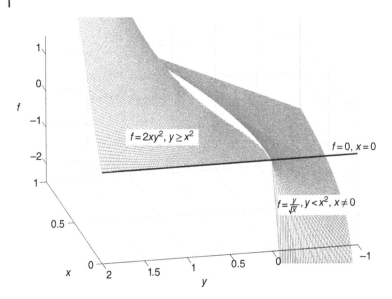

Figure 4.5 Remark 2 to Example 6, Remark to Example 16, function $f(x, y)$.

Now consider the function $F(y) = \int_0^1 f(x, y)dx$. It can be calculated separately on three intervals: the condition $y < 0$ implies $y < x^2$ and, therefore,

$$F(y) = \int_0^1 \frac{y}{\sqrt{x}}dx = 2\sqrt{x}y|_0^1 = 2y;$$

for $x \in [0, 1]$, the inequality $y > 1$ implies $y > x^2$ and, consequently,

$$F(y) = \int_0^1 2xy^2 dx = x^2 y^2|_0^1 = y^2;$$

finally, for $0 \le y \le 1$ one obtains

$$F(y) = \int_0^{\sqrt{y}} 2xy^2 dx + \int_{\sqrt{y}}^1 \frac{y}{\sqrt{x}}dx$$

$$= x^2 y^2|_0^{\sqrt{y}} + 2y\sqrt{x}|_{\sqrt{y}}^1 = y^3 + 2y - 2y^{5/4}.$$

The function $F(y) = \begin{cases} 2y, y < 0 \\ y^3 + 2y - 2y^{5/4}, 0 \le y \le 1 \\ y^2, y > 1 \end{cases}$ is continuous in each interval

$(-\infty, 0)$, $(0, 1)$, and $(1, +\infty)$ according to the rules of continuous functions. It is easy to check that it is also continuous at the points $y = 0$ and $y = 1$:

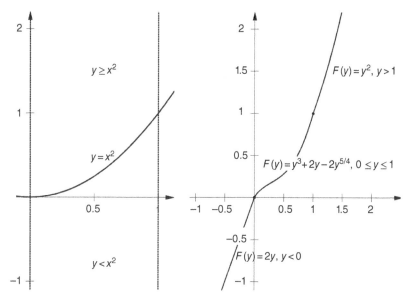

Figure 4.6 Remark 2 to Example 6, Remark to Example 16, domain of function $f(x, y)$ and function $F(y) = \int_0^1 f(x, y)dx$.

$$\lim_{y \to 0_-} F(y) = \lim_{y \to 0_-} 2y = 0 = F(0) = \lim_{y \to 0_+} (y^3 + 2y - 2y^{5/4}) = \lim_{y \to 0_+} F(y);$$

$$\lim_{y \to 1_-} F(y) = \lim_{y \to 1_-} (y^3 + 2y - 2y^{5/4}) = 1 = F(1) = \lim_{y \to 1_+} y^2 = \lim_{y \to 1_+} F(y).$$

Thus, $F(y)$ is continuous on $Y = \mathbb{R}$.

Remark 3. Another direction for strengthening the original statement is to maximize the number of discontinuity points: a function $f(x, y)$ is discontinuous at each point of $[a, b] \times Y$, but nevertheless $F(y) = \int_a^b f(x, y)dx$ is continuous on Y. The illustration of this situation can be given with the function $f(x, y) = \text{sgn}x \cdot D(y)$ defined on $\tilde{S} = [-1, 1] \times \mathbb{R}$ (recall that $D(y)$ is Dirichlet's function). At each point (x_0, y_0), $x_0 \neq 0$, this function is discontinuous just because $f(x_0, y) = \pm D(y)$ is discontinuous in y at each point. At the points $(0, y_0)$, the function $f(x, y)$ is discontinuous because in any neighborhood of $(0, y_0)$ there are the points with $x > 0, y \in \mathbb{Q}$, which lead to the partial limit equal to 1, different from the value of function $f(0, y_0) = 0$. Therefore, $f(x, y)$ is discontinuous at each point of $\tilde{S}$. Nevertheless, $F(y) = \int_{-1}^1 f(x, y)dx = D(y) \int_{-1}^1 \text{sgn}x \, dx = 0$, $\forall y \in \mathbb{R}$, that is, $F(y)$ is a continuous function on $\mathbb{R}$.

4.2 Differentiability

Example 7. Both $f(x, y)$ and $f_y(x, y)$ are continuous in y for any fixed x and also in x for any fixed y, but $F_y(y) = \frac{d}{dy} \int_a^b f(x, y) dx \neq \int_a^b f_y(x, y) dx$.

Solution

Let us show that the function $f(x, y) = \begin{cases} \frac{y^3}{x^3} e^{-y^2/x^2}, & x \neq 0 \\ 0, & x = 0 \end{cases}$ and its partial derivative $f_y(x, y) = \begin{cases} \left(\frac{3y^2}{x^3} - \frac{2y^4}{x^5} \right) e^{-y^2/x^2}, & x \neq 0 \\ 0, & x = 0 \end{cases}$ are continuous separately in each variable on $\mathbb{R}^2$. For $x \neq 0$, the continuity of $f(x, y)$ separately in x and in y follows directly from the arithmetic and composition rules. For $x = 0$, the continuity in y is evident $(f(0, y) = 0)$, and the continuity in x can be checked by the definition: for $y_0 = 0$, one has $\lim_{x \to 0^-} f(x, 0) = \lim_{x \to 0} 0 = 0 = f(0, 0)$, and for $y_0 \neq 0$ one gets

$$\lim_{x \to 0^-} f(x, y_0) = \lim_{x \to 0} \frac{y_0^3}{x^3} e^{-y_0^2/x^2} = \lim_{t \to \infty} \frac{y_0^3 t^3}{e^{y_0^2 t^2}} = \lim_{t \to \infty} \frac{3y_0^3 t^2}{2ty_0^2 e^{y_0^2 t^2}}$$

$$= \frac{3}{2} \lim_{t \to \infty} \frac{y_0 t}{e^{y_0^2 t^2}} = \frac{3}{2} \lim_{t \to \infty} \frac{y_0}{2ty_0^2 e^{y_0^2 t^2}} = 0 = f(0, y_0)$$

(the change of the variable $x = \frac{1}{t}$ and l'Hospital's rule were applied). Similar considerations show the continuity of $f_y(x, y)$ separately in x and in y. Hence, the conditions of the statement hold.

At the same time, note that $f(x, y)$ and $f_y(x, y)$ are discontinuous at the origin as functions of two variables, since along the path $y = x$ we have

$$\lim_{x \to 0^-} f(x, x) = \lim_{x \to 0} \frac{x^3}{x^3} e^{-x^2/x^2} = e^{-1} \neq 0 = f(0, 0)$$

and

$$\lim_{x \to 0^-} f_y(x, x) = \lim_{x \to 0} \left(\frac{3x^2}{x^3} - \frac{2x^4}{x^5} \right) e^{-x^2/x^2} = \lim_{x \to 0} \frac{1}{x} e^{-1} = \infty.$$

This leads to the impossibility to change the order of integration and differentiation at zero. Indeed,

$$F(y) = \int_0^1 \frac{y^3}{x^3} e^{-y^2/x^2} dx = \frac{y}{2} e^{-y^2/x^2} \Big|_0^1 = \frac{y}{2} e^{-y^2}$$

and, consequently,

$$F_y(y) = \frac{1}{2} e^{-y^2} - y^2 e^{-y^2},$$

which gives $F_y(0) = \frac{1}{2}$. On the other hand, $f_y(x, 0) = 0$ for any x, so

$$\int_0^1 f_y(x, 0) dx = \int_0^1 0 \, dx = 0 \neq F_y(0).$$

Remark. Note that under the assumptions of Example 7, the derivative $F_y(y)$ may even not exist at some point $y_0 \in Y$, although $\int_a^b f_y(x,y)dx$ exists for $\forall y \in Y$. To this effect, consider the function $f(x,y) = \begin{cases} \frac{y^2}{x^3}e^{-y^2/x^2}, x \neq 0 \\ 0, x = 0 \end{cases}$ on $\mathbb{R}^2$.

Its partial derivative is $f_y(x,y) = \begin{cases} \left(\frac{2y}{x^3} - \frac{2y^3}{x^5}\right)e^{-y^2/x^2}, x \neq 0 \\ 0, x = 0 \end{cases}$. The proof of the continuity of $f(x,y)$ and $f_y(x,y)$ separately in each variable on $\mathbb{R}^2$ can be done in the same way as in Example 7. At the same time, note that both functions are discontinuous and unbounded (as functions of two variables) at $(0,0)$:

$$\lim_{x\to 0} f(x,x) = \lim_{x\to 0} \frac{1}{x}e^{-1} = \infty$$

and

$$\lim_{x\to 0} f_y(x, 2x) = \lim_{x\to 0} \frac{1}{x^2}\left(\frac{4x}{x} - \frac{16x^3}{x^3}\right)e^{-4} = \lim_{x\to 0} \frac{-12}{x^2}e^{-4} = -\infty.$$

Let us calculate the required integrals on $[0,1]$:

$$F(y) = \int_0^1 f(x,y)dx = \int_0^1 \frac{y^2}{x^3}e^{-y^2/x^2}dx$$
$$= \frac{1}{2}e^{-y^2/x^2}\Big|_0^1 = \frac{1}{2}e^{-y^2}, \forall y \neq 0,$$
$$F(0) = \int_0^1 f(x,0)dx = \int_0^1 0\, dx = 0,$$

$$\int_0^1 f_y(x,y)dx = \int_0^1 \frac{2y}{x^3}e^{-y^2/x^2}dx - \int_0^1 \frac{2y^3}{x^5}e^{-y^2/x^2}dx$$
$$= \int_0^1 \frac{2y}{x^3}e^{-y^2/x^2}dx - \frac{y}{x^2}e^{-y^2/x^2}\Big|_0^1 - \int_0^1 \frac{2y}{x^3}e^{-y^2/x^2}dx$$
$$= -y\,e^{-y^2} + \lim_{x\to 0}\frac{y}{x^2}e^{-y^2/x^2} = -y\,e^{-y^2}, \forall y \neq 0,$$
$$\int_0^1 f_y(x,0)dx = \int_0^1 0\, dx = 0.$$

Therefore,

$$\int_0^1 f_y(x,y)dx = -y\,e^{-y^2}, \forall y \in \mathbb{R},$$

while $F(y)$ is discontinuous at $y = 0$ $(\lim_{y\to 0} F(y) = \frac{1}{2} \neq 0 = F(0))$ and, consequently, is not differentiable at 0.

Example 8. Both $f(x, y)$ and $f_y(x, y)$ are continuous in $\mathbb{R}^2$ except at one point, but $F_y(y) = \frac{d}{dy} \int_a^b f(x, y)dx \neq \int_a^b f_y(x, y)dx$.

Solution

The function $f(x, y) = \begin{cases} \frac{y^3}{x^2 + y^4}, & x^2 + y^2 \neq 0 \\ 0, & x^2 + y^2 = 0 \end{cases}$ has partial derivative $f_y(x, y) = \frac{3x^2 y^2 - y^6}{(x^2 + y^4)^2}$ at each point $(x, y) \in \mathbb{R}^2 \setminus (0, 0)$, and both functions $f(x, y)$ and $f_y(x, y)$ are continuous on $\mathbb{R}^2 \setminus (0, 0)$ due to the arithmetic properties of continuous functions. At the origin, $f(x, y)$ is discontinuous and unbounded:

$$\lim_{x \to 0, y = 0} f(x, y) = \lim_{x \to 0} 0 = 0, \quad \lim_{y \to 0, x = 0} f(x, y) = \lim_{y \to 0} \frac{1}{y} = \infty.$$

The derivative $f_y(x, y)$ does not exist at $(0, 0)$ and is unbounded in its neighborhood: $\lim_{y \to 0, x = 0} f_y(x, y) = \lim_{y \to 0} \frac{-y^6}{y^8} = \infty$. Thus, the example conditions are satisfied.

Since $f(x, y)$ is continuous on $\mathbb{R}$ with respect to x for any fixed y, it is Riemann integrable in x on an arbitrary interval $[a, b]$. In particular, for any fixed $y \neq 0$ one has

$$F(y) = \int_0^1 f(x, y)dx = y^3 \int_0^1 \frac{1}{x^2 + y^4}dx = \frac{y^3}{y^2} \arctan \frac{x}{y^2} \Big|_0^1 = y \arctan \frac{1}{y^2},$$

and at $y = 0$ it is just $F(0) = \int_0^1 f(x, 0)dx = \int_0^1 0 \, dx = 0$. Therefore,

$$F'(y) = \arctan \frac{1}{y^2} - \frac{2y^2}{1 + y^4}, \quad y \neq 0$$

and

$$F'(0) = \lim_{y \to 0} \frac{F(y) - F(0)}{y} = \lim_{y \to 0} \arctan \frac{1}{y^2} = \frac{\pi}{2}.$$

At the same time, $f_y(x, 0) = 0$, $\forall x \neq 0$ and, consequently, is Riemann integrable on any $[a, b]$. However, the value of integral $\int_0^1 f_y(x, 0)dx = 0$ is different from $F_y(0) = \frac{\pi}{2}$.

Remark. It may happen that under the example conditions, $F_y(y)$ even does not exist at some points. In fact, consider the function $f(x, y) = \begin{cases} \frac{y^2}{x^2 + y^2}, & x^2 + y^2 \neq 0 \\ 0, & x^2 + y^2 = 0 \end{cases}$ that has partial derivative $f_y(x, y) = \frac{2x^2 y}{(x^2 + y^2)^2}$ at each point $(x, y) \in \mathbb{R}^2 \setminus (0, 0)$, and, according to the arithmetic rules of continuous functions, both $f(x, y)$ and $f_y(x, y)$ are continuous on $\mathbb{R}^2 \setminus (0, 0)$. At the origin, both functions are discontinuous:

$$\lim_{x \to 0, y = 0} f(x, y) = 0 \neq 1 = \lim_{y \to 0, x = 0} f(x, y), \quad \lim_{x \to 0, y = x} f_y(x, y) = \lim_{x \to 0} \frac{2x^3}{4x^4} = \infty$$

(besides the fact that $f_y(x, y)$ does not exist at the origin).

For $\forall y \neq 0$, the integral $\int_0^1 f_y(x, y)dx$ can be calculated using the integration by parts:

$$\int_0^1 f_y(x, y)dx = \int_0^1 \frac{2x^2 y}{(x^2 + y^2)^2}dx = -\frac{xy}{x^2 + y^2}\bigg|_0^1 + \int_0^1 \frac{y}{x^2 + y^2}dx$$

$$= -\frac{y}{1 + y^2} + \arctan\frac{x}{y}\bigg|_0^1 = -\frac{y}{1 + y^2} + \arctan\frac{1}{y},$$

and for $y = 0$ the result is immediate $\int_0^1 f_y(x, 0)dx = \int_0^1 0dx = 0$. Thus, $\int_0^1 f_y(x, y)dx$ exists for $\forall y \in \mathbb{R}$.

The integral $\int_0^1 f(x, y)dx$ also exists for any fixed y:

$$F(y) = \int_0^1 f(x, y)dx = y\arctan\frac{x}{y}\bigg|_0^1 = y\arctan\frac{1}{y}, \forall y \neq 0$$

and

$$F(0) = \int_0^1 f(x, 0)dx = \int_0^1 0\ dx = 0.$$

However, $F'(0)$ does not exist at $y = 0$:

$$\lim_{y \to 0_+} \frac{F(y) - F(0)}{y} = \lim_{y \to 0_+} \arctan\frac{1}{y} = \frac{\pi}{2} \neq -\frac{\pi}{2} = \lim_{y \to 0_-} \frac{F(y) - F(0)}{y}.$$

Remark to Examples 7 and 8. In both examples, the functions $f(x, y)$ and $f_y(x, y)$ are discontinuous at the origin, which made impossible the interchange of the operations of integration and partial derivative. However, it may happen that the interchange of the two operations is impossible even when $f(x, y)$ is continuous on $\mathbb{R}^2$, as shown in Example 9.

Example 9. A function $f(x, y)$ is continuous on $\mathbb{R}^2$ and $f_y(x, 0)$ exists for $\forall x \in \mathbb{R}$, but $F_y(0) = \frac{d}{dy}\int_a^b f(x, y)dx\big|_{y=0} \neq \int_a^b f_y(x, 0)dx$.

Solution
Consider the function defined on $\mathbb{R}^2$ as follows: if $y \geq 0$, then $f(x, y) =$

$$\begin{cases} x, x \in [0, \sqrt{y}] \\ -x + 2\sqrt{y}, x \in [\sqrt{y}, 2\sqrt{y}], \text{ and if } y < 0 \text{ then } f(x, y) = -f(x, |y|). \text{ To show that} \\ 0, \text{otherwise} \end{cases}$$

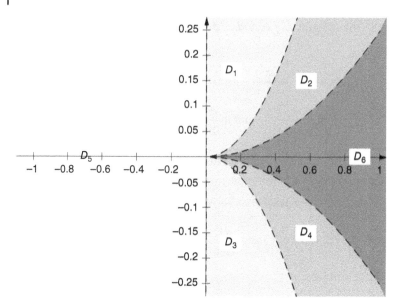

Figure 4.7 Example 9, parts of the domain of $f(x, y)$.

$f(x, y)$ is continuous on $\mathbb{R}^2$, we divide all the plane in six domains, which have no common interior point:

$$D_1 = \{(x, y) : \ y \geq 0, 0 \leq x \leq \sqrt{y}\},$$
$$D_2 = \{(x, y) : \ y \geq 0, \sqrt{y} \leq x \leq 2\sqrt{y}\},$$
$$D_3 = \{(x, y) : \ y \leq 0, 0 \leq x \leq \sqrt{|y|}\},$$
$$D_4 = \{(x, y) : \ y \leq 0, \sqrt{|y|} \leq x \leq 2\sqrt{|y|}\},$$
$$D_5 = \{(x, y) : \ x \leq 0\}, D_6 = \{(x, y) : \ x \geq 2\sqrt{|y|}\}.$$

Inside each of D_i the function is represented by a simple formula, which guarantees the continuity. Therefore, we should check the continuity on the boundaries of these domains.

For $y > 0$, there are three boundary curves—$x = 0$, $x = \sqrt{y}$, and $x = 2\sqrt{y}$—which we analyze separately as follows:

1) Any point $(0, y_0) \in D_1 \cap D_5$ can be approached by the points lying in D_1 or D_5; for the points of D_1, we have

$$\lim_{(x,y)\to(0,y_0)} f(x, y) = \lim_{(x,y)\to(0,y_0)} x = 0 = f(0, y_0),$$

and for those in D_5,

$$\lim_{(x,y)\to(0,y_0)} f(x, y) = \lim_{(x,y)\to(0,y_0)} 0 = 0 = f(0, y_0),$$

which means that $f(x, y)$ is continuous on $x = 0$;

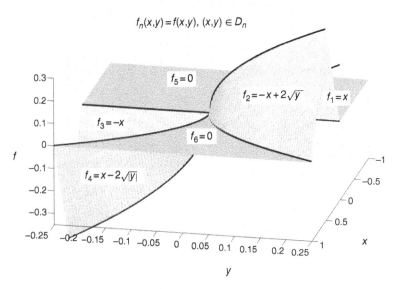

$$f_n(x,y) = f(x,y), \ (x,y) \in D_n$$

Figure 4.8 Example 9, function $f(x,y)$.

2) Any point $(x_0, x_0^2) \in D_1 \cap D_2$ can be approached by the points lying in D_1 or D_2; for the points of D_1, we have

$$\lim_{(x,y) \to (x_0, x_0^2)} f(x,y) = \lim_{(x,y) \to (x_0, x_0^2)} x = x_0 = f(x_0, x_0^2),$$

and for those in D_2,

$$\lim_{(x,y) \to (x_0, x_0^2)} f(x,y) = \lim_{(x,y) \to (x_0, x_0^2)} (-x + 2\sqrt{y}) = -x_0 + 2x_0 = x_0 = f(x_0, x_0^2),$$

that is, $f(x,y)$ is continuous on $x = \sqrt{y}$;

3) Any point $\left(x_0, \frac{x_0^2}{4}\right) \in D_2 \cap D_6$ can be approached by the points lying in D_2 or D_6; for D_2, we have

$$\lim_{(x,y) \to \left(x_0, \frac{x_0^2}{4}\right)} f(x,y) = \lim_{(x,y) \to \left(x_0, \frac{x_0^2}{4}\right)} (-x + 2\sqrt{y})$$

$$= -x_0 + 2\frac{x_0}{2} = 0 = f\left(x_0, \frac{x_0^2}{4}\right),$$

and for D_6 we get

$$\lim_{(x,y) \to \left(x_0, \frac{x_0^2}{4}\right)} f(x,y) = \lim_{(x,y) \to \left(x_0, \frac{x_0^2}{4}\right)} 0 = 0 = f\left(x_0, \frac{x_0^2}{4}\right),$$

which proves the continuity of $f(x,y)$ on $x = 2\sqrt{y}$.

Due to the symmetry with respect to x-axis, the function is also continuous for $y < 0$ on the boundaries of the domains D_3, D_4, D_5, and D_6. It remains to show the continuity at the origin. Approaching by the points of D_1 or D_3, we have

$$\lim_{(x,y)\to(0,0)} f(x,y) = \lim_{(x,y)\to(0,0)} (\pm x) = 0 = f(0,0).$$

For the points of D_2 or D_4, we obtain

$$\lim_{(x,y)\to(0,0)} f(x,y) = \lim_{(x,y)\to(0,0)} \pm(-x + 2\sqrt{|y|}) = 0 = f(0,0).$$

For D_5 or D_6, we have

$$\lim_{(x,y)\to(0,0)} f(x,y) = \lim_{(x,y)\to(0,0)} 0 = 0 = f(0,0).$$

Hence, $\lim_{(x,y)\to(0,0)} f(x,y) = 0 = f(0,0)$ for any path passing through $(0,0)$, that is, $f(x,y)$ is continuous at the origin.

Now let us find $f_y(x,0)$. Any point $(x_0,0)$, $x_0 \le 0$ lies in D_5 and, consequently, all the corresponding points (x_0,y) also lie in D_5. For any point $(x_0,0)$, $x_0 > 0$, located in D_6, there exists a small neighborhood such that all the corresponding points (x_0,y) also lie in D_6. In either case, the values of $f(x,y)$ at $(x_0,0)$ and (x_0,y) are zero, and therefore, $f_y(x_0,0) = \lim_{y\to 0} \frac{f(x_0,y)-f(x_0,0)}{y-0} = 0$ for any $x_0 \in \mathbb{R}$.

The next step is the calculation of $F(y) = \int_{-1}^{1} f(x,y)dx$ for $\forall y \in \left(-\frac{1}{4}, \frac{1}{4}\right)$. First, if $y \in \left(0, \frac{1}{4}\right)$, then

$$F(y) = \int_{-1}^{1} f(x,y)dx$$

$$= \int_{-1}^{0} 0\, dx + \int_{0}^{\sqrt{y}} x\, dx + \int_{\sqrt{y}}^{2\sqrt{y}} (-x + 2\sqrt{y})dx + \int_{2\sqrt{y}}^{1} 0\, dx$$

$$= \frac{x^2}{2}\Big|_{0}^{\sqrt{y}} + \left(-\frac{x^2}{2} + 2x\sqrt{y}\right)\Big|_{\sqrt{y}}^{2\sqrt{y}} = y.$$

Similarly, for $y \in \left(-\frac{1}{4}, 0\right)$, we get

$$F(y) = \int_{-1}^{1} f(x,y)dx$$

$$= \int_{-1}^{0} 0\, dx + \int_{0}^{\sqrt{|y|}} (-x)dx + \int_{\sqrt{|y|}}^{2\sqrt{|y|}} (x - 2\sqrt{|y|})dx + \int_{2\sqrt{|y|}}^{1} 0dx$$

$$= -\frac{x^2}{2}\Big|_{0}^{\sqrt{|y|}} + \left(\frac{x^2}{2} - 2x\sqrt{|y|}\right)\Big|_{\sqrt{|y|}}^{2\sqrt{|y|}} = -|y| = y.$$

Additionally,

$$F(0) = \int_{-1}^{1} f(x,0)dx = \int_{-1}^{1} 0\ dx = 0.$$

Thus, $F(y) = \int_{-1}^{1} f(x,y)dx = y$ and, consequently, $F_y(y) = 1$ for $\forall y \in \left(-\frac{1}{4}, \frac{1}{4}\right)$. In particular, $F_y(0) = \frac{d}{dy}\int_{-1}^{1} f(x,y)dx\big|_{y=0} = 1$, but $\int_{-1}^{1} f_y(x,0)dx = \int_{-1}^{1} 0\ dx = 0$.

Remark to Examples 7–9. The important condition to ensure the validity of differentiation under the sign of integral is the continuity of $f_y(x,y)$ in the considered domain, which was not satisfied in all these three examples. However, this condition is sufficient and not necessary as shown in Examples 10 and 11.

Example 10. Although $f(x,y)$ has infinitely many points of discontinuity on $[a,b] \times Y$, but $\frac{d}{dy}\int_a^b f(x,y)dx = \int_a^b f_y(x,y)dx$, $\forall y \in Y$.

Solution
Consider the function $f(x,y) = y + R(x)$ on $S = [0,1] \times \mathbb{R}$ (recall that $R(x)$ is the Riemann function defined as $R(x) = \begin{cases} \frac{1}{n}, x = \frac{m}{n} \in \mathbb{Q} \\ 0, x \in \mathbb{I} \end{cases}$, where m is integer, n is natural, and $\frac{m}{n}$ is in lowest terms). It is known that $R(x)$ is discontinuous at all rational points and continuous at all irrational points. It is also known that $R(x)$ is Riemann integrable on an arbitrary interval $[a,b]$ and $\int_a^b R(x)dx = 0$ (see explanations on integrability of $R(x)$ in Example 13 of Chapter 3). Using the last formula on $[0,1]$, one obtains

$$F(y) = \int_0^1 f(x,y)dx = \int_0^1 (y + R(x))dx$$
$$= \int_0^1 y\ dx + \int_0^1 R(x)dx = y, \forall y \in \mathbb{R},$$

that is, $F_y(y) = 1$, $\forall y \in \mathbb{R}$. On the other hand, $\int_0^1 f_y(x,y)dx = \int_0^1 1\ dx = 1$. Thus, $\frac{d}{dy}\int_0^1 f(x,y)dx = 1 = \int_0^1 f_y(x,y)dx$, $\forall y \in \mathbb{R}$.

Example 11. Although $f_y(x,y)$ is discontinuous at (x_0, y_0), but $\frac{d}{dy}\int_a^b f(x,y)dx\big|_{y=y_0} = \int_a^b f_y(x,y_0)dx$.

Solution
The function $f(x,y) = \begin{cases} \frac{y^3}{x^2+y^2}, x^2 + y^2 \neq 0 \\ 0, x^2 + y^2 = 0 \end{cases}$ possesses the partial derivative $f_y(x,y) = \begin{cases} \frac{3x^2y^2+y^4}{(x^2+y^2)^2}, x^2 + y^2 \neq 0 \\ 1, x^2 + y^2 = 0 \end{cases}$. The function is continuous on $\mathbb{R}^2$: for any

$(x, y) \neq (0, 0)$, it follows straightforward from the arithmetic properties, and for $(x, y) = (0, 0)$ the continuity follows from the fact that $f(0, 0) = 0$ and evaluation $|f(x, y)| = \frac{|y^3|}{x^2 + y^2} \leq |y| \underset{y \to 0}{\to} 0$. The derivative $f_y(x, y)$ is continuous on $\mathbb{R}^2 \setminus (0, 0)$ due to the arithmetic properties of continuous functions, and discontinuous at the origin, because $\lim_{x \to 0^-} f_y(x, 0) = 0 \neq 1 = \lim_{y \to 0^-} f_y(0, y)$.

Let us find the required integrals on $[0, 1]$. First,

$$F(y) = \int_0^1 f(x, y)dx = y^2 \arctan \frac{x}{y}\Big|_0^1 = y^2 \arctan \frac{1}{y}, \forall y \neq 0$$

and

$$F(0) = \int_0^1 f(x, 0)dx = \int_0^1 0 \, dx = 0.$$

Therefore,

$$F_y(y) = 2y \arctan \frac{1}{y} - \frac{y^2}{1 + y^2}, \forall y \neq 0$$

and

$$F_y(0) = \lim_{y \to 0} \frac{F(y) - F(0)}{y} = \lim_{y \to 0} y \arctan \frac{1}{y} = 0.$$

On the other hand,

$$\int_0^1 f_y(x, y)dx = \int_0^1 \frac{2x^2 y^2}{(x^2 + y^2)^2} dx + y^2 \int_0^1 \frac{1}{x^2 + y^2} dx$$

$$= -\frac{xy^2}{x^2 + y^2}\Big|_0^1 + \frac{2y^2}{y} \arctan \frac{x}{y}\Big|_0^1$$

$$= -\frac{y^2}{1 + y^2} + 2y \arctan \frac{1}{y}, \forall y \neq 0,$$

and

$$\int_0^1 f_y(x, 0)dx = \int_0^1 0dx = 0.$$

Thus,

$$\frac{d}{dy} \int_0^1 f(x, y)dx = \int_0^1 f_y(x, y)dx, \forall y \in \mathbb{R},$$

in particular,

$$F_y(0) = \frac{d}{dy} \int_0^1 f(x, y)dx\Big|_{y=0} = 0 = \int_0^1 f_y(x, 0)dx.$$

The counterexample with the function $f(x, y) = \begin{cases} \frac{y^4}{x^2}e^{-y^2/x}, & x > 0 \\ 0, & x = 0 \end{cases}$ defined on $X \times Y = [0, +\infty) \times \mathbb{R}$ shows that both the function and its partial derivative can be discontinuous at a chosen point, but still the integration and partial differentiation can be interchanged. Indeed, the given function is discontinuous at $(0, 0)$: choosing the path $x = y^2$, we get

$$\lim_{y \to 0} f(y^2, y) = \lim_{y \to 0} \frac{y^4}{y^4}e^{-y^2/y^2} = e^{-1} \neq 0 = f(0, 0).$$

The partial derivative in y is defined in the considered domain:

$$f_y(x, y) = \begin{cases} \left(\frac{4y^3}{x^2} - \frac{2y^5}{x^3}\right)e^{-y^2/x}, & x > 0 \\ 0, & x = 0 \end{cases},$$

but it is also discontinuous at $(0, 0)$: using the same path $x = y^2$, we obtain

$$\lim_{y \to 0} f_y(y^2, y) = \lim_{y \to 0} \left(\frac{4y^3}{y^4} - \frac{2y^5}{y^6}\right)e^{-y^2/y^2} = \lim_{y \to 0} \frac{2}{y}e^{-1} = \infty.$$

Let us calculate the integrals:

$$F(y) = \int_0^1 \frac{y^4}{x^2}e^{-y^2/x}dx = y^2 e^{-y^2/x}|_0^1 = y^2 e^{-y^2}$$

and, consequently

$$F_y(y) = (2y - 2y^3)e^{-y^2}, \forall y \in \mathbb{R}.$$

At the same time,

$$\int_0^1 f_y(x, 0)dx = \int_0^1 0dx = 0,$$

and for $\forall y \neq 0$

$$\int_0^1 f_y(x, y)dx = \int_0^1 \frac{4y^3}{x^2}e^{-y^2/x}dx - \int_0^1 \frac{2y^5}{x^3}e^{-y^2/x}dx$$

$$= \int_0^1 \frac{4y^3}{x^2}e^{-y^2/x}dx - \frac{2y^3}{x}e^{-y^2/x}\Big|_0^1 - \int_0^1 \frac{2y^3}{x^2}e^{-y^2/x}dx$$

$$= 2ye^{-y^2/x}\Big|_0^1 - \frac{2y^3}{x}e^{-y^2/x}\Big|_0^1 = (2y - 2y^3)e^{-y^2}.$$

To calculate the last integral, we have applied the integration by parts with $u = \frac{1}{x}$ and $dv = \frac{y^2}{x^2}e^{-y^2/x}dx$, and also l'Hospital's rule for the limit

$$\lim_{x \to 0_+} \frac{2y^3}{x}e^{-y^2/x} = \lim_{t \to +\infty} \frac{2y^3 t}{e^{y^2 t}} = \lim_{t \to +\infty} \frac{2y^3}{y^2 e^{y^2 t}} = 0.$$

Hence,

$$\frac{d}{dy} \int_0^1 f(x,y)dx = \int_0^1 f_y(x,y)dx, \forall y \in \mathbb{R},$$

in particular,

$$F_y(0) = \frac{d}{dy} \int_0^1 f(x,y)dx \bigg|_{y=0} = 0 = \int_0^1 f_y(x,0)dx,$$

although both $f(x,y)$ and $f_y(x,y)$ are discontinuous at $(0,0)$.

4.3 Integrability

Example 12. A function $f(x,y)$, defined on $[a,b] \times Y$, converges on $[a,b]$ to a function $\varphi(x)$ as y approaches y_0, and $f(x,y)$ is not Riemann integrable on $[a,b]$ for each fixed $y \in Y$, but $\varphi(x)$ is Riemann integrable on $[a,b]$.

(This is an analogue to Example 10 for a sequence in Chapter 3.)

Solution

The function $f(x,y) = \begin{cases} y, x \in \mathbb{Q} \\ 0, x \in \mathbb{I} \end{cases}$ considered on $[0,1] \times (0,1]$ with $y_0 = 0$ converges to 0: $\varphi(x) = \lim\limits_{y \to 0} f(x,y) = 0, \forall x \in [0,1]$. Therefore, the limit function is Riemann integrable on $[0,1]$: $\int_0^1 \varphi(x)dx = 0$. Moreover, the convergence is uniform on $[0,1]$, because $|f(x,y) - \varphi(x)| \leq y \xrightarrow[y \to 0]{} 0$. At the same time, the original function can be represented in the form $f(x,y) = y \cdot D(x)$, where $D(x)$ is Dirichlet's function, which is not Riemann integrable on any interval. Therefore, $f(x,y)$ is not Riemann integrable on $[0,1]$ for $\forall y \in (0,1]$.

Example 13. A function $f(x,y)$ is defined on $[a,b] \times [c,d]$ and one of the iterated integrals— $\int_c^d dy \int_a^b f(x,y)dx$ or $\int_a^b dx \int_c^d f(x,y)dy$—exists, but another does not exist.

Solution

Consider the function $f(x,y) = \begin{cases} \frac{2y^3 - 6x^2y}{(x^2+y^2)^3}, x^2 + y^2 \neq 0 \\ 0, x^2 + y^2 = 0 \end{cases}$ on $[0,1] \times [0,1]$. Recall that the integral value does not depend on the specification of function at a singular point; therefore, we can calculate each integral regardless of the second part in the function definition. The calculation of the first iterated integral is as follows. For each fixed $y \in [0,1]$, we get

$$\int_0^1 f(x,y)dx = \int_0^1 \frac{2y^3 - 6x^2 y}{(x^2 + y^2)^3} dx$$

$$= \int_0^1 \frac{2y(y^2 + x^2)}{(x^2 + y^2)^3} dx - \int_0^1 \frac{8yx^2}{(x^2 + y^2)^3} dx$$

$$= \int_0^1 \frac{2y}{(x^2 + y^2)^2} dx + \frac{2xy}{(x^2 + y^2)^2}\Big|_0^1 - \int_0^1 \frac{2y}{(x^2 + y^2)^2} dx$$

$$= \frac{2y}{(1 + y^2)^2}$$

(the integration by parts with $u = x$ and $dv = \frac{x}{(x^2+y^2)^3} dx$ was applied). This result is valid for $y = 0$, in which case the function $f(x, 0) = 0$ and the corresponding integral is also zero. Then

$$\int_0^1 dy \int_0^1 f(x,y)dx = \int_0^1 \frac{2y}{(1 + y^2)^2} dy = -\frac{1}{1 + y^2}\Big|_0^1 = \frac{1}{2}.$$

For the second iterated integral, we have at each fixed $x \in (0, 1]$:

$$G(x) = \int_0^1 f(x,y)dy = \int_0^1 \frac{2y^3 - 6x^2 y}{(x^2 + y^2)^3} dy$$

$$= \int_0^1 \frac{2y}{(x^2 + y^2)^2} dy - \int_0^1 \frac{8yx^2}{(x^2 + y^2)^3} dy$$

$$= -\frac{1}{x^2 + y^2}\Big|_0^1 + 2x^2 \frac{1}{(x^2 + y^2)^2}\Big|_0^1$$

$$= -\frac{1}{x^2 + 1} + \frac{2x^2}{(x^2 + 1)^2} - \frac{1}{x^2}.$$

For $x = 0$, the integral of $f(0, y) = \begin{cases} \frac{2}{y^3}, y \neq 0 \\ 0, y = 0 \end{cases}$ diverges

$$\int_0^1 f(0, y)dy = \int_0^1 \frac{2}{y^3} dy = -\frac{1}{y^2}\Big|_0^1 = +\infty,$$

so, formally, the last result is included in the integral formula for $x \in (0, 1]$, because the last term $\frac{1}{x^2}$ approaches infinity as x approaches 0, whereas the first two terms have finite limits. Again, the mere fact that the function is not defined at some point in the interval of integration does not affect the existence or the value of the integral. However, since $\lim_{x \to 0} G(x) = \infty$, the function $G(x)$ is not Riemann integrable on $[0, 1]$. Moreover, the integral $\int_0^1 G(x)dx$ does not exist even in the sense of the improper integral, because the functions $\frac{1}{x^2+1}$ and $\frac{2x^2}{(x^2+1)^2}$ are Riemann integrable on $[0, 1]$, whereas the improper integral $\int_0^1 \frac{1}{x^2} dx$ is divergent.

Note that $f(x,y)$ is continuous on $\mathbb{R}^2 \setminus (0,0)$. At the origin, the function is discontinuous, because $\lim_{y\to 0} f(0,y) = \lim_{y\to 0} \frac{2}{y^3} = \infty$.

Remark. Of course, one can construct a function discontinuous at only one point of the considered domain for which both iterated integrals do not exist. The function $f(x,y) = \begin{cases} \frac{2xy}{(x^2+y^2)^2}, & x^2+y^2 \neq 0 \\ 0, & x^2+y^2 = 0 \end{cases}$ considered on $[0,1]\times[0,1]$ is one of such examples. This function is continuous on $\mathbb{R}^2$ except at the origin, where $\lim_{x\to 0} f(x,x) = \lim_{x\to 0} \frac{2x^2}{4x^4} = \infty$. The first iterated integral is calculated as follows. For each fixed $y \in (0,1]$, we get

$$\int_0^1 f(x,y)dx = \int_0^1 \frac{2xy}{(x^2+y^2)^2}dx = -\frac{y}{1+y^2} + \frac{1}{y}$$

and at $y = 0$

$$\int_0^1 f(x,0)dx = 0.$$

Therefore,

$$\int_0^1 dy \int_0^1 f(x,y)dx = \int_0^1 -\frac{y}{1+y^2}dy + \int_0^1 \frac{1}{y}dy.$$

The first integral in the right-hand side exists

$$\int_0^1 -\frac{y}{1+y^2}dy = -\frac{1}{2}\ln(1+y^2)\Big|_0^1 = -\frac{1}{2}\ln 2,$$

but the second diverges $\int_0^1 \frac{1}{y}dy = \ln y|_0^1 = +\infty$, which implies the divergence of the first iterated integral. Due to the symmetry of the function, the same is true for the second iterated integral.

Example 14. A function $f(x,y)$ is defined on $[a,b]\times[c,d]$ and both iterated integrals $\int_c^d dy \int_a^b f(x,y)dx$ and $\int_a^b dx \int_c^d f(x,y)dy$ exist, but they assume different values.

Solution

Consider the function $f(x,y) = \begin{cases} \frac{x^2-y^2}{(x^2+y^2)^2}, & x^2+y^2 \neq 0 \\ 0, & x^2+y^2 = 0 \end{cases}$ on $[0,1]\times[0,1]$. The first iterated integral can be calculated as follows:

$$\int_0^1 f(x,y)dx = \int_0^1 \frac{x^2-y^2}{(x^2+y^2)^2}dx = \int_0^1 \frac{2x^2}{(x^2+y^2)^2}dx - \int_0^1 \frac{x^2+y^2}{(x^2+y^2)^2}dx$$

$$= -\frac{x}{x^2+y^2}\Big|_0^1 + \int_0^1 \frac{dx}{x^2+y^2} - \int_0^1 \frac{dx}{x^2+y^2} = -\frac{1}{1+y^2}$$

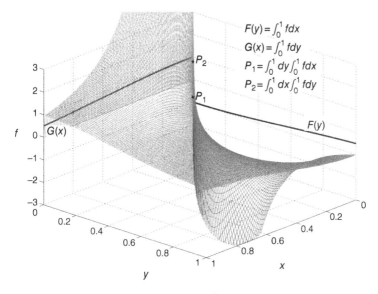

$$F(y) = \int_0^1 f\,dx$$
$$G(x) = \int_0^1 f\,dy$$
$$P_1 = \int_0^1 dy \int_0^1 f\,dx$$
$$P_2 = \int_0^1 dx \int_0^1 f\,dy$$

Figure 4.9 Example 14, function $f(x, y) = \begin{cases} \frac{x^2 - y^2}{(x^2+y^2)^2}, & x^2 + y^2 \neq 0 \\ 0, & x^2 + y^2 = 0 \end{cases}$.

(we applied the integration by parts with $u = x$ and $dv = \frac{2x}{(x^2+y^2)^2}dx$) and, consequently,

$$\int_0^1 dy \int_0^1 f(x, y)dx = -\int_0^1 \frac{dy}{1 + y^2} = -\arctan y\big|_0^1 = -\frac{\pi}{4}.$$

For the second iterated integral, we have

$$\int_0^1 f(x, y)dy = \int_0^1 \frac{x^2 - y^2}{(x^2 + y^2)^2}dy = \frac{y}{x^2 + y^2}\bigg|_0^1 = \frac{1}{x^2 + 1},$$

and

$$\int_0^1 dx \int_0^1 f(x, y)dy = \int_0^1 \frac{dx}{1 + x^2} = \arctan x\big|_0^1 = \frac{\pi}{4}.$$

Here, in the evaluation of both iterated integrals, we used again the fact that the result of calculation of integral does not depend on the definition of a function at a singular point. Thus, the iterated integrals exist, but have different values.

Note that like in Example 13 the function $f(x, y)$ is continuous on $\mathbb{R}^2 \setminus (0, 0)$ and has the essential discontinuity at the origin: $\lim_{y \to 0} f(0, y) = \lim_{y \to 0} \frac{-y^2}{y^4} = -\infty.$

Remark to Examples 13 and 14. These examples show that the discontinuity of $f(x,y)$ at only one point in the considered domain is sufficient to cause inequality of the iterated integrals or nonexistence of one of them (or even both). At the same time, the continuity of $f(x,y)$ on the considered domain is not necessary condition for the existence and equality of the iterated integrals as shown in the next example.

Example 15. A function $f(x,y)$ is defined on $[a,b] \times [c,d]$ and $\int_c^d dy \int_a^b f(x,y)dx = \int_a^b dx \int_c^d f(x,y)dy$, but $f(x,y)$ is discontinuous on $[a,b] \times [c,d]$.

Solution

The function $f(x,y) = \begin{cases} \frac{xy^2}{(x^2+y^2)^2}, & x^2 + y^2 \neq 0 \\ 0, & x^2 + y^2 = 0 \end{cases}$ is continuous on $\mathbb{R}^2 \setminus (0,0)$ due to the arithmetic properties of continuous functions, and it has the essential discontinuity at the origin, because $\lim_{x \to 0} f(x,x) = \lim_{x \to 0} \frac{x^3}{4x^4} = \infty$. Let us find the iterated integrals, taking again into account that the function values at isolated points do not affect the value of the Riemann integral. For the first iterated integral, we get

$$\int_0^1 f(x,y)dx = \int_0^1 \frac{xy^2}{(x^2+y^2)^2}dx = -\frac{y^2}{2}\frac{1}{x^2+y^2}\Big|_0^1$$

$$= -\frac{y^2}{2(1+y^2)} + \frac{1}{2} = \frac{1}{2(1+y^2)}$$

and, therefore,

$$\int_0^1 dy \int_0^1 f(x,y)dx = \frac{1}{2}\int_0^1 \frac{dy}{1+y^2} = \frac{1}{2}\arctan y\Big|_0^1 = \frac{\pi}{8}.$$

For the second iterated integral, we have

$$\int_0^1 f(x,y)dy = \int_0^1 \frac{xy^2}{(x^2+y^2)^2}dy = -\frac{xy}{2(x^2+y^2)}\Big|_0^1 + \frac{x}{2}\int_0^1 \frac{dy}{x^2+y^2}$$

$$= -\frac{x}{2(x^2+1)} + \frac{1}{2}\arctan\frac{y}{x}\Big|_0^1$$

$$= -\frac{x}{2(x^2+1)} + \frac{1}{2}\arctan\frac{1}{x}, \forall x \neq 0$$

(the integration by parts with $u = y$ and $dv = \frac{y}{(x^2+y^2)^2}dy$ was applied), and

$$\int_0^1 f(0,y)dy = \int_0^1 0\, dy = 0.$$

Therefore,

$$
\int_0^1 dx \int_0^1 f(x,y)dy = -\frac{1}{2}\int_0^1 \frac{xdx}{1+x^2} + \frac{1}{2}\int_0^1 \arctan\frac{1}{x}dx
$$

$$
= -\frac{1}{2}\int_0^1 \frac{xdx}{1+x^2} + \frac{x}{2}\arctan\frac{1}{x}\Big|_0^1 + \frac{1}{2}\int_0^1 \frac{xdx}{1+x^2}
$$

$$
= \frac{1}{2}\arctan 1 = \frac{\pi}{8}
$$

(again the integration by parts was applied, at this time with $u = \arctan\frac{1}{x}$ and $dv = dx$). Thus, both iterated integrals exist and are equal, despite the fact that the function is not continuous on $[0,1] \times [0,1]$.

Remark. It can be shown a stronger result that the existence and equality of the iterated integrals are not sufficient to guarantee the existence of the double integral: a function $f(x,y)$ is defined on $[a,b] \times [c,d]$ and both iterated integrals exist and are equal, but nevertheless the double integral does not exist.

For a counterexample, consider the function $f(x,y)$ defined on the unit square $S = [0,1] \times [0,1]$ as follows: $f(x,y) = \begin{cases} 1, (x,y) \in T \\ 0, (x,y) \notin T \end{cases}$, where T consists of the points of S such that $(x,y) = \left(\frac{m}{p}, \frac{n}{p}\right)$, $m,n,p \in \mathbb{N}$, where both fractions are in lowest terms (note that the denominator is the same). First, let us show that the iterated integral $\int_0^1 dy \int_0^1 f(x,y)dx$ exists. In fact, if we choose any irrational y_0, then $f(x,y_0) = 0$ and $\int_0^1 f(x,y_0)dx = 0$. On the other hand, for a rational $y_0 = \frac{n}{p}$ (with the fraction in lowest terms), $f(x,y_0)$ is different from zero only in the finite number of the points, because for a given natural p there are at most p numbers of the form $x = \frac{m}{p}$, $m \in \mathbb{N}$, in $[0,1]$. Therefore, again $\int_0^1 f(x,y_0)dx = 0$. Since the inner integral is zero for any y_0, the iterated integral exists and $\int_0^1 dy \int_0^1 f(x,y)dx = 0$. Due to the symmetry of the function and domain with respect to interchange of variables, the same result is true for the second iterated integral: $\int_0^1 dx \int_0^1 f(x,y)dy = 0$.

Now let us prove that the double integral $\iint_S f(x,y)dA$ does not exist. To this end, let us consider the special uniform partitions obtained by division of each side of S in p equal parts, where p is a prime number. Note that the diameter of such partitions approaches zero as p approaches infinity. Each elementary square

$$
R_{ij} = [x_{i-1}, x_i] \times [y_{j-1}, y_j], i,j = 1, \cdots, p
$$

contains the points $(c_i, d_j) \notin T$ and also the points $(c_i, d_j) \in T$. The possibility of the former choice is evident, since in any interval $[x_{i-1}, x_i]$ (or $[y_{j-1}, y_j]$) there are irrational points. The latter choice can be made by using the points of the upper right corner in each R_{ij}:

$$(c_i, d_j) = (x_i, y_j) = \left(\frac{i}{p}, \frac{j}{p}\right), i, j = 1, \cdots, p-1,$$

except for the elementary squares of the upper row ($j = p$) and the right column ($i = p$), for which we choose

$$(c_i, d_p) = \left(\frac{i}{p}, \frac{p-1}{p}\right), i = 1, \cdots, p-1,$$

$$(c_p, d_j) = \left(\frac{p-1}{p}, \frac{j}{p}\right), j = 1, \cdots, p-1,$$

and

$$(c_p, d_p) = \left(\frac{p-1}{p}, \frac{p-1}{p}\right).$$

The corresponding double Riemann sums

$$S(f; P) = \sum_{j=1}^{p} \sum_{i=1}^{p} f(c_i, d_j) \cdot \Delta x_i \Delta y_j$$

are equal 0 in the case $(c_i, d_j) \notin T$ and their partial limit is 0, but these sums are equal 1 if $(c_i, d_j) \in T$ and the corresponding partial limit is 1. Since two partial limits of the Riemann sums are different, there is no limit of the double Riemann sums as the partition diameter approaches 0, which means that the double integral does not exist.

Example 16. A function $f(x, y)$ has infinitely many discontinuity points in $[a, b] \times [c, d]$, but $\int_c^d dy \int_a^b f(x, y)dx = \int_a^b dx \int_c^d f(x, y)dy$.

Solution
Consider the function $f(x, y) = \text{sgn}(x - y)$ on $[0, 1] \times [-2, 2]$. It was shown in Example 6 that $F(y) = \int_0^1 f(x, y)dx = \begin{cases} 1, y < 0 \\ 1 - 2y, 0 \le y \le 1 \\ -1, y > 1 \end{cases}$ for $\forall y \in \mathbb{R}$. Therefore,

$$\int_{-2}^{2} dy \int_{0}^{1} f(x, y)dx = \int_{-2}^{2} F(y)dy$$

$$= \int_{-2}^{0} 1 \, dy + \int_{0}^{1} (1 - 2y)dy + \int_{1}^{2} (-1)dy$$

$$= y|_{-2}^{0} + (y - y^2)|_{0}^{1} - y|_{1}^{2} = 1.$$

On the other hand,

$$G(x) = \int_{-2}^{2} f(x,y)dy = \int_{-2}^{x} 1\, dy + \int_{x}^{2} (-1)dy = y|_{-2}^{x} - y|_{x}^{2} = 2x$$

and, consequently,

$$\int_{0}^{1} dx \int_{-2}^{2} f(x,y)dy = \int_{0}^{1} G(x)dx = \int_{0}^{1} 2x\, dx = x^2|_{0}^{1} = 1.$$

Thus, the two iterated integrals are equal. Nevertheless, $f(x,y)$ is discontinuous at each point of the line $y = x$, $x \in [0,1]$.

Remark. One can strengthen the above example to the following form: a function $f(x,y)$ is discontinuous and unbounded at infinitely many points in $[a,b] \times [c,d]$, but nevertheless $\int_{c}^{d} dy \int_{a}^{b} f(x,y)dx = \int_{a}^{b} dx \int_{c}^{d} f(x,y)dy$.

A counterexample is provided by the function $f(x,y) = \begin{cases} \frac{y}{\sqrt{x}}, y < x^2, x \neq 0 \\ 2xy^2, y \geq x^2 \\ 0, x = 0 \end{cases}$ on

$[0,1] \times [-2,2]$. In Remark 2 to Example 6, it was shown that

$$F(y) = \int_{0}^{1} f(x,y)dx = \begin{cases} 2y, & y < 0 \\ y^3 + 2y - 2y^{5/4}, & 0 \leq y \leq 1 \\ y^2, & y > 1 \end{cases}.$$

Therefore,

$$\int_{-2}^{2} dy \int_{0}^{1} f(x,y)dx = \int_{-2}^{2} F(y)dy$$

$$= \int_{-2}^{0} 2y\, dy + \int_{0}^{1} (y^3 + 2y - 2y^{5/4})dy + \int_{1}^{2} y^2 dy$$

$$= y^2|_{-2}^{0} + \left(\frac{y^4}{4} + y^2 - \frac{8}{9}y^{9/4} \right)\Big|_{0}^{1} + \frac{y^3}{3}\Big|_{1}^{2} = -\frac{47}{36}.$$

For another iterated integral, one gets

$$G(x) = \int_{-2}^{2} f(x,y)dy = \int_{-2}^{x^2} \frac{y}{\sqrt{x}}dy + \int_{x^2}^{2} 2xy^2 dy$$

$$= \frac{y^2}{2\sqrt{x}}\Big|_{-2}^{x^2} + \frac{2}{3}xy^3\Big|_{x^2}^{2} = \frac{1}{2}x^{7/2} - 2x^{-1/2} + \frac{16}{3}x - \frac{2}{3}x^7, \forall x \neq 0$$

and

$$G(0) = \int_{-2}^{2} f(0, y) dy = \int_{-2}^{2} 0 \, dy = 0.$$

Then

$$\int_{0}^{1} dx \int_{-2}^{2} f(x, y) dy = \int_{0}^{1} G(x) dx$$

$$= \int_{0}^{1} \left(\frac{1}{2} x^{7/2} - 2x^{-1/2} + \frac{16}{3} x - \frac{2}{3} x^{7} \right) dx$$

$$= \left(\frac{1}{9} x^{9/2} - 4x^{1/2} + \frac{8}{3} x^{2} - \frac{1}{12} x^{8} \right) \Big|_{0}^{1} = -\frac{47}{36}.$$

Thus, the two iterated integrals are equal. However, as shown in Remark 2 to Example 6, $f(x, y)$ is discontinuous and unbounded at each point of the segment $x = 0$, $y \in [-2, 0]$.

Exercises

1 Show that the function $f(x, y) = \frac{4x^3}{y^2}(1 - x^4)^{1/y^2}$ defined on $X \times Y = [0, 1] \times (0, 1]$ with $y_0 = 0$ fulfills all the conditions of the statement in Example 1, but $\int_{0}^{1} \lim_{y \to y_0} f(x, y) dx \neq \lim_{y \to y_0} \int_{0}^{1} f(x, y) dx$. Check the character of the convergence of $f(x, y)$ on $X = [0, 1]$ as y approaches y_0. Do the same for $f(x, y) = \frac{2xy^2}{(x^2+y^2)^2}$ defined on $[0, 1] \times (0, 1]$ with $y_0 = 0$.

2 Verify whether the function $f(x, y) = \frac{2xy}{x^4+y^2}$ satisfies the conditions of Examples 1 and 2 on $X \times Y = [0, 1] \times (0, 1]$ with $y_0 = 0$. What about the relation between the integrals $\int_{0}^{1} \lim_{y \to y_0} f(x, y) dx$ and $\lim_{y \to y_0} \int_{0}^{1} f(x, y) dx$? Analyze the nature of the convergence of $f(x, y)$ on $X = [0, 1]$ as y approaches y_0 and monotonicity of $f(x, y)$ in y on $Y = (0, 1]$.

3 Show that the functions
   a) $f(x, y) = \frac{2xy}{x^4+y^2}$, $X \times Y = [-1, 1] \times (0, 1]$, $y_0 = 0$,
   b) $f(x, y) = \arctan \frac{x}{y}$, $X \times Y = [0, 1] \times (0, +\infty)$, $y_0 = 0$,
   c) $f(x, y) = \frac{2xy^3}{(x^2+y^2)^2}$, $X \times Y = [0, 1] \times (0, 1]$, $y_0 = 0$
   provide counterexamples to Example 3. Verify which conditions are violated in the results (the theorem and corollary) on passing to the limit under the integral sign.

4 Prove that the following statement is false: "if $f(x, y)$ is continuous on $[a, b] \times Y$, then $\int_{a}^{b} \lim_{y \to y_0} f(x, y) dx = \lim_{y \to y_0} \int_{a}^{b} f(x, y) dx$."

(Hint: use the counterexamples with one of the following functions:
a) $f(x,y) = \frac{1}{y^2}e^{-x/y^2}$, $X \times Y = [0,1] \times (0,1]$, $y_0 = 0$,

b) $f(x,y) = \frac{\cos x}{y^2}e^{-x/y^2}$, $X \times Y = \left[0, \frac{\pi}{2}\right] \times (0,1]$, $y_0 = 0$.
Check the character of the convergence of $f(x,y)$ on $X = [a,b]$ as y approaches y_0.)

5   Verify whether the following statement is false: "if $f(x,y)$ is continuous on $[a,b] \times Y$, and $\int_a^b \lim_{y \to y_0} f(x,y)dx = \lim_{y \to y_0} \int_a^b f(x,y)dx$, then there exists a finite limit $\lim_{y \to y_0} f(x,y)dx$ for any fixed $x \in [a,b]$."

(Hint: construct a counterexample using the function $f(x,y) = \frac{1}{y}e^{-x/y^2}$, $X \times Y = [0,1] \times (0,1]$, $y_0 = 0$. Investigate the character of the convergence of $f(x,y)$ on $X = [0,1]$ as y approaches y_0. Another counterexample: try $f(x,y) = \text{sgn}ye^{-x/y^2}$, $X \times Y = [0,1] \times [-1,0) \cup (0,1]$, $y_0 = 0$.)

6   Verify whether the following statement is false: "if $f(x,y)$ is discontinuous at each point of $[a,b] \times Y$, then $\int_a^b \lim_{y \to y_0} f(x,y)dx \neq \lim_{y \to y_0} \int_a^b f(x,y)dx$."
(Hint: construct a counterexample using the function $f(x,y) = D(y)e^{-x/y}$, $X \times Y = [0,1] \times (0,1]$, $y_0 = 0$, where $D(y)$ is Dirichlet's function.)

7   Verify whether the following statement is false: "if $f(x,y)$ is continuous on $[a,b] \times Y$, there exists $\varphi(x) = \lim_{y \to y_0} f(x,y)dx$ for $\forall x \in [a,b]$ and $\int_a^b \lim_{y \to y_0} f(x,y)dx = \lim_{y \to y_0} \int_a^b f(x,y)dx$, then $f(x,y)$ converges to $\varphi(x)$ uniformly on $[a,b]$."
(Hint: consider one of the following functions:
a) $f(x,y) = \frac{x}{y^2}e^{-x/y^2}$, $X \times Y = [0,1] \times (0,1]$, $y_0 = 0$,

b) $f(x,y) = \frac{\sin x}{y}e^{-x/y}$, $X \times Y = \left[0, \frac{\pi}{2}\right] \times (0,1]$, $y_0 = 0$,
to construct a counterexample.)

8   Show that each of the following functions

a) $f(x,y) = \begin{cases} \frac{2xy(x^2+2)}{x^4+y^2}, & x^2 + y^2 \neq 0 \\ 0, & x^2 + y^2 = 0 \end{cases}$

b) $f(x,y) = \begin{cases} \frac{2x}{y^2}e^{-x^2/y^2}, & y \neq 0 \\ 0, & y = 0 \end{cases}$

c) $f(x,y) = \begin{cases} \frac{2xy^2}{(x^2+y^2)^2}, & x^2 + y^2 \neq 0 \\ 0, & x^2 + y^2 = 0 \end{cases}$

provides a counterexample to Examples 4 and 5. Consider $[a,b] = [0,1]$.

9 For the function $f(x, y) = \begin{cases} 2x, y < x^2 \\ 2xy^2, y \geq x^2 \end{cases}$, find such a set $X \times Y$ that can be used to provide a counterexample to Example 6.

10 Show that the statement in Remark 1 to Example 6 can be exemplified by

the function $f(x, y) = \begin{cases} \frac{2xy^{8/3}}{(x^2+y^2)^2}, x^2 + y^2 \neq 0 \\ 0, x^2 + y^2 = 0 \end{cases}$ considered on $[0, 1] \times \mathbb{R}$, and

the statement in Remark 2—by the function $f(x, y) = \begin{cases} \frac{2y}{\sqrt[3]{x}}, y < x, x \neq 0 \\ 3x^2y, y \geq x \\ 0, x = 0 \end{cases}$

considered on $[0, 1] \times \mathbb{R}$.

11 Use the following functions to construct counterexamples to Examples 7 and 8:

a) $f(x, y) = \begin{cases} \frac{y^5}{x^3}e^{-y^4/x^2}, x \neq 0 \\ 0, x = 0 \end{cases}$ on $X \times Y = \mathbb{R}^2$,

b) $f(x, y) = \begin{cases} \frac{y\sqrt{y}}{x^3}e^{-y/x^2}, x \neq 0 \\ 0, x = 0 \end{cases}$ on $X \times Y = \mathbb{R} \times [0, +\infty)$,

c) $f(x, y) = \begin{cases} \frac{2x}{y}e^{-x^2/y^2}, y \neq 0 \\ 0, y = 0 \end{cases}$ on $X \times Y = \mathbb{R}^2$,

d) $f(x, y) = \begin{cases} \frac{2xy^3}{x^4+y^4}, x^2 + y^2 \neq 0 \\ 0, x^2 + y^2 = 0 \end{cases}$ on $X \times Y = \mathbb{R}^2$,

e) $f(x, y) = \begin{cases} \frac{2xy^2}{x^4+y^4}, x^2 + y^2 \neq 0 \\ 0, x^2 + y^2 = 0 \end{cases}$ on $X \times Y = \mathbb{R}^2$,

f) $f(x, y) = \begin{cases} \frac{y^3}{x^2}e^{-y^2/x}, x > 0 \\ 0, x = 0 \end{cases}$ on $X \times Y = [0, +\infty) \times \mathbb{R}$,

g) $f(x, y) = \begin{cases} \frac{y^2}{x^2}e^{-y^2/x}, x > 0 \\ 0, x = 0 \end{cases}$ on $X \times Y = [0, +\infty) \times \mathbb{R}$,

h) $f(x, y) = \begin{cases} \frac{y^2}{x^3}e^{-y^2/x^2}, x \neq 0 \\ 0, x = 0 \end{cases}$ on $X \times Y = \mathbb{R}^2$.

In all cases, consider $[a, b] = [0, 1]$.

12 Show that the following functions

a) $f(x, y) = \begin{cases} \frac{2xy^4}{x^4+y^4}, x^2 + y^2 \neq 0 \\ 0, x^2 + y^2 = 0 \end{cases}$ on $X \times Y = \mathbb{R}^2$,

b) $f(x, y) = \begin{cases} \frac{y^6}{x^3}e^{-y^4/x^2}, x \neq 0 \\ 0, x = 0 \end{cases}$ on $X \times Y = \mathbb{R}^2$

are counterexamples to Example 11. Consider $[a, b] = [0, 1]$.

13 Use the functions

a) $f(x, y) = \begin{cases} \frac{3xy^2 - x^3}{(x^2 + y^2)^3}, & x^2 + y^2 \neq 0 \\ 0, & x^2 + y^2 = 0 \end{cases}$ on $X \times Y = [0, 1]^2$,

b) $f(x, y) = \begin{cases} \frac{x^3 - xy^2}{(x^2 + y^2)^3}, & x^2 + y^2 \neq 0 \\ 0, & x^2 + y^2 = 0 \end{cases}$ on $X \times Y = [0, 1]^2$,

c) $f(x, y) = \begin{cases} \frac{2x - y}{(x+y)^4}, & x + y > 0 \\ 0, & (x, y) = (0, 0) \end{cases}$ on $X \times Y = [0, 1]^2$,

d) $f(x, y) = \begin{cases} \frac{x^2 y - y^3}{(x^2 + y^2)^3}, & x^2 + y^2 \neq 0 \\ 0, & x^2 + y^2 = 0 \end{cases}$ on $X \times Y = [0, 1] \times [0, 1]$,

e) $f(x, y) = \begin{cases} \frac{x - 2y}{(x+y)^4}, & x + y > 0 \\ 0, & (x, y) = (0, 0) \end{cases}$ on $X \times Y = [0, 1] \times [0, 1]$

to construct counterexamples to Example 13.

14 Prove that the following statement is false: "if $f(x, y)$ is continuous on $[a, b] \times [c, d]$, except at one point, then there exists at least one of the integrals $\int_c^d dy \int_a^b f(x, y) dx$ and $\int_a^b dx \int_c^d f(x, y) dy$."

(Hint: construct a counterexample with the function $f(x, y) = \begin{cases} \frac{4x - y}{(x+y)^4}, & x + y > 0 \\ 0, & (x, y) = (0, 0) \end{cases}$ on $X \times Y = [0, 1] \times [0, 1]$. Another function is

$f(x, y) = \begin{cases} \frac{x - y}{(x+y)^4}, & x + y > 0 \\ 0, & (x, y) = (0, 0) \end{cases}$ on $X \times Y = [0, 1] \times [0, 1]$.)

15 Show that the function $f(x, y) = \begin{cases} \frac{y^2 - 2xy}{(x+y)^4}, & x + y > 0 \\ 0, & (x, y) = (0, 0) \end{cases}$ on $X \times Y = [0, 1]^2$ is a counterexample to Example 14. Do the same for the function $f(x, y) = \begin{cases} \frac{x^2 + 2xy - 2y^2}{(x+y)^4}, & x + y > 0 \\ 0, & (x, y) = (0, 0) \end{cases}$ on $X \times Y = [0, 1]^2$.

16 Use the following functions to provide counterexamples to Example 15:

a) $f(x, y) = \begin{cases} \frac{2xy}{x^4 + y^2}, & x^2 + y^2 \neq 0 \\ 0, & x^2 + y^2 = 0 \end{cases}$ on $X \times Y = [0, 1]^2$,

b) $f(x, y) = \begin{cases} \frac{2x^3}{y^3} e^{-x^2/y^2}, & y \neq 0 \\ 0, & y = 0 \end{cases}$ on $X \times Y = [0, 1]^2$,

c) $f(x, y) = \begin{cases} \frac{x^2 - 4xy + y^2}{(x+y)^4}, & x + y > 0 \\ 0, & x^2 + y^2 = 0 \end{cases}$ on $X \times Y = [0, 1]^2$,

d) $f(x,y) = \begin{cases} \frac{2x-y}{(x+y)^2}, x+y>0 \\ 0, x^2+y^2=0 \end{cases}$ on $X \times Y = [0,1]^2$,

e) $f(x,y) = \begin{cases} \frac{2xy^3}{(x^2+y^2)^2}, x^2+y^2 \neq 0 \\ 0, x^2+y^2=0 \end{cases}$ on $X \times Y = [0,1]^2$,

f) $f(x,y) = \begin{cases} \frac{x-3y}{(x+y)^2}, x+y>0 \\ 0, (x,y)=(0,0) \end{cases}$ on $X \times Y = [0,1]^2$.

17 Show that the function $f(x,y) = \begin{cases} 2x, y<x^2 \\ 2xy^2, y \geq x^2 \end{cases}$ on $[0,1] \times [-2,2]$ provides one more counterexample to Example 16.

18 Show that the function $f(x,y) = \begin{cases} \frac{2y}{\sqrt[3]{x}}, y<x, x \neq 0 \\ 3x^2y, y \geq x \\ 0, x=0 \end{cases}$ on $[0,1] \times [-2,2]$ is one more counterexample to Remark in Example 16.

Further Reading

B.M. Budak and S.V. Fomin, *Multiple Integrals, Field Theory and Series*, Mir Publisher, Moscow, 1978.

G.M. Fichtengolz, *Differential- und Integralrechnung, Vol.1–3*, V.E.B. Deutscher Verlag Wiss., Berlin, 1968.

V.A. Ilyin and E.G. Poznyak, *Fundamentals of Mathematical Analysis, Vol.1,2*, Mir Publisher, Moscow, 1982.

V.A. Zorich, *Mathematical Analysis I, II*, Springer, Berlin, 2004.

CHAPTER 5

Improper Integrals Depending on a Parameter

5.1 Pointwise, Absolute, and Uniform Convergence

Example 1. An integral $\int_a^{+\infty} f(x,y)dx$ converges on a set Y, but it does not converge uniformly on this set.

Solution
Consider $F(y) = \int_0^{+\infty} ye^{-xy}dx$ on $Y = (0,+\infty)$. This integral converges on $(0,+\infty)$: $F(y) = \int_0^{+\infty} ye^{-xy}dx = -e^{-xy}|_0^{+\infty} = 1$. To show that the convergence is not uniform, let us evaluate the remainder of the integral for $\forall A > 0$ and $y_A = \frac{1}{A}$:

$$\int_A^{+\infty} y_A e^{-xy_A}dx = -e^{-xy_A}|_A^{+\infty} = e^{-A\cdot 1/A} = e^{-1} \underset{A\to+\infty}{\nrightarrow} 0.$$

Example 2. An integral $\int_a^{+\infty} f(x,y)dx$ converges on a (finite or infinite) interval $Y = (c,d)$ and converges uniformly on any interval $[c_1, d_1] \subset (c,d)$, but it does not converge uniformly on (c,d).

Solution
Consider the integral of Example 1: $F(y) = \int_0^{+\infty} ye^{-xy}dx$ on $Y = (0,+\infty)$. It was shown that it converges on $Y = (0,+\infty)$, but nonuniformly. At the same time, the convergence is uniform on an interval $[a,+\infty)$ for any $a > 0$. Indeed, for $A > 0$ and simultaneously for all $y \in [a,+\infty)$, we get

$$\int_A^{+\infty} ye^{-xy}dx = -e^{-xy}|_A^{+\infty} = e^{-Ay} \leq e^{-Aa} \underset{A\to+\infty}{\to} 0.$$

Example 3. An integral $\int_a^{+\infty} f(x,y)dx$ converges on a (finite or infinite) interval $Y = (c,d)$, but it does not converge uniformly on an interval $[c_1, d_1] \subset (c,d)$.

Counterexamples on Uniform Convergence: Sequences, Series, Functions, and Integrals, First Edition.
Andrei Bourchtein and Ludmila Bourchtein.
© 2017 John Wiley & Sons, Inc. Published 2017 by John Wiley & Sons, Inc.
Companion website: www.wiley.com/go/bourchtein/counterexamples_on_uniform_convergence

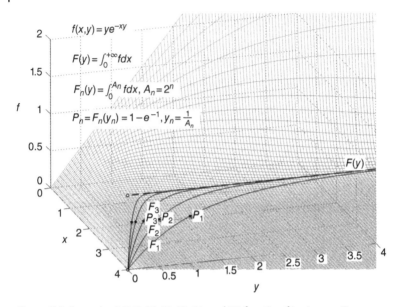

Figure 5.1 Examples 1, 2, 9, 10, 25, 28, 31, and 33, function $f(x, y) = ye^{-xy}$.

Solution

Consider $F(y) = \int_0^{+\infty} \frac{\sin xy}{x}\,dx$ on $Y = \mathbb{R}$. Note that the point $x = 0$ is not singular for the function $f(x, y) = \frac{\sin xy}{x}$, because for $y = 0$ we have $f(x, 0) = 0$, and for $y \neq 0$ we get $\lim_{x \to 0} f(x, y) = \lim_{x \to 0} \frac{\sin xy}{xy} \cdot y = y$, that is, at $x = 0$ the discontinuity is removable for any $y \in \mathbb{R}$. Let us show that the integral converges on $\mathbb{R}$. For $y = 0$, we have $F(0) = \int_0^{+\infty} 0\,dx = 0$; for $y > 0$, changing the variable by the formula $xy = t$, we obtain

$$F(y) = \int_0^{+\infty} \frac{\sin xy}{x}\,dx = \int_0^{+\infty} \frac{\sin t}{t}\,dt = \frac{\pi}{2}$$

(the last integral is a known result); and for $y < 0$, we take $y = -|y|$ and make a similar substitution $x|y| = t$ to get

$$F(y) = -\int_0^{+\infty} \frac{\sin x|y|}{x}\,dx = -\int_0^{+\infty} \frac{\sin t}{t}\,dt = -\frac{\pi}{2}.$$

Hence, $F(y) = \begin{cases} -\pi/2, y < 0 \\ 0, y = 0 \\ \pi/2, y > 0 \end{cases} = \frac{\pi}{2}\operatorname{sgn} y, \forall y \in \mathbb{R}.$

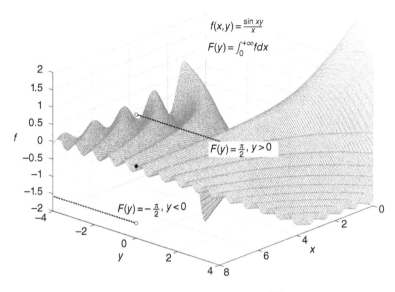

$$f(x,y) = \frac{\sin xy}{x}$$

$$F(y) = \int_0^{+\infty} f \, dx$$

$$F(y) = \frac{\pi}{2}, \, y > 0$$

$$F(y) = -\frac{\pi}{2}, \, y < 0$$

Figure 5.2 Examples 3, 5, 25, and 27, function $f(x,y) = \frac{\sin xy}{x}$.

Now we consider an arbitrary interval $Y_1 = [c_1, d_1]$ such that $c_1 < 0 < d_1$ and prove that the convergence on Y_1 is not uniform. To assess the integral remainder, for any $A > \frac{1}{\sqrt{d_1}}$ we choose $y_A = \frac{1}{A^2} \in [c_1, d_1]$ and obtain

$$\int_A^{+\infty} \frac{\sin x y_A}{x} \, dx = \int_{A y_A}^{+\infty} \frac{\sin t}{t} \, dt$$

$$= \int_{1/A}^{+\infty} \frac{\sin t}{t} \, dt \xrightarrow[A \to +\infty]{} \int_0^{+\infty} \frac{\sin t}{t} \, dt = \frac{\pi}{2} \neq 0,$$

that is, the convergence is not uniform on $[c_1, d_1]$.

Note that the convergence is uniform on any interval $[c, +\infty), c > 0$ according to Dirichlet's theorem: the integral $\int_0^A \sin xy \, dx$ is uniformly bounded, since

$$\left| \int_0^A \sin xy \, dx \right| = \left| \left(-\frac{1}{y} \cos xy \right) \Big|_0^{|A|} \right| \leq \frac{2}{|y|} \leq \frac{2}{c},$$

for $\forall A > 0$ and $\forall y \in [c, +\infty)$, and the function $\frac{1}{x}$ approaches zero, as $x \to +\infty$, monotonically and independently of y.

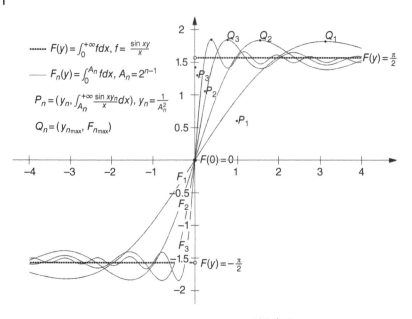

$$----\ F(y) = \int_0^{+\infty} f dx, \ f = \frac{\sin xy}{x}$$

$$----\ F_n(y) = \int_0^{A_n} f dx, \ A_n = 2^{n-1}$$

$$P_n = (y_n, \int_{A_n}^{+\infty} \frac{\sin xy_n}{x} dx), \ y_n = \frac{1}{A_n^2}$$

$$Q_n = (y_{n_{max}}, F_{n_{max}})$$

Figure 5.3 Examples 3, 5, 25, and 27, integral $F(y) = \int_0^{+\infty} \frac{\sin xy}{x} dx$.

Example 4. An integral $F(y) = \int_a^{+\infty} f(x,y) dx$ converges absolutely on a (finite or infinite) interval $Y = (c, d)$, but it does not converge uniformly on this interval.

Solution

Consider $F(y) = \int_0^{+\infty} e^{-xy} \sin x dx$ on $Y = (0, +\infty)$. Since $|f(x,y)| = |e^{-xy} \sin x| \le e^{-xy}$, $\forall x \in [0, +\infty)$, and the integral $\int_0^{+\infty} e^{-xy} dx = -\frac{1}{y} e^{-xy}\Big|_0^{+\infty} = \frac{1}{y}$ converges on $(0, +\infty)$, it follows that the original integral converges absolutely on $Y = (0, +\infty)$.

However, we will see that the convergence is not uniform. The antiderivative of $f(x,y)$ can be found by integrating by parts twice:

$$\int e^{-xy} \sin x dx = -e^{-xy} \cos x - y \int e^{-xy} \cos x dx$$

$$= -e^{-xy} \cos x - y e^{-xy} \sin x - y^2 \int e^{-xy} \sin x dx$$

($u = e^{-xy}$, $dv = \sin x dx$ is used the first time and $u = e^{-xy}$, $dv = \cos x dx$ - the second). Therefore,

$$\int e^{-xy} \sin x dx = \frac{-e^{-xy} \cos x - y e^{-xy} \sin x}{1 + y^2}.$$

Then the evaluation of the remainder goes as follows. For any $A > 0$ and corresponding $y_A = \frac{1}{A} \in (0, +\infty)$, we obtain

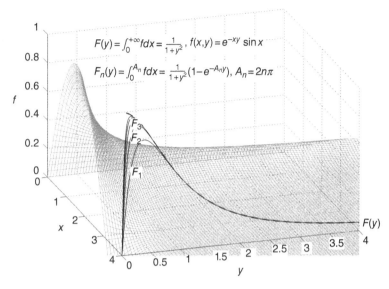

$$F(y) = \int_0^{+\infty} f dx = \frac{1}{1+y^2}, \; f(x,y) = e^{-xy}\sin x$$

$$F_n(y) = \int_0^{A_n} f dx = \frac{1}{1+y^2}(1-e^{-A_n y}), \; A_n = 2n\pi$$

Figure 5.4 Examples 4, 26 (second counterexample), and 34, function $f(x,y) = e^{-xy}\sin x$.

$$\int_A^{+\infty} e^{-xy_A}\sin x\, dx = -e^{-xy_A}\frac{\cos x + y_A \sin x}{1+y_A^2}\bigg|_A^{+\infty} = e^{-1}\left(\frac{\cos A}{1+\frac{1}{A^2}} + \frac{\frac{1}{A}\sin A}{1+\frac{1}{A^2}}\right)$$

(the second term disappears since $\lim_{x\to+\infty} e^{-xy_A}\frac{\cos x + y_A \sin x}{1+y_A^2} = 0$). Note now that $\lim_{A\to+\infty} \frac{\frac{1}{A}\sin A}{1+\frac{1}{A^2}} = 0$, whereas $\lim_{A\to+\infty} \frac{\cos A}{1+\frac{1}{A^2}}$ does not exist (for the sequence $A_n = 2n\pi \underset{n\to\infty}{\to} +\infty$, we have $\lim_{n\to+\infty} \frac{\cos A_n}{1+\frac{1}{A_n^2}} = 1$, but for $A_k = (2k+1)\pi \underset{k\to\infty}{\to} +\infty$ the result is different $\lim_{k\to+\infty} \frac{\cos A_k}{1+\frac{1}{A_k^2}} = -1$). Concluding, there is no limit of $\int_A^{+\infty} e^{-xy_A}\sin x\, dx$ as A approaches infinity, which means that the integral $\int_0^{+\infty} e^{-xy}\sin x\, dx$ does not converge uniformly on $Y = (0, +\infty)$.

Example 5. An integral $\int_a^{+\infty} f(x, y) dx$ converges uniformly on an interval Y, but it does not converge absolutely on this interval.

Solution
In Example 3, it was shown that $\int_0^{+\infty} \frac{\sin xy}{x} dx$ converges uniformly on any interval $Y = [c, +\infty)$, $c > 0$. This implies the uniform convergence of the integral $F(y) = \int_{\pi/2}^{+\infty} \frac{\sin xy}{x} dx$ on $Y = [1, +\infty)$.

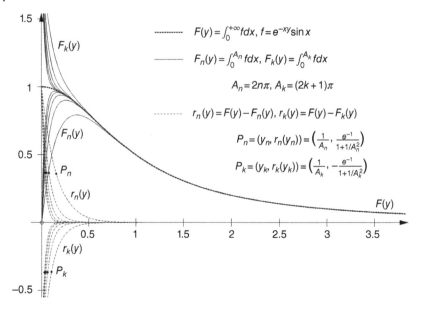

Figure 5.5 Examples 4, 26 (second counterexample), and 34, integral $F(y) = \int_0^{+\infty} e^{-xy} \sin x\, dx$.

Let us analyze the behavior of $\int_{\pi/2}^{+\infty} \frac{|\sin xy|}{x}dx$. Note that $\int_{\pi/2}^{+\infty} \frac{\cos 2xy}{x}dx$ converges (even uniformly) on $Y = [1, +\infty)$ by Dirichlet's theorem:

$$\left| \int_{\pi/2}^{A} \cos 2xy\, dx \right| = \left| \frac{1}{2y} \sin 2xy \right|_{\pi/2}^{A} \right|$$

$$\leq \frac{1}{2}|\sin 2Ay - \sin \pi y| \leq 1, \forall A > \frac{\pi}{2}, \forall y \in Y,$$

and $\frac{1}{x}$ approaches 0 monotonically as x approaches infinity.

Together with the inequality $|\sin xy| \geq \sin^2 xy = \frac{1}{2}(1 - \cos 2xy)$ and divergence of $\int_{\pi/2}^{+\infty} \frac{1}{2x}dx$, it proves the divergence of $\int_{\pi/2}^{+\infty} \frac{|\sin xy|}{x}dx$, that is, the original integral converges uniformly but not absolutely.

Example 6. An integral $\int_a^{+\infty} f(x, y)dx$ converges uniformly and absolutely on an interval Y, but $\int_a^{+\infty} |f(x, y)|dx$ does not converge uniformly on this interval.

Solution
Consider $F(y) = \int_1^{+\infty} \frac{\cos x}{x^y}dx$ on $Y = (1, +\infty)$. Since $|\cos x| \leq 1$ and the integral $\int_1^{+\infty} \frac{1}{x^y}dx = \frac{x^{1-y}}{1-y}\Big|_1^{+\infty} = \frac{1}{y-1}$ converges for $\forall y \in Y$, it follows that the integral

$\int_1^{+\infty} \frac{|\cos x|}{x^y} dx$ also converges, that is, the original integral converges absolutely for $\forall y \in Y$. The uniformity of the convergence follows from Dirichlet's theorem: first,

$$\left| \int_1^A \cos x \, dx \right| = |\sin x|_1^A| = |\sin A - \sin 1| \le 2, \forall A \in [1, +\infty), \forall y \in (1, +\infty)$$

and, second, the function $\frac{1}{x^y}$ is continuous and monotone in x on $[1, +\infty)$ for any fixed $y \in Y$ and also approaches 0, as x approaches infinity, uniformly on Y, because $0 < \frac{1}{x^y} < \frac{1}{x} \underset{x \to +\infty}{\longrightarrow} 0$ simultaneously for all $y \in Y$.

Now let us show that $\int_1^{+\infty} \frac{|\cos x|}{x^y} dx$ does not converge uniformly on Y. In effect, it is sufficient to note that $|\cos x| \ge \cos^2 x = \frac{1}{2}(1 + \cos 2x)$ and the integral $\int_1^{+\infty} \frac{\cos 2x}{x^y} dx$ converges uniformly on Y (it can be shown in the same way as for the original integral), whereas the integral $\int_1^{+\infty} \frac{1}{x^y} dx$ converges on Y, but not uniformly. The last can be seen by evaluating the remainder with $\forall A > 1$ and $y_A = 1 + \frac{1}{A} \in Y$:

$$\int_A^{+\infty} \frac{1}{x^{y_A}} dx = \left. \frac{x^{1-y_A}}{1 - y_A} \right|_A^{+\infty} = \frac{A^{1-y_A}}{y_A - 1} = A^{1-1/A} \underset{A \to +\infty}{\longrightarrow} +\infty.$$

Example 7. An integral $\int_a^{+\infty} f(x, y) dx$ converges uniformly and absolutely on Y, but there is no bound of $f(x, y)$ such that $|f(x, y)| \le \varphi(x), \forall y \in Y$, where the improper integral $\int_a^{+\infty} \varphi(x) dx$ converges.

Solution

The function $f(x, y) = \begin{cases} \frac{1}{x}, x \in [y, y + 1] \\ 0, \text{otherwise} \end{cases}$ defined on $[1, +\infty) \times [1, +\infty)$ is nonnegative, so the convergence and absolute convergence of the integral $\int_1^{+\infty} f(x, y) dx$ is the same thing for this function. Direct calculations

$$\int_1^{+\infty} f(x, y) dx = \int_y^{y+1} \frac{1}{x} dx = \ln x|_y^{y+1} = \ln \frac{y+1}{y}$$

show that the integral converges for each $y \in [1, +\infty)$. Noting that for each fixed y it holds $f(x, y) \le \frac{1}{x}$ and $f(x, y)$ is 0 outside the interval $[y, y + 1]$ of the unit length, one can prove the uniformity of the convergence on $Y = [1, +\infty)$ applying the Cauchy criterion: for $\forall \epsilon > 0$, one can choose $A_\epsilon = \frac{1}{\epsilon}$ and arbitrary $A_2 > A_1 > A_\epsilon$ to obtain $\int_{A_1}^{A_2} f(x, y) dx < \frac{1}{A_\epsilon} = \epsilon$ simultaneously for all $y \in [1, +\infty)$.

At the same time, the function $f(x, y)$ does not admit majoration on $Y = [1, +\infty)$ in the form $|f(x, y)| \le \varphi(x)$ such that the improper integral

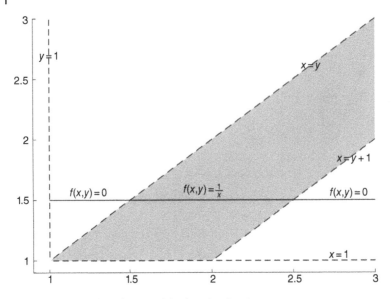

Figure 5.6 Example 7, domain of the function $f(x, y)$.

$\int_a^{+\infty} \varphi(x)dx$ converges. In fact, the function definition can be rewritten in the form $f(x, y) = \begin{cases} \frac{1}{x}, y \in [x - 1, x] \\ 0, \text{otherwise} \end{cases}$. Therefore, for each fixed x, the lowest upper bound of $|f(x, y)|$ is the value $\frac{1}{x}$ that $f(x, y)$ assumes on $[x - 1, x]$. Thus, the smallest function $\varphi(x)$ for which $|f(x, y)| \le \varphi(x), \forall y \in Y$ is $\varphi(x) = \frac{1}{x}$. However, the improper integral $\int_1^{+\infty} \frac{1}{x}dx$ diverges.

Remark. The converse general statement is true and represents the Weierstrass test for improper integrals.

Example 8. Suppose functions $f(x, y)$ and $g(x, y)$ are positive and continuous on $[a, +\infty) \times Y$, and $\lim\limits_{x \to +\infty} \frac{f(x,y)}{g(x,y)} = 1$, $\forall y \in Y$. One of the integrals $\int_a^{+\infty} f(x, y)dx$ or $\int_a^{+\infty} g(x, y)dx$ converges uniformly on Y, but another converges nonuniformly.

Solution
Consider $f(x, y) = \frac{2y^2}{x^4+y^4}$ and $g(x, y) = \frac{2y^2}{x^4+y^2}$ on $[1, +\infty) \times \mathbb{R}$. Both functions are positive and continuous on the chosen domain and $\lim\limits_{x \to +\infty} \frac{f(x,y)}{g(x,y)} = \lim\limits_{x \to +\infty} \frac{x^4+y^2}{x^4+y^4} = 1$, $\forall y \in \mathbb{R}$. Since $f(x, y) = \frac{2y^2}{x^4+y^4} = \frac{1}{x^2} \frac{2x^2y^2}{x^4+y^4} \le \frac{1}{x^2}$, $\forall y \in \mathbb{R}$, and $\int_1^{+\infty} \frac{1}{x^2}dx$ converges, it

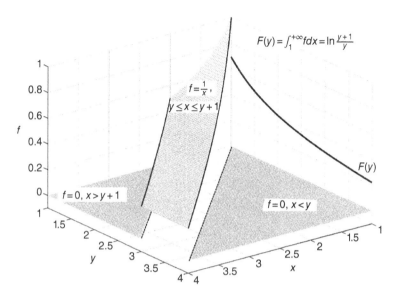

$$F(y) = \int_1^{+\infty} f \, dx = \ln \frac{y+1}{y}$$

Figure 5.7 Example 7, function $f(x, y) = \begin{cases} \frac{1}{x}, x \in [y, y+1] \\ 0, \text{otherwise} \end{cases}$.

follows that the integral $\int_1^{+\infty} f(x, y) dx = \int_1^{+\infty} \frac{2y^2}{x^4 + y^4} dx$ converges uniformly on $\mathbb{R}$. The second integral converges pointwise on $\mathbb{R}$, because for any fixed $y \in \mathbb{R}$ it holds $|g(x, y)| = \frac{2y^2}{x^4 + y^2} \le \frac{2y^2}{x^4}$ and the integral $\int_1^{+\infty} \frac{1}{x^4} dx$ converges. However, this convergence is nonuniform on $\mathbb{R}$ according to the Cauchy criterion: for $\forall A > 1$, one can choose $B = A + 1$ and $y_A = (A + 1)^2$ to obtain

$$\left| \int_A^{A+1} g(x, y_A) dx \right| = \int_A^{A+1} \frac{2y_A^2}{x^4 + y_A^2} dx > \frac{2y_A^2}{(A+1)^4 + y_A^2} (A + 1 - A)$$

$$= \frac{2(A+1)^4}{2(A+1)^4} = 1 \underset{A \to +\infty}{\nrightarrow} 0.$$

Remark. The following comparison theorem for the pointwise convergence of improper integrals is true: if for any fixed $y \in Y$, the functions $f(x, y)$ and $g(x, y)$ are positive on $[a, +\infty)$ and $\lim\limits_{x \to +\infty} \frac{f(x,y)}{g(x,y)} = C = const \neq 0$ (in particular, $C = 1$), then both integrals $\int_a^{+\infty} f(x, y) dx$ and $\int_a^{+\infty} g(x, y) dx$ are pointwise convergent or divergent. Example 8 shows that this property cannot be extended to uniform convergence.

5.2 Convergence of the Sum and Product

Example 9. Improper integrals $\int_a^{+\infty} f(x,y)dx$ and $\int_a^{+\infty} g(x,y)dx$ converge nonuniformly on Y, but the integral $\int_a^{+\infty} f(x,y)+g(x,y)dx$ converges uniformly on Y.

Solution
The idea of a trivial counterexample is simple. One can choose $f(x,y)$, which gives rise to a nonuniformly convergent integral and then use $g(x,y) = -f(x,y)$. Just to not stick with the zero function $f(x,y)+g(x,y)$, one can choose any uniformly convergent integral and use the function $h(x,y)$ in such an integral to define $g(x,y) = h(x,y) - f(x,y)$. One of such examples with nonzero function $h(x,y)$ is provided below.

Consider the functions $f(x,y) = ye^{-xy}$ and $g(x,y) = (y^2 - y)e^{-xy}$ on $[0,+\infty) \times (0,+\infty)$. The first improper integral converges

$$\int_0^{+\infty} ye^{-xy}dx = -e^{-xy}|_0^{+\infty} = 1, \forall y \in (0,+\infty),$$

but this convergence is not uniform on $(0,+\infty)$, since for $y_A = \frac{1}{A}$ one has

$$\int_A^{+\infty} y_A e^{-xy_A}dx = -e^{-xy_A}|_A^{+\infty} = e^{-1} \underset{A\to+\infty}{\not\to} 0.$$

Similarly,

$$\int_0^{+\infty} (y^2 - y)e^{-xy}dx = -(y-1)e^{-xy}|_0^{+\infty} = y-1, \forall y \in (0,+\infty),$$

but for $y_A = \frac{1}{A}$,

$$\int_A^{+\infty} (y_A^2 - y_A)e^{-xy_A}dx = -(y_A - 1)e^{-xy_A}|_A^{+\infty}$$

$$= \left(\frac{1}{A} - 1\right)e^{-1} \underset{A\to+\infty}{\to} -e^{-1} \neq 0,$$

that is, the second improper integral converges nonuniformly on $(0,+\infty)$.
The integral of $f(x,y)+g(x,y)$ also converges on $(0,+\infty)$ (due to the arithmetic properties of improper integrals or direct calculations):

$$\int_0^{+\infty} y^2 e^{-xy}dx = -ye^{-xy}|_0^{+\infty} = y, \forall y \in (0,+\infty).$$

However, contrary to the first two integrals, this convergence is uniform on $(0, +\infty)$. This can be shown by evaluating the residual $\int_A^{+\infty} y^2 e^{-xy} dx = y e^{-Ay}$. Note that the critical point equation for the last function $(y e^{-Ay})_y = (1 - Ay) e^{-Ay} = 0$ gives the only critical point $y_A = \frac{1}{A}$, which is also the maximum point in y. Since the function $y e^{-Ay}$ is positive for $\forall y \in (0, +\infty)$, $\forall A > 0$, the following evaluation holds

$$\int_A^{+\infty} y^2 e^{-xy} dx = y e^{-Ay} \leq \sup_{(0,+\infty)} y e^{-Ay} = \frac{1}{A} e^{-1} \xrightarrow[A \to +\infty]{} 0,$$

which means the uniform convergence.

Remark 1. An example of nonuniformly convergent integral $\int_a^{+\infty} f(x, y) + g(x, y) dx$ when both improper integrals $\int_a^{+\infty} f(x, y) dx$ and $\int_a^{+\infty} g(x, y) dx$ converges nonuniformly on Y can also be provided. As a trivial counterexample, one can choose $f(x, y) = g(x, y)$. In a bit more elaborated example, one can pick up any function $h(x, y)$, which generates uniformly convergent integral and define $g(x, y) = f(x, y) + h(x, y)$. For instance, $f(x, y) = y e^{-xy}$ and $g(x, y) = y e^{-xy} + y^2 e^{-xy}$ is one of such counterexamples.

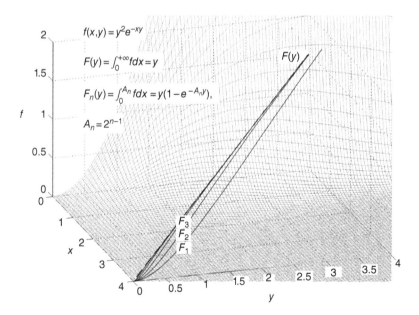

Figure 5.8 Examples 9 and 10, function $f(x, y) = y^2 e^{-xy}$.

Remark 2. The same results are true for the difference of nonuniformly convergent integrals.

Example 10. Improper integrals $\int_a^{+\infty} f(x,y)dx$ and $\int_a^{+\infty} g(x,y)dx$ converge nonuniformly on Y, but the integral $\int_a^{+\infty} f(x,y)g(x,y)dx$ converges uniformly on Y.

Solution
It was shown in Example 9 that the improper integral $\int_0^{+\infty} ye^{-xy}dx$ converges nonuniformly on $(0, +\infty)$ to 1. In a similar way, it can be shown that the integral $\int_0^{+\infty} \sqrt{y}e^{-xy}dx$ is also nonuniformly convergent on $(0, +\infty)$:

$$\int_0^{+\infty} \sqrt{y}e^{-xy}dx = -\frac{1}{\sqrt{y}}e^{-xy}\Big|_0^{+\infty} = \frac{1}{\sqrt{y}}, \forall y \in (0, +\infty),$$

and for $y_A = \frac{1}{A}$

$$\int_A^{+\infty} \sqrt{y_A}e^{-xy_A}dx = \frac{e^{-Ay_A}}{\sqrt{y_A}} = \sqrt{A}e^{-1} \xrightarrow[A \to +\infty]{} +\infty.$$

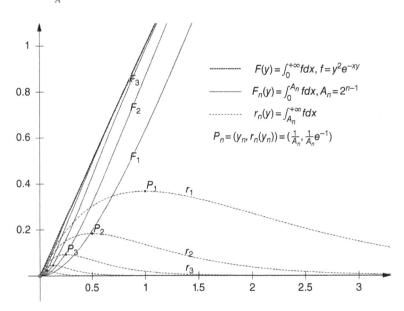

Figure 5.9 Examples 9 and 10, integral $F(y) = \int_0^{+\infty} y^2 e^{-xy}dx$.

Nevertheless, the integral

$$\int\limits_{0}^{+\infty} f(x,y)g(x,y)dx = \int\limits_{0}^{+\infty} y^{3/2}e^{-2xy}dx$$

converges uniformly on $(0,+\infty)$. The convergence can be verified by the direct calculation:

$$\int\limits_{0}^{+\infty} y^{3/2}e^{-2xy}dx = -\frac{1}{2}\sqrt{y}e^{-2xy}\Big|_{0}^{+\infty} = \frac{1}{2}\sqrt{y}, \forall y \in (0,+\infty).$$

To evaluate the character of the convergence, note that $y_A = \frac{1}{4A}$ is the only maximum point in y of the function $\sqrt{y}e^{-2Ay}$, which is seen from the critical point equation $(\sqrt{y}e^{-2Ay})_y = \left(\frac{1}{2\sqrt{y}} - 2A\sqrt{y}\right)e^{-2Ay} = 0$. Since the function $\sqrt{y}e^{-2Ay}$ is positive for $\forall y \in (0,+\infty)$, $\forall A > 0$, the following evaluation for the integral residual holds:

$$\int\limits_{A}^{+\infty} y^{3/2}e^{-2xy}dx = \frac{1}{2}\sqrt{y}e^{-2Ay} \leq \frac{1}{2}\sqrt{y_A}e^{-2Ay_A} = \frac{1}{2}\frac{1}{\sqrt{4A}}e^{-1/2} \underset{A\to+\infty}{\to} 0,$$

which means the uniform convergence.

Remark 1. The following example also takes place: an integral $\int_a^{+\infty} f(x,y)dx$ converges nonuniformly on Y, but the integral $\int_a^{+\infty} f^2(x,y)dx$ converges uniformly on Y. For instance, it was shown in Example 9 that $\int_0^{+\infty} ye^{-xy}dx$ converges nonuniformly on $(0,+\infty)$. However, the integral $\int_0^{+\infty} y^2e^{-2xy}dx$ converges uniformly on the same interval that can be shown in the same way as for $\int_0^{+\infty} y^2e^{-xy}dx$ in Example 9.

Remark 2. It is also not true that if $\int_a^{+\infty} f(x,y)dx$ and $\int_a^{+\infty} g(x,y)dx$ converge nonuniformly on Y, then the integral $\int_a^{+\infty} f(x,y)g(x,y)dx$ converges uniformly on Y. For a counterexample, consider the functions $f(x,y) = g(x,y) = \sqrt{y}e^{-xy}$. It is shown in the last example, that $\int_0^{+\infty} \sqrt{y}e^{-xy}dx$ converges nonuniformly on $(0,+\infty)$, and the nonuniform convergence of the integral $\int_0^{+\infty} f(x,y)g(x,y)dx = \int_0^{+\infty} ye^{-2xy}dx$ can be shown in the same way as in Example 9.

Example 11. Improper integrals $\int_a^{+\infty} f(x,y)dx$ and $\int_a^{+\infty} g(x,y)dx$ converge uniformly on Y, but the integral $\int_a^{+\infty} f(x,y)g(x,y)dx$ converges nonuniformly on Y.

Solution

Consider $f(x,y) = g(x,y) = \frac{\sin x}{\sqrt{x}}e^{-xy}$ on $[1,+\infty) \times (0,+\infty)$. The improper integral $\int_1^{+\infty} \frac{\sin x}{\sqrt{x}}e^{-xy}dx$ converges uniformly on $(0,+\infty)$ by Abel's theorem, since the integral $\int_1^{+\infty} \frac{\sin x}{\sqrt{x}}dx$ converges (uniformly, because it does not depend on y) and the function e^{-xy} is monotone in $x \in [1,+\infty)$ for any fixed $y \in (0,+\infty)$ and is also uniformly bounded: $|e^{-xy}| \leq 1$, $\forall x \in [1,+\infty)$, $\forall y \in (0,+\infty)$.

However, the integral

$$\int_1^{+\infty} f(x,y)g(x,y)dx = \int_1^{+\infty} \frac{\sin^2 x}{x}e^{-2xy}dx$$

$$= \frac{1}{2}\int_1^{+\infty} \frac{1}{x}e^{-2xy}dx - \frac{1}{2}\int_1^{+\infty} \frac{\cos 2x}{x}e^{-2xy}dx$$

converges nonuniformly on $(0,+\infty)$, since the first integral in the right-hand side converges nonuniformly and the second - uniformly on $(0,+\infty)$. In fact, the uniform convergence of the second integral follows from Abel's theorem: the integral $\int_1^{+\infty} \frac{\cos 2x}{x}dx$ converges uniformly on $(0,+\infty)$ and e^{-2xy} is monotone in $x \in [1,+\infty)$ for any fixed $y \in (0,+\infty)$ and is also uniformly bounded ($|e^{-2xy}| \leq 1$ on $[1,+\infty) \times (0,+\infty)$). The first integral converges since $0 < \frac{1}{x}e^{-2xy} \leq e^{-2xy}$, $\forall x \in [1,+\infty)$ and the integral $\int_1^{+\infty} e^{-2xy}dx$ converges. To prove that the convergence is not uniform, for any $A > 1$ and corresponding $B = 2A$ and $y_A = \frac{1}{2A}$, we perform the following evaluations:

$$\left| \int_A^B \frac{1}{x}e^{-2xy}dx \right| = \int_A^{2A} \frac{1}{x}e^{-2xy_A}dx \geq \frac{1}{2A}\int_A^{2A} e^{-2xy_A}dx$$

$$= \frac{1}{4Ay_A}(e^{-2Ay_A} - e^{-4Ay_A}) = \frac{1}{2}(e^{-1} - e^{-2}) \underset{A \to +\infty}{\not\to} 0.$$

According to the Cauchy criterion, it means that the convergence is nonuniform on $(0,+\infty)$.

Remark 1. The given counterexample can also be used for the following example: an integral $\int_a^{+\infty} f(x,y)dx$ converges uniformly on Y, but the integral $\int_a^{+\infty} f^2(x,y)dx$ does not converge uniformly on Y.

Remark 2. One can also show that the following statement is false: if $\int_a^{+\infty} f(x,y)dx$ and $\int_a^{+\infty} g(x,y)dx$ converge uniformly on Y, then the integral $\int_a^{+\infty} f(x,y)g(x,y)dx$ converges nonuniformly on Y. For a counterexample, consider the functions $f(x,y) = g(x,y) = \frac{\sin x}{x}e^{-xy}$ on $[1,+\infty) \times (0,+\infty)$. The integral $\int_1^{+\infty} \frac{\sin x}{x} e^{-xy} dx$ converges uniformly on $(0, +\infty)$ by Abel's theorem: the integral $\int_1^{+\infty} \frac{\sin x}{x} dx$ converges (uniformly, because it does not depend on y) and the function e^{-xy} is monotone in $x \in [1, +\infty)$ for any fixed $y \in (0, +\infty)$ and is also uniformly bounded—$|e^{-xy}| \leq 1$ on $[1, +\infty) \times (0, +\infty)$. The integral

$$\int_a^{+\infty} f^2(x, y) dx = \int_1^{+\infty} \frac{\sin^2 x}{x^2} e^{-2xy} dx$$

also converges uniformly by the Weierstrass test: the evaluation $0 \leq \frac{\sin^2 x}{x^2} e^{-2xy} \leq \frac{1}{x^2}$ holds for $\forall x \in [1, +\infty)$, $\forall y \in (0, +\infty)$ and the majorant integral $\int_1^{+\infty} \frac{1}{x^2} dx$ converges.

Remark 3. The following general statement for the sum and difference is true: if improper integrals $\int_a^{+\infty} f(x, y) dx$ and $\int_a^{+\infty} g(x, y) dx$ converge uniformly on Y, then the integral $\int_a^{+\infty} f(x, y) \pm g(x, y) dx$ also converges uniformly on Y.

Example 12. An improper integral $\int_a^{+\infty} f(x, y) dx$ converges uniformly and an integral $\int_a^{+\infty} g(x, y) dx$ converges nonuniformly on Y, but the integral $\int_a^{+\infty} f(x, y) g(x, y) dx$ converges uniformly on Y.

Solution
Consider $f(x, y) = \frac{\sin x}{x}$ and $g(x, y) = e^{-xy}$ on $[1, +\infty) \times (0, +\infty)$. The improper integral $\int_1^{+\infty} \frac{\sin x}{x} dx$ converges uniformly: it does not depend on y and its convergence is guaranteed by Dirichlet's theorem, since $\left| \int_1^A \sin x dx \right| \leq 2$, $\forall A > 1$ and $\frac{1}{x}$ converges monotonically to 0 as x approaches $+\infty$. On the other hand, the integral $\int_1^{+\infty} e^{-xy} dx$ converges nonuniformly on $(0, +\infty)$, which can be proved in the same way as for similar integrals in Example 9: the direct calculation

$$\int_1^{+\infty} e^{-xy} dx = -\frac{1}{y} e^{-xy} \Big|_1^{+\infty} = \frac{e^{-y}}{y}, \forall y \in (0, +\infty)$$

shows the convergence, and the residual evaluation with $y_A = \frac{1}{A}$

$$\int_A^{+\infty} e^{-xy_A} dx = \frac{e^{-Ay_A}}{y_A} = Ae^{-1} \underset{A \to +\infty}{\to} +\infty$$

reveals that this convergence is not uniform on $(0, +\infty)$. At the same time, the integral

$$\int_1^{+\infty} f(x,y)g(x,y)dx = \int_1^{+\infty} \frac{\sin x}{x} e^{-xy} dx$$

converges uniformly on $(0, +\infty)$ (see Remark 2 to Example 11 for details).

Remark 1. An example with the opposite conclusion also takes place: an improper integral $\int_a^{+\infty} f(x,y)dx$ converges uniformly and an integral $\int_a^{+\infty} g(x,y)dx$ converges nonuniformly on Y, but the integral $\int_a^{+\infty} f(x,y)g(x,y)dx$ converges nonuniformly on Y. It can be illustrated using the functions $f(x,y) = \frac{\sin x}{x}$ and $g(x,y) = xe^{-xy}$ on $[1, +\infty) \times (0, +\infty)$. In fact, as shown above, $\int_1^{+\infty} \frac{\sin x}{x} dx$ converges uniformly. The convergence of the second integral can be proved by integrating by parts:

$$\int_1^{+\infty} xe^{-xy}dx = -\frac{x}{y}e^{-xy}\Big|_1^{+\infty} + \frac{1}{y}\int_1^{+\infty} e^{-xy}dx = \frac{1}{y}e^{-y} + \frac{1}{y^2}e^{-y}, \forall y \in (0, +\infty),$$

but the evaluation of the integral residual with $y_A = \frac{1}{A}$

$$\int_A^{+\infty} xe^{-xy_A}dx = \left(\frac{A}{y_A} + \frac{1}{y_A^2}\right)e^{-Ay_A} = 2A^2e^{-1} \underset{A\to+\infty}{\to} +\infty$$

reveals that this convergence is not uniform on $(0, +\infty)$. At the same time, the integral

$$\int_1^{+\infty} f(x,y)g(x,y)dx = \int_1^{+\infty} \sin xe^{-xy}dx$$

converges nonuniformly on $(0, +\infty)$, as was shown in Example 4.

Remark 2. The following general statement is true for the sum and difference: if an improper integral $\int_a^{+\infty} f(x,y)dx$ converges uniformly and $\int_a^{+\infty} g(x,y)dx$ converges nonuniformly on Y, then the integral $\int_a^{+\infty} f(x,y) \pm g(x,y)dx$ converges nonuniformly on Y.

Example 13. Improper integrals $\int_a^{+\infty} f(x,y)dx$ and $\int_a^{+\infty} g(x,y)dx$ diverge on Y, but nevertheless the integral $\int_a^{+\infty} f(x,y)g(x,y)dx$ converges uniformly on Y.

Solution

Consider $f(x,y) = y \sin xy$ and $g(x,y) = \frac{1}{xy}$ on $[1,+\infty) \times [1,+\infty)$. For the first improper integral, we have

$$\int\limits_{1}^{+\infty} y \sin xy\, dx = -\cos xy\big|_{1}^{+\infty} = \cos y - \lim_{x \to +\infty} \cos xy.$$

The last limit does not exist for any $y_0 \in [1,+\infty)$: choosing the sequence $x_n = \frac{2n\pi}{y_0} \underset{n \to +\infty}{\to} +\infty$, we get $\lim\limits_{n \to +\infty} \cos x_n y_0 = 1$, while another sequence $x_m = \frac{(2m+1)\pi}{y_0} \underset{m \to +\infty}{\to} +\infty$ gives different result $\lim\limits_{m \to +\infty} \cos x_m y_0 = -1$. There-fore, the first integral diverges for $\forall y \in [1,+\infty)$. The second integral diverges as well:

$$\int\limits_{1}^{+\infty} \frac{1}{xy}\, dx = \frac{1}{y} \ln x \bigg|_{1}^{+\infty} \underset{x \to +\infty}{\to} +\infty, \forall y \in [1,+\infty).$$

Nevertheless, the integral

$$\int\limits_{1}^{+\infty} f(x,y)g(x,y)\, dx = \int\limits_{1}^{+\infty} \frac{\sin xy}{x}\, dx$$

converges uniformly on $[1,+\infty)$ by Dirichlet's theorem: first, the integral of $\sin xy$ is uniformly bounded

$$\left| \int\limits_{1}^{A} \sin xy\, dx \right| = \left| -\frac{1}{y} \cos xy \right|_{1}^{A} \le \frac{2}{y} \le 2, \forall A > 1, \forall y \in [1,+\infty),$$

and, second, the function $\frac{1}{x}$ converges monotonically (and uniformly on $[1,+\infty)$) to 0 as $x \to +\infty$.

Example 14. Improper integrals $\int_{a}^{+\infty} f(x,y)dx$ and $\int_{a}^{+\infty} g(x,y)dx$ converge uniformly on Y, but the integral $\int_{a}^{+\infty} f(x,y)g(x,y)dx$ diverges on Y.

Solution

Consider $f(x,y) = g(x,y) = \frac{\sin xy}{\sqrt{x}}$ on $[1,+\infty) \times [1,+\infty)$. The uniform con-vergence of the improper integral $\int_{1}^{+\infty} \frac{\sin xy}{\sqrt{x}} dx$ on $[1,+\infty)$ can be shown by applying Dirichlet's theorem: the integral $\left| \int_{1}^{A} \sin xy\, dx \right|$ is uniformly bounded on $A > 1$ and $y \in [1,+\infty)$ (see details in Example 13), and the function $\frac{1}{\sqrt{x}}$ converges monotonically (and uniformly on $[1,+\infty)$) to 0 as $x \to +\infty$.

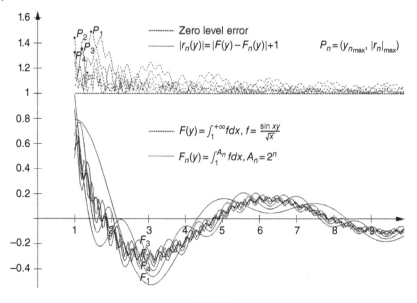

Figure 5.10 Example 14, integral $F(y) = \int_1^{+\infty} \frac{\sin xy}{\sqrt{x}} dx$.

However, the integral

$$\int\limits_1^{+\infty} f(x,y)g(x,y)dx = \int\limits_1^{+\infty} \frac{\sin^2 xy}{x}dx = \frac{1}{2}\int\limits_1^{+\infty} \frac{1}{x}dx - \frac{1}{2}\int\limits_1^{+\infty} \frac{\cos 2xy}{x}dx$$

diverges for $\forall y \in [1, +\infty)$. In fact, the second integral in the right-hand side converges uniformly on $[1, +\infty)$ according to Dirichlet's theorem: first, the integral of $\cos 2xy$ is uniformly bounded

$$\left| \int\limits_1^A \cos 2xy dx \right| = \left| \frac{1}{2y} \sin 2xy \right|_1^A \right|$$

$$\le \frac{1}{2y} |\sin 2Ay - \sin 2y| \le 1, \forall A > 1, \forall y \in [1, +\infty),$$

and, second, the function $\frac{1}{x}$ converges monotonically (and uniformly on $[1, +\infty)$) to 0 as $x \to +\infty$. At the same time, the first integral in the right-hand side diverges for $\forall y \in [1, +\infty)$. It implies the divergence of $\int_1^{+\infty} \frac{\sin^2 xy}{x}dx$ for $\forall y \in [1, +\infty)$.

Remark to Examples 10–14. Examples 10–14 show that the information on the nature of convergence of integrals $\int_a^{+\infty} f(x,y)dx$ and $\int_a^{+\infty} g(x,y)dx$

does not allow us to make a conclusion about the behavior of the integral $\int_a^{+\infty} f(x,y)g(x,y)dx$.

5.3 Dirichlet's and Abel's Theorems

Remark to Examples 15–17. In the following three examples, we analyze the conditions of Dirichlet's theorem on uniform convergence of improper integrals. It is shown that an omission of any of the conditions of the theorem can lead to nonuniform convergence of $\int_a^{+\infty} f(x,y)g(x,y)dx$.

Example 15. For each fixed $y \in Y$, integral $\int_a^A f(x,y)dx$ is bounded and function $g(x,y)$ is monotone in $x \in [a, +\infty)$; additionally, $g(x,y)$ converges uniformly to 0 on Y as $x \to +\infty$, but the improper integral $\int_a^{+\infty} f(x,y)g(x,y)dx$ does not converge uniformly on Y.

Solution
Consider the functions $f(x,y) = \sin xy$ and $g(x,y) = \frac{1}{x}$ on $X \times Y = [1, +\infty) \times (0, +\infty)$. The estimate

$$\left| \int_1^A f(x,y)dx \right| = \left| \int_1^A \sin xy\, dx \right| = \left| \left(-\frac{1}{y} \cos xy \right) \Big|_1^A \right| = \frac{1}{y} |\cos y - \cos Ay| \le \frac{2}{y}$$

shows that the integral $\int_1^A f(x,y)dx$ is bounded for each $y \in Y$. The second function $g(x,y) = \frac{1}{x}$ is monotone in x and converges (uniformly) to 0 as x approaches ∞. In Example 3, it was proved that $\int_0^{+\infty} \frac{\sin xy}{x}dx$ converges on $\mathbb{R}$, which implies the convergence of $\int_1^{+\infty} \frac{\sin xy}{x}dx$ on $(0, +\infty)$. Let us show that this convergence is nonuniform on $(0, +\infty)$. Using the same reasoning as in Example 3, we evaluate the integral remainder, for any $A > 1$ and $y_A = \frac{1}{A^2} \in (0, +\infty)$:

$$\int_A^{+\infty} \frac{\sin xy_A}{x}dx = \int_{Ay_A}^{+\infty} \frac{\sin t}{t}dt$$

$$= \int_{1/A}^{+\infty} \frac{\sin t}{t}dt \xrightarrow[A\to+\infty]{} \int_0^{+\infty} \frac{\sin t}{t}dt = \frac{\pi}{2} \ne 0,$$

that is, the convergence is not uniform on $(0, +\infty)$.

The convergence fails to be uniform due to the violation of the condition of uniform boundedness of the integrals $\int_a^A f(x, y)dx$ on Y: for $\forall A > 1$, choosing $y_A = \frac{\pi}{2A} > 0$ one gets

$$\left| \int_1^A f(x, y_A)dx \right| = \left| \int_1^A \sin xy_A dx \right| = \frac{1}{y_A} |\cos y_A - \cos Ay_A|$$

$$= \frac{2A}{\pi} \left| \cos \frac{\pi}{2A} \right| \underset{A \to +\infty}{\to} +\infty,$$

that is, $\int_1^A \sin xydx$ is not uniformly bounded on $(0, +\infty)$.

Example 16. The integral $\int_a^A f(x, y)dx$ is uniformly bounded on Y, and $g(x, y)$ converges uniformly to 0 on Y as $x \to +\infty$, but the improper integral $\int_a^{+\infty} f(x, y)g(x, y)dx$ does not converge uniformly on Y.

Solution
Consider the functions $f(x, y) = \frac{\sin x}{\sqrt{x}}$ and $g(x, y) = \frac{\sin x}{\sqrt{x}}e^{-xy}$ on $X \times Y = [1, +\infty) \times (0, +\infty)$. Since $\left| \int_1^A \sin xdx \right| = |-\cos x|_1^A| \le 2$, $\forall A > 1$ and $\frac{1}{\sqrt{x}}$ converges monotonically to 0 as $x \to +\infty$, it follows from Dirichlet's theorem for improper integrals (without a parameter) that $\int_1^{+\infty} \frac{\sin x}{\sqrt{x}}dx$ converges, and this convergence is uniform on $\mathbb{R}$ (and, in particular, on $(0, +\infty)$), because $f(x, y)$ does not depend on y. Therefore, the integral $\int_1^A \frac{\sin x}{\sqrt{x}}dx$ is uniformly bounded on $(0, +\infty)$. The function $g(x, y)$ converges uniformly on $(0, +\infty)$ to 0, since $\forall y \in (0, +\infty)$ one has

$$|g(x, y)| = \frac{|\sin x|}{\sqrt{x}}e^{-xy} \le \frac{1}{\sqrt{x}} \underset{x \to +\infty}{\to} 0.$$

It remains to prove that $\int_1^{+\infty} f(x, y)g(x, y)dx$ does not converge uniformly on $(0, +\infty)$. Due to the equality $\sin^2 x = \frac{1-\cos 2x}{2}$, the last integral can be written as follows:

$$\int_1^{+\infty} f(x, y)g(x, y)dx = \frac{1}{2}\int_1^{+\infty} \frac{1}{x}e^{-xy}dx - \frac{1}{2}\int_1^{+\infty} \frac{\cos 2x}{x}e^{-xy}dx.$$

The second integral in the right-hand side converges uniformly on $(0, +\infty)$ according to Dirichlet's theorem. Indeed, the integral $\int_1^A \cos 2xdx$ is uniformly bounded on $(0, +\infty)$: $\left| \int_1^A \cos 2xdx \right| = \left| \frac{1}{2}\sin 2x \right|_1^A | \le 1$, and the function $\frac{e^{-xy}}{x}$ is monotone in x and converges uniformly on $(0, +\infty)$ to 0: $\left| \frac{e^{-xy}}{x} \right| \le \frac{1}{x} \underset{x \to +\infty}{\to} 0$. The first integral in the right-hand side converges for each fixed $y \in (0, +\infty)$ due to the comparison theorem: $0 < \frac{e^{-xy}}{x} \leq e^{-xy}$, $\forall x \geq 1$ and $\int_1^{+\infty} e^{-xy} dx = -\frac{1}{y} e^{-xy} \Big|_1^{+\infty} = \frac{e^{-y}}{y}$ converges for $\forall y \in (0, +\infty)$. However, the convergence of the first integral is nonuniform: according to the Cauchy criterion, for any $A > 1$ one can choose $B = 2A$ and $y_A = \frac{1}{2A} \in (0, +\infty)$ such that

$$\int_A^{2A} \frac{e^{-xy_A}}{x} dx > \frac{1}{2A} e^{-2Ay_A}(2A - A) = \frac{1}{2} e^{-1} \underset{A \to +\infty}{\nrightarrow} 0.$$

Since the first integral converges nonuniformly and the second—uniformly, it follows that $\int_1^{+\infty} f(x, y)g(x, y)dx$ converges nonuniformly on $(0, +\infty)$. Note that this result is caused by the fact that $g(x, y)$ is not monotone in x, that is, the corresponding condition of monotonicity in Dirichlet's theorem is violated.

Example 17. The integral $\int_a^A f(x, y)dx$ is uniformly bounded on Y, and the function $g(x, y)$ is monotone in x and converges to 0 for $\forall y \in Y$ as $x \to +\infty$, but the improper integral $\int_a^{+\infty} f(x, y)g(x, y)dx$ does not converge uniformly on Y.

Solution
Consider the functions $f(x, y) = y \sin xy$ and $g(x, y) = \frac{1}{xy}$ on $X \times Y = [1, +\infty) \times (0, +\infty)$. The following evaluation holds simultaneously for $\forall A > 1$ and $\forall y \in Y$: $\left| \int_1^A y \sin xy dx \right| = |-\cos xy|_1^A| \leq 2$, which means the uniform boundedness of the integral $\int_1^A f(x, y)dx$ on $(0, +\infty)$. For each fixed $y \in (0, +\infty)$, the function $g(x, y) = \frac{1}{xy}$ is decreasing in x on $[1, +\infty)$ and $\lim_{x \to +\infty} \frac{1}{xy} = 0$, but the last convergence is not uniform on $(0, +\infty)$ since for $\forall x > 1$ there exists $y_x = \frac{1}{x}$ such that $g(x, y_x) = \frac{1}{xy_x} = 1 \underset{x \to +\infty}{\nrightarrow} 0$. Thus, the condition of the uniform convergence of $g(x, y)$ in Dirichlet's theorem is not satisfied. This leads to nonuniform convergence of $\int_1^{+\infty} f(x, y)g(x, y)dx = \int_1^{+\infty} \frac{\sin xy}{x} dx$, which was already shown in Example 15.

Remark 1 to Examples 15–17. The violation of one of the conditions of Dirichlet's theorem can lead not only to nonuniform convergence of $\int_a^{+\infty} f(x, y)g(x, y)dx$, but even to divergence. For instance, for the functions $f(x, y) = y \sin xy$ and $g(x, y) = \frac{\sin xy}{x}$ considered on $X \times Y = [1, +\infty) \times (0, +\infty)$, the only condition violated is the monotonicity of $g(x, y)$ in x. Indeed, it was shown in Example 17 that the integral $\int_1^A f(x, y)dx = \int_1^A y \sin xydx$ is uniformly bounded on $(0, +\infty)$, and the uniform convergence of $g(x, y)$ to 0 follows immediately from the estimate $|g(x, y)| = \frac{|\sin xy|}{x} \leq \frac{1}{x} \underset{x \to +\infty}{\to} 0$. Nevertheless, the integral $\int_1^{+\infty} f(x,y)g(x,y)dx = \int_1^{+\infty} \frac{y}{x}\sin^2 xy dx$ is divergent that can be proved as follows. Represent this integral in the following form:

$$\int_1^{+\infty} \frac{y}{x}\sin^2 xy dx = \frac{1}{2}\int_1^{+\infty}\frac{y}{x}dx - \frac{1}{2}\int_1^{+\infty}\frac{y\cos 2xy}{x}dx.$$

Note that the second integral in the right-hand side converges (even uniformly on $(0,+\infty)$) by Dirichlet's theorem: the integral $\int_1^A y\cos 2xy dx$ is uniformly bounded since $\left|\int_1^A y\cos 2xy dx\right| = \left|\frac{1}{2}\sin 2xy\Big|_1^A\right| \le 1$ simultaneously for $\forall A > 1$, $\forall y \in (0,+\infty)$, and the function $\frac{1}{x}$ converges monotonically and uniformly on $(0,+\infty)$ to 0. At the same time, the first integral in the right-hand side $\int_1^{+\infty} y\frac{1}{x}dx$ diverges for each fixed $y \in (0,+\infty)$. Therefore, the integral $\int_1^{+\infty}\frac{y}{x}\sin^2 xy dx$ diverges.

Remark 2 to Examples 15–17.   As shown in Examples 15–17 and Remark 1, the violation of one of the conditions of Dirichlet's theorem can lead to nonuniform convergence and divergence of the integral $\int_a^{+\infty} f(x,y)g(x,y)dx$. At the same time, these conditions are sufficient, but not necessary: the integral $\int_a^{+\infty} f(x,y)g(x,y)dx$ still can be uniformly convergent if one of the conditions or even all the conditions are not satisfied. This situation is illustrated in the next example.

Example 18. All the conditions of Dirichlet's theorem are violated, but nevertheless the improper integral $\int_a^{+\infty} f(x,y)g(x,y)dx$ converges uniformly on Y.

Solution
For the functions $f(x,y) = \frac{1}{x^2 y}$ and $g(x,y) = \frac{y\cos xy}{x}$ defined on $X \times Y = [1,+\infty) \times (0,+\infty)$, all the conditions of Dirichlet's theorem are violated. In fact, according to the evaluation

$$\left|\int_1^A f(x,y)dx\right| = \left|\int_1^A \frac{1}{x^2 y}dx\right| = \left|-\frac{1}{xy}\Big|_1^A\right| = \frac{1}{y}\left(1 - \frac{1}{A}\right) \le \frac{1}{y}$$

the integrals $\int_1^A f(x,y)dx$ are bounded for each fixed $y \in (0,+\infty)$, but the boundedness is not uniform on $(0,+\infty)$, since for $\forall A > 1$ there exists $y_A = \frac{1}{A} \in (0,+\infty)$ such that

$$\left|\int_1^A f(x,y_A)dx\right| = \frac{1}{y_A}\left(1 - \frac{1}{A}\right) = A - 1 \underset{A\to+\infty}{\longrightarrow} +\infty.$$

Further, the functions $g(x, y)$ are not monotone in x and, although $g(x, y)$ converges to 0 as x approaches infinity for each fixed y, this convergence is not uniform, since for the sequence of the points (x_k, y_k), $x_k = y_k = \sqrt{2k\pi}$, $\forall k \in \mathbb{N}$ one has $x_k \underset{k\to+\infty}{\to} +\infty$, while $g(x_k, y_k) = \frac{x_k \cos 2k\pi}{x_k} = 1 \underset{k\to+\infty}{\nrightarrow} 0$. Even so, the integral $\int_1^{+\infty} f(x, y)g(x, y)dx = \int_1^{+\infty} \frac{\cos xy}{x^3}dx$ converges uniformly on $(0, +\infty)$ due to the Weierstrass test: $\left|\frac{\cos xy}{x^3}\right| \le \frac{1}{x^3}$ and $\int_1^{+\infty} \frac{1}{x^3}dx$ converges.

Remark to Examples 19–22. In the next four examples, we analyze in a similar way the conditions of Abel's theorem on uniform convergence of improper integrals. Examples 19–21 analyze what can happen if one of the conditions is violated, and Example 22 shows that all the conditions are sufficient but not necessary for the uniform convergence of $\int_a^{+\infty} f(x, y)g(x, y)dx$.

Example 19. For each fixed $y \in Y$, integral $\int_a^{+\infty} f(x, y)dx$ is convergent and function $g(x, y)$ is monotone in $x \in [a, +\infty)$; additionally, $g(x, y)$ is uniformly bounded on $[a, +\infty) \times Y$, but the improper integral $\int_a^{+\infty} f(x, y)g(x, y)dx$ does not converge uniformly on Y.

Solution
Consider the functions $f(x, y) = ye^{-xy}$ and $g(x, y) = e^{-xy}$ on $X \times Y = [0, +\infty) \times (0, +\infty)$. Since

$$\int_0^{+\infty} f(x, y)dx = \int_0^{+\infty} ye^{-xy}dx = -e^{-xy}|_0^{+\infty} = 1,$$

the integral $\int_0^{+\infty} f(x, y)dx$ converges for each $y \in (0, +\infty)$. However, this convergence is not uniform on $(0, +\infty)$, because for $\forall A > 0$ choosing $y_A = \frac{1}{A} \in (0, +\infty)$, one gets

$$\int_A^{+\infty} f(x, y_A)dx = -e^{-xy_A}|_A^{+\infty} = e^{-1} \underset{A\to+\infty}{\nrightarrow} 0.$$

The conditions for $g(x, y) = e^{-xy}$ are satisfied: it is decreasing in x for each fixed $y \in (0, +\infty)$ and is uniformly bounded $|g(x, y)| = e^{-xy} \le 1$, $\forall(x, y) \in [0, +\infty) \times (0, +\infty)$. The convergence of the integral of the product can be proved by direct calculation:

$$\int_0^{+\infty} f(x, y)g(x, y)dx = \int_0^{+\infty} ye^{-2xy}dx = -\frac{1}{2}e^{-2xy}\bigg|_0^{+\infty} = \frac{1}{2},$$

but this convergence is not uniform, which is seen by choosing $y_A = \frac{1}{2A} \in (0, +\infty)$ for $\forall A > 0$ and arriving to

$$\int\limits_A^{+\infty} f(x, y_A)g(x, y_A)dx = -\frac{1}{2}e^{-2xy_A}\Big|_A^{+\infty} = \frac{1}{2}e^{-1} \underset{A \to +\infty}{\nrightarrow} 0.$$

The last result is caused by nonuniform convergence of $\int_0^{+\infty} f(x, y)dx$, which violates one of the conditions of Abel's theorem.

Example 20. The integral $\int_a^{+\infty} f(x, y)dx$ is uniformly convergent and $g(x, y)$ is uniformly bounded on Y, but the improper integral $\int_a^{+\infty} f(x, y)g(x, y)dx$ does not converge uniformly on Y.

Solution
The functions $f(x, y) = \frac{\sin x}{x}$ and $g(x, y) = e^{-xy} \sin x$ defined on $X \times Y = [1, +\infty) \times (0, +\infty)$ satisfy the conditions of this example. In fact, the convergence of $\int_1^{+\infty} f(x, y)dx = \int_1^{+\infty} \frac{\sin x}{x}dx$ can be proved in the same way as that of $\int_1^{+\infty} \frac{\sin x}{\sqrt{x}}dx$ in Example 16, and this convergence is uniform since $f(x, y)$ does not depend on y. Also, the function $g(x, y)$ is uniformly bounded on $[1, +\infty) \times (0, +\infty)$ since $|g(x, y)| = |e^{-xy} \sin x| \leq 1$ for $\forall(x, y) \in [1, +\infty) \times (0, +\infty)$. However, the integral $\int_1^{+\infty} f(x, y)g(x, y)dx = \int_1^{+\infty} \frac{\sin^2 x}{x}e^{-xy}dx$ converges nonuniformly as was shown in Example 16. The last integral fails to converge uniformly because of the violation of the monotonicity condition for $g(x, y)$ in Abel's theorem.

Example 21. The integral $\int_a^{+\infty} f(x, y)dx$ converges uniformly on Y, and the function $g(x, y)$ is monotone in x and bounded for each fixed $y \in Y$, but the improper integral $\int_a^{+\infty} f(x, y)g(x, y)dx$ does not converge uniformly on Y.

Solution
Consider the functions $f(x, y) = y^2 e^{-xy}$ and $g(x, y) = \frac{1}{y}e^{-xy}$ on $X \times Y = [0, +\infty) \times (0, +\infty)$. Straightforward calculation

$$\int\limits_0^{+\infty} f(x, y)dx = \int\limits_0^{+\infty} y^2 e^{-xy}dx = -ye^{-xy}|_0^{+\infty} = y$$

shows the convergence of the integral $\int_0^{+\infty} f(x, y)dx$. Note that the function $h(y) = ye^{-Ay}$ for any fixed $A > 0$ has the critical point equation $h_y(y) = (1 - yA)e^{-Ay} = 0$ with the only critical point $y_A = \frac{1}{A}$, and this point is the maximum one, since $h_y(y) > 0$ for $y < y_A$ and $h_y(y) < 0$ for $y > y_A$. Therefore, the following evaluation of the residual

$$\left| \int_A^{+\infty} f(x,y)dx \right| = -ye^{-xy}\big|_A^{+\infty} = ye^{-Ay} \leq \max_{y\in(0,+\infty)} ye^{-Ay}$$

$$= y_A e^{-Ay_A} = \frac{1}{A}e^{-1} \xrightarrow[A\to+\infty]{} 0$$

proves the uniform convergence of $\int_0^{+\infty} f(x,y)dx$ on $(0,+\infty)$. Further, for each fixed $y \in (0,+\infty)$, the function $g(x,y) = \frac{1}{y}e^{-xy}$ is monotone (decreasing) and bounded: $|g(x,y)| = \frac{1}{y}e^{-xy} \leq \frac{1}{y}$. However, the boundedness is not uniform on $Y = (0,+\infty)$ since choosing $y_x = \frac{1}{x} \in (0,+\infty)$ for $\forall x \in (0,+\infty)$, one gets $g(x,y_x) = xe^{-1} \xrightarrow[x\to+\infty]{} +\infty$.

Let us turn to the integral of the product:

$$\int_0^{+\infty} f(x,y)g(x,y)dx = \int_0^{+\infty} ye^{-2xy}dx = -\frac{1}{2}e^{-2xy}\Big|_0^{+\infty} = \frac{1}{2},$$

that is, the integral is convergent for each $y \in (0,+\infty)$. However, this convergence is nonuniform, since for $\forall A > 0$ and $y_A = \frac{1}{2A}$, it follows the following evaluation of the residual:

$$\left| \int_A^{+\infty} f(x,y_A)g(x,y_A)dx \right| = \int_A^{+\infty} y_A e^{-2xy_A}dx$$

$$= -\frac{1}{2}e^{-2xy_A}\Big|_A^{+\infty} = \frac{1}{2}e^{-1} \xrightarrow[A\to+\infty]{} 0.$$

In this example, the nonuniformity of the convergence of $\int_0^{+\infty} f(x,y)g(x,y)dx$ was caused by weakening one of the conditions in Abel's theorem: the function $g(x,y)$ is bounded for each $y \in Y$, but it is not bounded uniformly on Y.

Remark 1 to Examples 19–21. In the same way as for Dirichlet's theorem, it may happen that the violation of only one condition in Abel's theorem will result in the divergence of $\int_a^{+\infty} f(x,y)g(x,y)dx$. For instance, the functions $f(x,y) = \frac{y\sin xy}{x}$ and $g(x,y) = \sin xy$ defined on $X \times Y = [1,+\infty) \times (0,+\infty)$ illustrate this situation. First, the integral $\int_1^{+\infty} f(x,y)dx = \int_1^{+\infty} \frac{y\sin xy}{x}dx$ converges uniformly on $(0,+\infty)$ by Dirichlet's theorem: $\left|\int_1^A y\sin xydx\right| = |-\cos xy|_1^A| \leq 2$, that is, the integral $\int_1^A f(x,y)dx$ is uniformly bounded on $(0,+\infty)$, the function $\frac{1}{x}$ decreases on $[1,+\infty)$ and converges uniformly on $(0,+\infty)$ to 0 as x approaches $+\infty$. Further, the function $g(x,y)$ is uniformly bounded on $[1,+\infty) \times (0,+\infty)$: $|g(x,y)| = |\sin xy| \leq 1$, $\forall(x,y) \in [1,+\infty) \times (0,+\infty)$. Thus, the only condition of Abel's theorem, which is not satisfied, is the monotonicity of $g(x, y)$. Nevertheless, as shown in Remark 1 to Examples 15–17, the integral $\int_1^{+\infty} f(x, y)g(x, y)dx = \int_1^{+\infty} \frac{y}{x}\sin^2 xy dx$ is divergent at each $y \in (0, +\infty)$.

Remark 2 to Examples 19–21. It was shown above that the violation of one of the conditions of Abel's theorem can lead to nonuniform convergence and even divergence of the integral $\int_a^{+\infty} f(x, y)g(x, y)dx$. At the same time, the conditions of Abel's theorem are sufficient, but not necessary, as it is illustrated in Example 22.

Example 22. All the conditions of Abel's theorem are violated, but nevertheless the improper integral $\int_a^{+\infty} f(x, y)g(x, y)dx$ converges uniformly on Y.

Solution
For the functions $f(x, y) = \frac{1}{x^2 y}$ and $g(x, y) = y\cos xy$ defined on $X \times Y = [1, +\infty) \times (0, +\infty)$, all the conditions of Abel's theorem are violated. In fact, the direct calculation

$$\int_1^{+\infty} f(x, y)dx = \int_1^{+\infty} \frac{1}{x^2 y}dx = -\frac{1}{xy}\Big|_1^{+\infty} = \frac{1}{y}$$

shows that the integral $\int_1^{+\infty} f(x, y)dx$ converges, but this convergence is nonuniform on $(0, +\infty)$, since for $\forall A > 1$ and the corresponding $y_A = \frac{1}{A} \in (0, +\infty)$ one gets

$$\int_A^{+\infty} f(x, y_A)dx = \int_A^{+\infty} \frac{1}{x^2 y_A}dx = \frac{1}{A y_A} = 1 \underset{A \to +\infty}{\nrightarrow} 0.$$

Further, the function $g(x, y) = y\cos xy$ is not monotone in x. Finally, $g(x, y)$ is bounded for each fixed $y \in (0, +\infty)$: $|g(x, y)| = |y\cos xy| \leq y$, but this boundedness is not uniform on $(0, +\infty)$ since for the sequence of the points (x_k, y_k), $x_k = y_k = \sqrt{2k\pi}$, $\forall k \in \mathbb{N}$ one has $x_k \underset{k \to +\infty}{\to} +\infty$, while $g(x_k, y_k) = \sqrt{2k\pi}\cos 2k\pi \underset{k \to +\infty}{\to} +\infty$. Nevertheless, the integral $\int_1^{+\infty} f(x, y)g(x, y)dx = \int_1^{+\infty} \frac{\cos xy}{x^2}dx$ converges uniformly on $(0, +\infty)$ by the Weierstrass test: $\left|\frac{\cos xy}{x^2}\right| \leq \frac{1}{x^2}$, $\forall y \in (0, +\infty)$ and $\int_1^{+\infty} \frac{1}{x^2}dx$ converges.

5.4 Existence of the Limit and Continuity

Example 23. A function $f(x, y)$ is continuous in x on $[a, +\infty)$ for any fixed $y \in Y$ and converges uniformly on $[a, +\infty)$ to a function $\varphi(x)$, as y approaches y_0, and also both improper integrals $\int_a^{+\infty} f(x,y)dx$ and $\int_a^{+\infty} \varphi(x)dx$ are convergent, but $\lim\limits_{y\to y_0} \int_a^{+\infty} f(x,y)dx \neq \int_a^{+\infty} \lim\limits_{y\to y_0} f(x,y)dx.$

Solution

Consider the function $f(x,y) = \begin{cases} \frac{1}{x^3 y} e^{-\frac{1}{2x^2 y}}, x > 0 \\ 0, x = 0 \end{cases}$ on $X \times Y = [0,+\infty) \times (0,1]$

and $y_0 = 0$. This function is continuous in x on X for any fixed $y \in Y$. For $x > 0$, it follows from the arithmetic properties and composition rule of continuous functions, and for $x = 0$ it can be shown as follows:

$$\lim_{x\to 0} f(x,y) = \lim_{x\to 0} \frac{1}{x^3 y} e^{-\frac{1}{2x^2 y}} = \lim_{t\to+\infty} \frac{1}{y} \frac{t^3}{e^{t^2/2y}}$$

$$= \frac{1}{y} \lim_{t\to+\infty} \frac{3t^2}{\frac{1}{2y} 2te^{t^2/2y}} = 3 \lim_{t\to+\infty} \frac{1}{\frac{1}{y} te^{t^2/2y}} = 0$$

(the change of the variable $t = \frac{1}{x}$ was applied with subsequent application of l'Hospital's rule).

Next, let us show that $f(x,y)$ converges to $\varphi(x) \equiv 0$ as y approaches 0: for $x = 0$, the result is immediate—$f(0,y) = 0$, $\forall y \in Y$—and for other x, it can be seen from the evaluation

$$\lim_{y\to 0} f(x,y) = \lim_{y\to 0} \frac{1}{x^3 y} e^{-\frac{1}{2x^2 y}} = \lim_{t\to+\infty} \frac{2}{x} \frac{t}{e^t} = \lim_{t\to+\infty} \frac{2}{x} \frac{1}{e^t} = 0$$

(here, the change of the variable $t = \frac{1}{2x^2 y}$ was followed by l'Hospital's rule). Moreover, $f(x,y)$ converges to $\varphi(x) \equiv 0$ uniformly on X due to the following argument. Since $f(x,y) > 0$ for $x > 0$, $f(0,y) = 0$, and $\lim\limits_{x\to+\infty} f(x,y) = 0$, the maximum $\max\limits_{x\in[0,+\infty)} |f(x,y) - \varphi(x)| = \max\limits_{x\in[0,+\infty)} f(x,y)$ is attained at an interior point of $[0,+\infty)$ for each fixed $y \in Y$. The corresponding partial derivative $f_x(x,y) = \frac{1-3x^2 y}{x^6 y^2} e^{-\frac{1}{2x^2 y}}$ vanishes at the points where $1 - 3x^2 y = 0$, or $x = \pm\frac{1}{\sqrt{3y}}$. The only critical point in $[0,+\infty)$ is $x_y = \frac{1}{\sqrt{3y}}$, which therefore is the maximum point. Then, for all $x \in [0,+\infty)$, we have

$$|f(x,y) - \varphi(x)| \leq \max_{x\in[0,+\infty)} f(x,y) = f(x_y,y)$$

$$= \frac{3y\sqrt{3y}}{y} e^{-3y/2y} = 3\sqrt{3y}e^{-3/2} \xrightarrow[y\to 0]{} 0,$$

which means the uniform convergence on $X = [0,+\infty)$.

Finally, note that both improper integrals exist:

$$\int_0^{+\infty} \lim_{y\to 0} f(x,y)dx = \int_0^{+\infty} 0 dx = 0$$

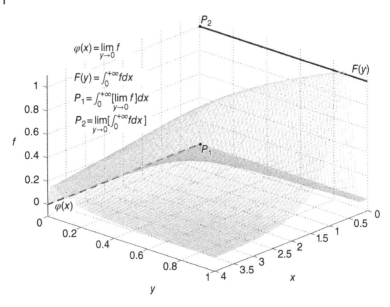

Figure 5.11 Example 23, function $f(x, y) = \begin{cases} \frac{1}{x^3 y} e^{-\frac{1}{2x^2 y}}, x > 0 \\ 0, x = 0 \end{cases}$.

and

$$\int\limits_{0}^{+\infty} f(x, y)dx = \int\limits_{0}^{+\infty} \frac{1}{x^3 y} e^{-\frac{1}{2x^2 y}} dx = e^{-\frac{1}{2x^2 y}} \Big|_{0}^{+\infty} = 1,$$

but

$$\lim_{y \to 0} \int\limits_{0}^{+\infty} f(x, y)dx = 1 \neq 0 = \int\limits_{0}^{+\infty} \lim_{y \to 0^-} f(x, y)dx.$$

Remark. In the given counterexample, the interchange of the limit and integral is impossible because the convergence of the integral $\int_0^{+\infty} f(x, y)dx$ is not uniform on $Y = (0, 1]$. The last fact can be verified through the evaluation of the remainder: for $\forall A > 1$ one can choose $y_A = \frac{1}{2A^2} \in Y$ to obtain

$$\int\limits_{A}^{+\infty} f(x, y_A)dx = e^{-\frac{1}{2x^2 y_A}} \Big|_{A}^{+\infty} = 1 - e^{-\frac{1}{2A^2 y_A}} = 1 - e^{-1} \underset{A \to +\infty}{\not\to} 0.$$

Thus, the example shows that the condition of the uniform convergence of $\int_a^{+\infty} f(x, y)dx$ on Y is important for the possibility of the calculation of the limit under the sign of an improper integral. However, this condition is sufficient but not necessary as shown in the next example.

Example 24. A function $f(x, y)$ is continuous in x on $[a, +\infty)$ for any fixed $y \in Y$ and converges uniformly on $[a, +\infty)$ to $\varphi(x)$, as y approaches y_0, and $\lim_{y \to y_0} \int_a^{+\infty} f(x, y)dx = \int_a^{+\infty} \lim_{y \to y_0} f(x, y)dx$, but $\int_a^{+\infty} f(x, y)dx$ converges nonuniformly on Y.

Solution
Consider the function $f(x, y) = \left(\frac{x}{y^2} - \frac{1}{y} \right) e^{-x/y}$ on $X \times Y = [1, +\infty) \times [1, +\infty)$ and $y_0 = +\infty$. This function is continuous in x on X for any fixed $y \in Y$ (just apply the arithmetic and composition rules for continuous functions). Also $f(x, y)$ converges on X to $\varphi(x) \equiv 0$: $\lim_{y \to +\infty} \left(\frac{x}{y^2} - \frac{1}{y} \right) e^{-x/y} = 0$, $\forall x \in X$. To show that this convergence is uniform, let us find the extrema with respect to x of $f(x, y) - \varphi(x) = f(x, y)$. The critical points are found from the equation $f_x(x, y) = \frac{2y - x}{y^3} e^{-x/y} = 0$, which results in $x_y = 2y$. If $x_y < 2y$ then $f_x(x, y) > 0$, and if $x_y > 2y$ then $f_x(x, y) < 0$, that is, $x_y = 2y$ is the maximum point. Note also that $f(x, y)$ is strictly increasing on $[1, 2y)$ and has negative values when $x < y$ and positive when $x > y$. Therefore,

$$\max_{x \in [1, +\infty)} |f(x, y) - \varphi(x)| = \max\{|f(1, y)|, |f(x_y, y)|\}.$$

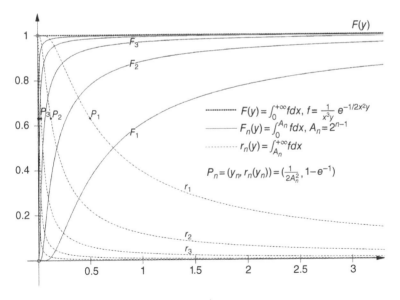

Figure 5.12 Example 23, integral $F(y) = \int_0^{+\infty} f(x, y)dx$.

For the last two functions of y, we have

$$\lim_{y \to +\infty} |f(1, y)| = \lim_{y \to +\infty} \left(\frac{1}{y} - \frac{1}{y^2} \right) e^{-1/y} = 0$$

and

$$\lim_{y \to +\infty} |f(x_y, y)| = \lim_{y \to +\infty} \left(\frac{2y}{y^2} - \frac{1}{y} \right) e^{-2y/y} = \lim_{y \to +\infty} \frac{1}{y} e^{-2} = 0.$$

Thus, simultaneously for all $x \in X$, we get

$$|f(x, y) - \varphi(x)| \leq \max_{x \in [1, +\infty)} |f(x, y)| = \max\{|f(1, y)|, |f(x_y, y)|\} \underset{y \to +\infty}{\to} 0,$$

which implies the uniform convergence of $f(x, y)$ on X.

Finally, let us check the condition of interchanging the integral and limit. The first integral is trivial:

$$\int_1^{+\infty} \lim_{y \to +\infty} f(x, y) dx = \int_1^{+\infty} 0 dx = 0.$$

For the second, we have

$$\int_1^{+\infty} f(x, y) dx = -\frac{x}{y} e^{-x/y} \Big|_1^{+\infty} = \frac{1}{y} e^{-1/y},$$

where the indeterminate form at $x \to +\infty$ was evaluated using l'Hospital's rule:

$$\lim_{x \to +\infty} -\frac{x}{y} e^{-x/y} = \lim_{x \to +\infty} -\frac{x/y}{e^{x/y}} = \lim_{x \to +\infty} -\frac{1}{e^{x/y}} = 0.$$

Therefore,

$$\lim_{y \to +\infty} \int_1^{+\infty} f(x, y) dx = \lim_{y \to +\infty} \frac{1}{y} e^{-1/y} = 0 = \int_1^{+\infty} \lim_{y \to +\infty} f(x, y) dx.$$

Hence, all the conditions of the statement are satisfied.

Nevertheless, the improper integral $\int_1^{+\infty} f(x, y) dx$ converges nonuniformly on Y: for $\forall A > 1$, choosing $y_A = A \in Y$ we obtain

$$\int_A^{+\infty} f(x, y_A) dx = -\frac{x}{y_A} e^{-x/y_A} \Big|_A^{+\infty} = \frac{A}{y_A} e^{-A/y_A} = e^{-1} \underset{A \to +\infty}{\not\to} 0.$$

Example 25. A function $f(x, y)$ is continuous in two variables on $X \times Y = [a, +\infty) \times [c, d]$ and the improper integral $F(y) = \int_a^{+\infty} f(x, y)dx$ converges on Y, but $F(y)$ is discontinuous on Y.

Solution
The function $f(x, y) = ye^{-xy}$ is continuous on $X \times Y = [0, +\infty) \times [0, +\infty)$ and the improper integral converges on Y: $F(0) = \int_0^{+\infty} 0dx = 0$ and for $y \neq 0$ we have $F(y) = \int_0^{+\infty} ye^{-xy}dx = -e^{-xy}|_0^{+\infty} = 1$. However, $F(y)$ is discontinuous at $y = 0$.

Note that the improper integral $\int_0^{+\infty} f(x, y)dx$ converges nonuniformly on Y: for $\forall A > 0$, there exists $y_A = \frac{1}{A} \in Y$ such that

$$\int_A^{+\infty} y_A e^{-xy_A} dx = -e^{-xy_A}|_A^{+\infty} = e^{-Ay_A} = e^{-1} \underset{A \to +\infty}{\nrightarrow} 0.$$

Another counterexample is $f(x, y) = \begin{cases} \frac{\sin xy}{x}, & x \neq 0 \\ y, & x = 0 \end{cases}$ considered on $X \times Y = [0, +\infty) \times \mathbb{R}$. This function is continuous at every (x, y), $x \neq 0$ due to the arithmetic rules and at every $(0, y)$ because $\lim_{x \to 0} f(x, y) = \lim_{x \to 0} \frac{\sin xy}{xy} \cdot y = y = f(0, y)$. Besides, the improper integral converges (albeit nonuniformly) on Y: $F(y) = \int_0^{+\infty} \frac{\sin xy}{x} dx = \frac{\pi}{2}\text{sgn}y$ (see calculations for a similar function in Example 3). However, $F(y)$ is discontinuous at $y = 0$.

Remark. The condition of the uniform convergence is relevant for the continuity of an improper integral depending on a parameter, but it is not necessary as shown in Example 26.

Example 26. A function $f(x, y)$ is continuous in two variables on $X \times Y$, where $X = [a, +\infty)$ and Y is an interval, and $F(y) = \int_a^{+\infty} f(x, y)dx$ is continuous on Y, but the improper integral $\int_a^{+\infty} f(x, y)dx$ does not converge uniformly on Y.

Solution
The function $f(x, y) = \frac{1}{x^y}$ is continuous on $X \times Y = [1, +\infty) \times (1, +\infty)$. Also, the function

$$F(y) = \int_1^{+\infty} \frac{1}{x^y} dx = \frac{x^{1-y}}{1-y}\Big|_1^{+\infty} = \frac{1}{y-1}$$

is continuous on $Y = (1, +\infty)$. However, the improper integral $\int_1^{+\infty} f(x, y)dx$ converges nonuniformly on Y: for $\forall A > 1$, one can choose $y_A = 1 + \frac{1}{A} \in Y$ to obtain

$$\int_A^{+\infty} \frac{dx}{x^{y_A}} = \frac{x^{1-y_A}}{1 - y_A}\Big|_A^{+\infty} = A^{1-1/A} \underset{A \to +\infty}{\to} +\infty.$$

Another counterexample employs the function of Example 4: $f(x, y) = e^{-xy} \sin x$ considered on $X \times Y = [0, +\infty) \times (0, +\infty)$. Evidently, this function is continuous on $\mathbb{R}^2$, and in particular on $[0, +\infty) \times (0, +\infty)$. Using the calculations made in Example 4 for $y > 0$, we obtain

$$F(y) = \int_0^{+\infty} e^{-xy} \sin x dx = -e^{-xy} \frac{\cos x + y \sin x}{1 + y^2}\Big|_0^{+\infty} = \frac{1}{1 + y^2}.$$

This function is continuous on $(0, +\infty)$, but it was shown in Example 4 that the integral $\int_0^{+\infty} e^{-xy} \sin x dx$ does not converge uniformly on $Y = (0, +\infty)$.

Remark. Note that the continuity of $F(y)$ on Y does not guarantee the validity of the interchange between the limit and the improper integral. In fact, in the last counterexample the function $F(y) = \frac{1}{1+y^2}$, and in particular, $\lim_{y \to 0_+} F(y) = \lim_{y \to 0_+} \int_0^{+\infty} e^{-xy} \sin x dx = 1$. However, the improper integral $\int_0^{+\infty} \lim_{y \to 0_+} e^{-xy} \sin x dx = \int_0^{+\infty} \sin x dx$ is divergent.

5.5 Differentiability

Example 27. Both $f(x, y)$ and $f_y(x, y)$ are continuous on $X \times Y$, where $X = [a, +\infty)$ and Y is an interval, and $\int_a^{+\infty} f(x, y)dx$ converges uniformly on Y, but $\left(\int_a^{+\infty} f(x, y)dx\right)_y \neq \int_a^{+\infty} f_y(x, y)dx$ on Y.

Solution

In Examples 3 and 25, it was shown that $f(x, y) = \begin{cases} \frac{\sin xy}{x}, x \neq 0 \\ y, x = 0 \end{cases}$ is continuous on $\mathbb{R}^2$. Its partial derivative $f_y(x, y) = \begin{cases} \cos xy, x \neq 0 \\ 1, x = 0 \end{cases}$ is also continuous on $\mathbb{R}^2$. In particular, both $f(x, y)$ and $f_y(x, y)$ are continuous on $X \times Y = [0, +\infty) \times [1, +\infty)$. Besides, in Example 3 it was proved that $F(y) = \int_a^{+\infty} f(x, y)dx = \int_0^{+\infty} \frac{\sin xy}{x}dx$ converges uniformly on any interval $[c, +\infty)$, $c > 0$, in particular, on $Y = [1, +\infty)$. Since $F(y) = \frac{\pi}{2}$ for $\forall y \in [1, +\infty)$

(see calculations in Example 3), it follows that $F_y(y) = 0$. However, it is not possible to calculate the derivative inside the integral, because

$$\int_0^{+\infty} f_y(x, y)dx = \int_0^{+\infty} \cos xy dx = \frac{1}{y} \int_0^{+\infty} \cos t \, dt$$

diverges.

Example 28. Both $f(x, y)$ and $f_y(x, y)$ are continuous on $X \times Y$, where $X = [a, +\infty)$ and Y is an interval, and both integrals $\int_a^{+\infty} f(x, y)dx$ and $\int_a^{+\infty} f_y(x, y)dx$ converge on Y, but $\left(\int_a^{+\infty} f(x, y)dx \right)_y \neq \int_a^{+\infty} f_y(x, y)dx$ on Y.

Solution
Let us consider $f(x, y) = \left(\frac{y}{x} + \frac{1}{x^2} \right) e^{-xy}$ on $X \times Y = [1, +\infty) \times [0, +\infty)$. Both $f(x, y)$ and $f_y(x, y) = -ye^{-xy}$ are continuous on $X \times Y$. Also, both improper integrals converge on $Y = [0, +\infty)$, which can be checked by straightforward calculations:

$$F(y) = \int_1^{+\infty} f(x, y)dx = \int_1^{+\infty} \left(\frac{y}{x} + \frac{1}{x^2} \right) e^{-xy} dx$$

$$= -\frac{1}{x} e^{-xy} \Big|_1^{+\infty} = e^{-y}, \forall y \in (0, +\infty),$$

$$F(0) = \int_1^{+\infty} f(x, 0)dx = \int_1^{+\infty} \frac{1}{x^2} dx = -\frac{1}{x} \Big|_1^{+\infty} = 1,$$

$$G(y) = \int_1^{+\infty} f_y(x, y)dx = \int_1^{+\infty} -ye^{-xy} dx$$

$$= e^{-xy} \Big|_1^{+\infty} = -e^{-y}, \forall y \in (0, +\infty),$$

$$G(0) = \int_1^{+\infty} f_y(x, 0)dx = \int_1^{+\infty} 0 dx = 0.$$

Nevertheless, $F_y(y) = -e^{-y}$ and, consequently, $F_y(0) = -1 \neq 0 = G(0)$.

Note that the improper integral $\int_1^{+\infty} f_y(x, y)dx$ converges nonuniformly on $Y = [0, +\infty)$ because for $\forall A > 1$ there exists $y_A = \frac{1}{A} \in Y$ such that the remainder does not approach 0:

$$\int_A^{+\infty} f_y(x, y_A)dx = e^{-xy_A} \Big|_A^{+\infty} = -e^{-Ay_A} = -e^{-1} \underset{A \to +\infty}{\nrightarrow} 0.$$

At the same time, the second improper integral $\int_1^{+\infty} f(x,y)dx$ converges uniformly on $Y = [0, +\infty)$. Indeed, for $y \in (0, +\infty)$ one gets

$$\int_A^{+\infty} f(x,y)dx = -\frac{1}{x}e^{-xy}\Big|_A^{+\infty} = \frac{1}{A}e^{-Ay},$$

and for $y = 0$

$$\int_A^{+\infty} f(x,0)dx = \int_A^{+\infty} \frac{1}{x^2}dx = -\frac{1}{x}\Big|_A^{+\infty} = \frac{1}{A}.$$

Therefore, the evaluation $\left|\int_A^{+\infty} f(x,y)dx\right| \leq \frac{1}{A} \underset{A \to \infty}{\to} 0$ holds simultaneously for $\forall y \in [0, +\infty)$.

Remark. This example shows that the uniform convergence of $\int_a^{+\infty} f_y(x,y)dx$ is an important condition to interchange the calculation of partial derivative and improper integral. At the same time, this condition is not necessary as shown in the following example.

Example 29. Both $f(x,y)$ and $f_y(x,y)$ are continuous on $X \times Y$, where $X = [a, +\infty)$ and Y is an interval, the integrals $\int_a^{+\infty} f(x,y)dx$ and $\int_a^{+\infty} f_y(x,y)dx$ converge on Y and $\int_a^{+\infty} f_y(x,y)dx = \left(\int_a^{+\infty} f(x,y)dx\right)_y$, $\forall y \in Y$, but the improper integral $\int_a^{+\infty} f_y(x,y)dx$ converges nonuniformly on Y.

Solution
Consider the function of Example 28: $f(x,y) = \left(\frac{y}{x} + \frac{1}{x^2}\right)e^{-xy}$ but on the different domain $X \times Y = [1, +\infty) \times (0, +\infty)$. It was shown in Example 28 that $\int_1^{+\infty} f_y(x,y)dx$ does not converge uniformly on $[0, +\infty)$, and for exactly the same reason it does not converge uniformly on $(0, +\infty)$. However, all the statement conditions hold: $f(x,y)$ and $f_y(x,y) = -ye^{-xy}$ are continuous on $X \times Y$; both improper integrals converge:

$$F(y) = \int_a^{+\infty} f(x,y)dx = e^{-y}, G(y) = \int_a^{+\infty} f_y(x,y)dx = -e^{-y}, \forall y \in (0, +\infty);$$

and finally $F_y(y) = -e^{-y} = G(y)$, $\forall y \in (0, +\infty)$.

Remark. In the presented counterexample, the integral $\int_a^{+\infty} f(x,y)dx = \int_1^{+\infty}\left(\frac{y}{x} + \frac{1}{x^2}\right)e^{-xy}dx$ converges uniformly on Y (as was shown in Example 28). However, it may happen that the equality $\int_a^{+\infty} f_y(x,y)dx = \left(\int_a^{+\infty} f(x,y)dx\right)_y$

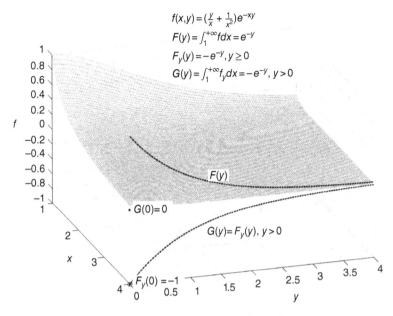

$$f(x,y) = \left(\tfrac{y}{x} + \tfrac{1}{x^2}\right)e^{-xy}$$
$$F(y) = \int_1^{+\infty} f\,dx = e^{-y}$$
$$F_y(y) = -e^{-y}, y \geq 0$$
$$G(y) = \int_1^{+\infty} f_y\,dx = -e^{-y}, y > 0$$

$F(y)$

$\cdot\, G(0) = 0$

$G(y) = F_y(y), y > 0$

$F_y(0) = -1$

Figure 5.13 Examples 28 and 29, function $f(x,y) = \left(\frac{y}{x} + \frac{1}{x^2}\right)e^{-xy}$.

holds for $\forall y \in Y$ even when both integrals $\int_a^{+\infty} f(x,y)dx$ and $\int_a^{+\infty} f_y(x,y)dx$ converge nonuniformly on Y. For such a case, let us consider the function $f(x,y) = ye^{-xy}$ on $X \times Y = [1,+\infty) \times (0,+\infty)$, which is continuous together with its derivative $f_y(x,y) = (1 - xy)e^{-xy}$ on $\mathbb{R}^2$ and, in particular, on $X \times Y$. The convergence of both integrals $F(y) = \int_1^{+\infty} f(x,y)dx$ and $G(y) = \int_1^{+\infty} f_y(x,y)dx$ on $(0,+\infty)$ can be verified by the direct calculations:

$$F(y) = \int_1^{+\infty} ye^{-xy}dx = -e^{-xy}|_1^{+\infty} = e^{-y}, \forall y \in (0,+\infty),$$

$$G(y) = \int_1^{+\infty} (1 - xy)e^{-xy}dx = xe^{-xy}|_1^{+\infty} = -e^{-y}, \forall y \in (0,+\infty).$$

Therefore, $F_y(y) = -e^{-y} = G(y), \forall y \in (0,+\infty)$. However, for $\forall A > 1$ and $y_A = \frac{1}{A}$ one gets

$$\left| \int_A^{+\infty} f(x,y_A)dx \right| = |-e^{-xy_A}|_A^{+\infty}| = e^{-Ay_A} = e^{-1} \underset{A \to +\infty}{\nrightarrow} 0$$

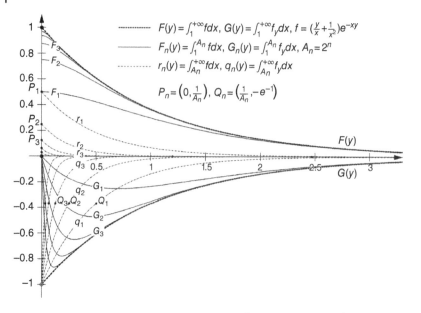

Figure 5.14 Examples 28 and 29, integrals $F(y) = \int_1^{+\infty} f(x,y)dx$, $G(y) = \int_1^{+\infty} f_y(x,y)dx$.

and

$$\left| \int\limits_A^{+\infty} f_y(x,y_A)dx \right| = \left| xe^{-xy_A} \big|_A^{+\infty} \right| = Ae^{-Ay_A} = Ae^{-1} \underset{A \to +\infty}{\longrightarrow} +\infty,$$

which means that both integrals converge nonuniformly on $(0,+\infty)$.

5.6   Integrability

Example 30. A function $f(x,y)$ is continuous on $[a,+\infty) \times [c,d]$ and the integral $\int_a^{+\infty} f(x,y)dx$ converges on $[c,d]$, but $\int_c^d dy \int_a^{+\infty} f(x,y)dx \neq \int_a^{+\infty} dx \int_c^d f(x,y)dy$.

Solution
The function $f(x,y) = (2xy - 2x^3y^3)e^{-x^2y^2}$ is continuous on $\mathbb{R}^2$, and in particular, on $X \times Y = [0,+\infty) \times [0,1]$. The improper integral converges on $[0,1]$ according to the direct calculations:

$$F(y) = \int\limits_0^{+\infty} (2xy - 2x^3y^3)e^{-x^2y^2}dx = x^2ye^{-x^2y^2} \big|_0^{+\infty} = 0.$$

However, the integrals assume different values:

$$\int_0^1 dy \int_0^{+\infty} f(x,y)dx = \int_0^1 F(y)dy = \int_0^1 0 dy = 0,$$

while

$$\int_0^{+\infty} dx \int_0^1 f(x,y)dy = \int_0^{+\infty} xy^2 e^{-x^2 y^2}|_0^1 dx$$

$$= \int_0^{+\infty} xe^{-x^2} dx = -\frac{1}{2}e^{-x^2}|_0^{+\infty} = \frac{1}{2}.$$

Note that the convergence of $F(y) = \int_0^{+\infty}(2xy - 2x^3 y^3)e^{-x^2 y^2} dx$ is not uniform on $[0, 1]$. Indeed, for $\forall A > 1$ and $y_A = \frac{1}{A} \in [0, 1]$ the remainder evaluation gives

$$\left| \int_A^{+\infty}(2xy_A - 2x^3 y_A^3)e^{-x^2 y_A^2} dx \right| = \left| x^2 y_A e^{-x^2 y_A^2}|_A^{+\infty} \right|$$

$$= A^2 y_A e^{-A^2 y_A^2} = Ae^{-1} \xrightarrow[A\to+\infty]{} +\infty.$$

Remark. The required equality will be satisfied if the convergence of $F(y) = \int_a^{+\infty} f(x,y)dx$ would be uniform on $[c, d]$. However, Example 31 shows that the last condition is not necessary.

Example 31. A function $f(x, y)$ is continuous on an infinite strip $[a, +\infty) \times [c, d]$ and $\int_c^d dy \int_a^{+\infty} f(x,y)dx = \int_a^{+\infty} dx \int_c^d f(x,y)dy$, but the improper integral $\int_a^{+\infty} f(x,y)dx$ converges nonuniformly on $[c, d]$.

Solution
The function $f(x, y) = ye^{-xy}$ is continuous on $\mathbb{R}^2$, and consequently, on $X \times Y = [1, +\infty) \times [0, 1]$. The improper integral converges on $[0, 1]$:

$$F(y) = \int_1^{+\infty} ye^{-xy}dx = -e^{-xy}|_1^{+\infty} = e^{-y}, \forall y \in (0, 1]$$

and $F(0) = \int_1^{+\infty} 0 dx = 0$. However, this convergence is not uniform: for $\forall A > 1$ and $y_A = \frac{1}{A} \in [0, 1]$, we get

$$\int_A^{+\infty} y_A e^{-xy_A}dx = -e^{-xy_A}|_A^{+\infty} = e^{-1} \xrightarrow[A\to+\infty]{} 0.$$

Let us check if it is possible to change the order of integration. On the one hand,

$$\int_0^1 dy \int_1^{+\infty} f(x,y)dx = \int_0^1 F(y)dy = \int_0^1 e^{-y}dy = 1 - e^{-1}.$$

On the other hand,

$$\int_1^{+\infty} dx \int_0^1 f(x,y)dy = \int_1^{+\infty} dx \int_0^1 ye^{-xy}dy$$

$$= \int_1^{+\infty}\left[-\frac{y}{x}e^{-xy}\Big|_0^1 + \int_0^1 \frac{1}{x}e^{-xy}dy\right]dx$$

$$= \int_1^{+\infty}\left[-\frac{1}{x}e^{-x} - \frac{1}{x^2}e^{-xy}\Big|_0^1\right]dx$$

$$= \int_1^{+\infty} -\frac{1}{x}e^{-x}dx - \int_1^{+\infty}\frac{e^{-x}}{x^2}dx + \int_1^{+\infty}\frac{1}{x^2}dx$$

$$= \frac{e^{-x}}{x}\Big|_1^{+\infty} + \int_1^{+\infty}\frac{e^{-x}}{x^2}dx - \int_1^{+\infty}\frac{e^{-x}}{x^2}dx - \frac{1}{x}\Big|_1^{+\infty}$$

$$= 1 - e^{-1}$$

(the integration by parts was applied for integration in y with $u = y$, $dv = e^{-xy}dy$ and for integration in x with $u = \frac{1}{x}$, $dv = e^{-x}dx$). Therefore, both integrals give the same result.

Example 32. A function $f(x,y)$ is continuous on $[a, +\infty) \times [c, +\infty)$, and the improper integrals $\int_c^{+\infty} dy \int_a^{+\infty} f(x,y)dx$ and $\int_a^{+\infty} dx \int_c^{+\infty} f(x,y)dy$ converge, but they assume different values.

Solution
The function $f(x,y) = \frac{x-y}{(x+y)^3}$ is continuous on $X \times Y = [1, +\infty) \times [2, +\infty)$. Let us evaluate the required improper integrals. For the first integral, we have

$$F(y) = \int_1^{+\infty}\frac{x-y}{(x+y)^3}dx = \int_1^{+\infty}\frac{x+y}{(x+y)^3}dx - 2\int_1^{+\infty}\frac{y}{(x+y)^3}dx$$

$$= \left(-\frac{1}{x+y} + \frac{y}{(x+y)^2}\right)\Big|_1^{+\infty} = -\frac{x}{(x+y)^2}\Big|_1^{+\infty} = \frac{1}{(1+y)^2},$$

and therefore,

$$\int_2^{+\infty} F(y)dy = \int_2^{+\infty} \frac{1}{(1+y)^2}dy = -\frac{1}{1+y}\Big|_2^{+\infty} = \frac{1}{3}.$$

The second integral gives

$$G(x) = \int_2^{+\infty} \frac{x-y}{(x+y)^3}dy = \int_2^{+\infty} \frac{2x}{(x+y)^3}dy - \int_2^{+\infty} \frac{x+y}{(x+y)^3}dy$$

$$= \left(-\frac{x}{(x+y)^2} + \frac{1}{x+y}\right)\Big|_2^{+\infty} = \frac{y}{(x+y)^2}\Big|_2^{+\infty} = -\frac{2}{(x+2)^2},$$

and then

$$\int_1^{+\infty} G(x)dx = \int_1^{+\infty} \frac{-2}{(x+2)^2}dx = \frac{2}{x+2}\Big|_1^{+\infty} = -\frac{2}{3}.$$

Thus, both integrals are convergent, but different.

Comparing the conditions of this example with the conditions of Theorem 1 (on integration with respect to parameter on infinite interval), one can see that the former does not require that $f(x, y)$ keeps the same sign over the considered domain. It is used in the counterexample and leads to different results for the integrals.

On the other hand, comparing the conditions of this example with the conditions of Theorem 2 (on integration with respect to parameter on infinite interval), one can note that the former does not require the convergence of the integrals of the absolute values. Applying the Weierstrass test, one can show that in the used counterexample the integral $F(y)$ converges uniformly on $Y = [2, +\infty)$: $|f(x, y)| = \frac{|x-y|}{(x+y)^3} < \frac{1}{x^2}$, $\forall(x, y) \in [1, +\infty) \times [2, +\infty)$, and $\int_1^{+\infty} \frac{1}{x^2}dx$ converges. Using the same reasoning, one can prove that the integral $G(x)$ converges uniformly on $X = [1, +\infty)$. However, both integrals $\int_2^{+\infty} dy \int_1^{+\infty} \frac{|x-y|}{(x+y)^3}dx$ and $\int_1^{+\infty} dx \int_2^{+\infty} \frac{|x-y|}{(x+y)^3}dy$ diverge. Indeed,

$$\int_1^{+\infty} \frac{|x-y|}{(x+y)^3}dx = \int_1^y \frac{y-x}{(x+y)^3}dx + \int_y^{+\infty} \frac{x-y}{(x+y)^3}dx$$

$$= \frac{x}{(x+y)^2}\Big|_1^y - \frac{x}{(x+y)^2}\Big|_y^{+\infty} = \frac{1}{2y} - \frac{1}{(1+y)^2},$$

and then

$$\int_2^{+\infty} dy \int_1^{+\infty} \frac{|x-y|}{(x+y)^3} dx = \int_2^{+\infty} \frac{1}{2y} dy - \int_2^{+\infty} \frac{1}{(1+y)^2} dy,$$

where the second integral converges, but the first diverges, implying in the divergence of the integral $\int_2^{+\infty} dy \int_1^{+\infty} |f(x,y)| dx$. Similar evaluations can be made for the second integral:

$$\int_1^{+\infty} dx \int_2^{+\infty} \frac{|x-y|}{(x+y)^3} dy$$

$$= \int_1^2 dx \int_2^{+\infty} \frac{y-x}{(x+y)^3} dy$$

$$+ \int_2^{+\infty} dx \left[\int_2^x \frac{x-y}{(x+y)^3} dy + \int_x^{+\infty} \frac{y-x}{(x+y)^3} dy \right]$$

$$= \int_1^2 \left(-\frac{1}{x+y} + \frac{x}{(x+y)^2} \right) \Big|_2^{+\infty} dx$$

$$+ \int_2^{+\infty} \left(\frac{1}{x+y} - \frac{x}{(x+y)^2} \right) \Big|_2^x + \left(-\frac{1}{x+y} + \frac{x}{(x+y)^2} \right) \Big|_x^{+\infty} dx$$

$$= \int_1^2 \left(\frac{1}{x+2} - \frac{x}{(x+2)^2} \right) dx$$

$$+ \int_2^{+\infty} \left(\frac{2}{2x} - \frac{2}{4x} - \frac{1}{x+2} + \frac{x}{(x+2)^2} \right) dx$$

$$= -\frac{2}{x+2} \Big|_1^2 + \frac{2}{x+2} \Big|_2^{+\infty} + \int_2^{+\infty} \frac{1}{2x} dx,$$

where the last integral diverges, resulting in the divergence of the integral $\int_1^{+\infty} dx \int_2^{+\infty} |f(x,y)| dy$.

Another counterexample can be provided with the function $f(x,y) = \frac{x^2-y^2}{(x^2+y^2)^2}$ considered on $X \times Y = [1,+\infty) \times [1,+\infty)$.

Remark. In these counterexamples, the improper integrals have different values because neither the conditions of Theorem 1 nor those of Theorem 2 are satisfied. At the same time, the conditions in Theorems 1 and 2 are sufficient, but not necessary for equality of the improper integrals.

Example 33. A function $f(x, y)$ is continuous and nonnegative on $[a, +\infty) \times [c, +\infty)$ and $\int_c^{+\infty} dy \int_a^{+\infty} f(x,y)dx = \int_a^{+\infty} dx \int_c^{+\infty} f(x,y)dy$, but at least one of the functions $F(y) = \int_a^{+\infty} f(x,y)dx$ and $G(x) = \int_c^{+\infty} f(x,y)dy$ is discontinuous on $[c, +\infty)$ or $[a, +\infty)$, respectively.

Solution
The function $f(x, y) = ye^{-xy}$ is continuous and nonnegative on $X \times Y = [1, +\infty) \times [0, +\infty)$. For $F(y)$, we obtain

$$F(y) = \int_1^{+\infty} ye^{-xy}dx = -e^{-xy}|_1^{+\infty} = e^{-y}, \forall y \in (0, +\infty)$$

and $F(0) = \int_1^{+\infty} 0 dx = 0$, that is, $F(y)$ is discontinuous at $y = 0$. Therefore, one of the conditions of Theorem 1 is not satisfied (this also implies that the integral $F(y)$ does not converge uniformly on $[0, d], \forall d > 0$, that is, one of the conditions of Theorem 2 is not satisfied either). Nevertheless, both iterated improper integrals exist and coincide: for the first integral, we have

$$\int_0^{+\infty} dy \int_1^{+\infty} f(x,y)dx = \int_0^{+\infty} F(y)dy = \int_0^{+\infty} e^{-y}dy = 1,$$

and for the second, we get

$$G(x) = \int_0^{+\infty} ye^{-xy}dy = -\frac{y}{x}e^{-xy}|_0^{+\infty} - \frac{1}{x^2}e^{-xy}|_0^{+\infty} = \frac{1}{x^2}$$

and

$$\int_1^{+\infty} dx \int_0^{+\infty} f(x,y)dy = \int_1^{+\infty} G(x)dx = \int_1^{+\infty} \frac{1}{x^2}dx = -\frac{1}{x}\Big|_1^{+\infty} = 1.$$

Example 34. A function $f(x, y)$ is continuous on $[a, +\infty) \times [c, +\infty)$, and both improper integrals $F(y) = \int_a^{+\infty} f(x,y)dx$ and $G(x) = \int_c^{+\infty} f(x,y)dy$ do not converge uniformly, but $\int_c^{+\infty} dy \int_a^{+\infty} f(x,y)dx = \int_a^{+\infty} dx \int_c^{+\infty} f(x,y)dy$.

Solution
The function $f(x, y) = e^{-xy} \sin x$ is continuous on $X \times Y = [0, +\infty) \times [0, +\infty)$, but does not maintain the sign there. For this reason, we check the

conditions of Theorem 2 only. Using the result found in Example 4: $\int e^{-xy} \sin x dx = -\frac{\cos x + y \sin x}{1+y^2} e^{-xy}$, we immediately obtain

$$F(y) = \int_0^{+\infty} e^{-xy} \sin x dx = -\frac{\cos x + y \sin x}{1 + y^2} e^{-xy} \Big|_0^{+\infty} = \frac{1}{1 + y^2}, \forall y \in (0, +\infty).$$

Since the integral $F(0) = \int_0^{+\infty} f(x, 0) dx = \int_0^{+\infty} \sin x dx$ diverges, let us consider an auxiliary interval $Y_1 = [c, +\infty)$, $\forall c > 0$ on which the integral $F(y)$ converges to $\frac{1}{1+y^2}$. Then

$$\int_c^{+\infty} F(y) dy = \int_c^{+\infty} \frac{1}{1 + y^2} dy = \arctan y \Big|_c^{+\infty} = \frac{\pi}{2} - \arctan c, \forall c > 0.$$

Finally, passing to the limit, we obtain the following result for the first iterated integral:

$$\int_0^{+\infty} dy \int_0^{+\infty} f(x, y) dx = \int_0^{+\infty} F(y) dy$$

$$= \lim_{c \to 0_+} \int_c^{+\infty} F(y) dy = \lim_{c \to 0_+} \left(\frac{\pi}{2} - \arctan c\right) = \frac{\pi}{2}.$$

For the second integral, we have

$$G(x) = \int_0^{+\infty} e^{-xy} \sin x dy = -\frac{\sin x}{x} e^{-xy} \Big|_0^{+\infty} = \frac{\sin x}{x}, \forall x \in (0, +\infty)$$

and

$$G(0) = \int_0^{+\infty} f(0, y) dy = \int_0^{+\infty} 0 dy = 0.$$

The function $G(x)$ has a discontinuity (removable) at $x = 0$, but it is continuous for $\forall x > 0$ and bounded on $[0, +\infty)$ (since $\left|\frac{\sin x}{x}\right| \le 1, \forall x \in (0, +\infty)$). Therefore, $G(x)$ is integrable and

$$\int_0^{+\infty} dx \int_0^{+\infty} f(x, y) dy = \int_0^{+\infty} G(x) dx = \int_0^{+\infty} \frac{\sin x}{x} dx = \frac{\pi}{2}.$$

Thus, two iterated integrals coincide.

Nevertheless, the conditions of Theorem 2 are not satisfied. Indeed, in Example 4, it was shown that $F(y) = \int_0^{+\infty} e^{-xy} \sin x dx$ does not converge uniformly on $Y = (0, +\infty)$. By the same reasoning, it does not converge uniformly on $Y_1 = [0, 1]$ either. Further, $G(x) = \int_0^{+\infty} e^{-xy} \sin x dy$ also does not converge uniformly on $X_1 = [0, 1]$, since for any $A > 1$ and corresponding $x_A = \frac{1}{A} \in [0, 1]$ we obtain

$$\left| \int_A^{+\infty} e^{-x_A y} \sin x_A dy \right| = \left| -\frac{\sin x_A}{x_A} e^{-x_A y} \right|_A^{+\infty} = \left| e^{-1} \frac{\sin 1/A}{1/A} \right| \xrightarrow[A \to +\infty]{} e^{-1} \neq 0.$$

Besides, both integrals $\int_0^{+\infty} dy \int_0^{+\infty} |f(x, y)| dx$ and $\int_0^{+\infty} dx \int_0^{+\infty} |f(x, y)| dy$ are divergent. For the first integral, the evaluation goes as follows. Since $|\sin x| \geq \sin^2 x = \frac{1}{2}(1 - \cos 2x)$, it follows that

$$\int_0^{+\infty} dy \int_0^{+\infty} e^{-xy} |\sin x| dx \geq \frac{1}{2} \int_0^{+\infty} dy \int_0^{+\infty} e^{-xy} dx$$

$$- \frac{1}{2} \int_0^{+\infty} dy \int_0^{+\infty} e^{-xy} \cos 2x dx.$$

The second integral in the right-hand side converges (one can carry out similar calculation as for $\int_0^{+\infty} dy \int_0^{+\infty} e^{-xy} \sin x dx$), but the first integral diverges, because for $\forall y > 0$ one gets

$$\int_0^{+\infty} dy \int_0^{+\infty} e^{-xy} dx = \int_0^{+\infty} \left. -\frac{1}{y} e^{-xy} \right|_0^{+\infty} dy = \int_0^{+\infty} \frac{1}{y} dy.$$

Therefore, the integral $\int_0^{+\infty} dy \int_0^{+\infty} e^{-xy} |\sin x| dx$ diverges. Using the analogous approach for the second iterated integral, we obtain

$$\int_0^{+\infty} e^{-xy} |\sin x| dy = \frac{|\sin x|}{x} \geq \frac{1}{2x}(1 - \cos 2x), \forall x > 0, \int_0^{+\infty} f(0, y) dy = 0,$$

and for $\forall x > 0$

$$\int_0^{+\infty} dx \int_0^{+\infty} e^{-xy} |\sin x| dy = \int_0^{+\infty} \frac{|\sin x|}{x} dx$$

$$= \int_0^1 \frac{|\sin x|}{x} dx + \int_1^{+\infty} \frac{|\sin x|}{x} dx$$

$$\geq \int_0^1 \frac{|\sin x|}{x} dx + \frac{1}{2} \int_1^{+\infty} \frac{dx}{x} - \frac{1}{2} \int_1^{+\infty} \frac{\cos 2x}{x} dx.$$

The first and third integrals in the right-hand side are convergent, whereas the second is divergent. Thus, the integral $\int_0^{+\infty} dx \int_0^{+\infty} e^{-xy}|\sin x|dy$ also diverges.

Exercises

1 Show that the following integrals converge on the given intervals and verify the character of the convergence:

a) $\int_0^{+\infty} e^{-xy}\cos x\,dx$, $Y = (0,+\infty)$

b) $\int_{-\infty}^{+\infty} \frac{\cos xy}{x^2+1}dx$, $Y = \mathbb{R}$

c) $\int_0^{+\infty} \frac{dx}{(x-y)^2+1}$, $Y = \mathbb{R}$

d) $\int_1^{+\infty} \frac{\cos x}{x^p}e^{-xy}dx$, $p > 0$, $Y = [0,+\infty)$

e) $\int_0^{+\infty} \frac{\sin x}{\sqrt[4]{x}}e^{-xy}dx$, $Y = [0,+\infty)$

f) $\int_0^{+\infty} \sqrt{y}e^{-x^2y}dx$, $Y = [0,+\infty)$

g) $\int_1^{+\infty} \frac{\sin x}{x^y}dx$, $Y = (0,+\infty)$

h) $\int_0^{+\infty} e^{-xy}\sin xy dx$, $Y = (0,+\infty)$.

2 Show that the integrals a), c), g), and h) in Exercise 1 are counterexamples to Example 2.

3 Use $\int_0^{+\infty} y^2 e^{-xy^2}dx$, $Y = \mathbb{R}$ as a counterexample to Example 3.

4 Verify whether the integrals a) and h) of Exercise 1 are counterexamples to Example 4.

5 Verify whether the integral $\int_1^{+\infty} \frac{\sin x}{x^y}dx$ considered on $Y = [10^{-1},1]$ provides a counterexample to Example 5.

6 Show that the integral $\int_1^{+\infty} \frac{\sin x}{x^y}dx$ converges absolutely and uniformly on $Y = (1,+\infty)$, but the integral $\int_1^{+\infty} \frac{|\sin x|}{x^y}dx$ does not converge uniformly on $Y = (1,+\infty)$. It provides one more counterexample to Example 6.

7 Show that the functions $f(x,y) = \frac{2xy}{x^4+y^4}$ and $g(x,y) = \frac{2xy}{x^4+y^2}$ defined on $X \times Y = [1,+\infty) \times [1,+\infty)$ provide one more counterexample to Example 8.

8 Check if the following statement is true: "if $\int_a^{+\infty} f(x,y)+g(x,y)dx$ converges uniformly on Y, then the integrals $\int_a^{+\infty} f(x,y)dx$ and $\int_a^{+\infty} g(x,y)dx$ also converge uniformly on Y." (Hint: use the functions $f(x,y) = 2xy^2e^{-x^2y^2}$ and $g(x,y) = (2xy^3 - 2xy^2)e^{-x^2y^2}$ on $X = [1,+\infty)$, $Y = \mathbb{R}$. Compare with Example 9.)

9 Check if the following statement is true: "if $\int_a^{+\infty} f(x,y) + g(x,y)dx$ converges nonuniformly on Y, then the integrals $\int_a^{+\infty} f(x,y)dx$ and $\int_a^{+\infty} g(x,y)dx$ also converge nonuniformly on Y." (Hint: use the functions $f(x,y) = y^3 e^{-xy^2}$ and $g(x,y) = y^2 e^{-xy^2}$ on $X = [0,+\infty)$, $Y = \mathbb{R}$. Compare with Example 9.)

10 Verify if the integrals $\int_X f(x,y)dx$, $\int_X g(x,y)dx$ and $\int_X f(x,y)g(x,y)dx$ converge on Y, and the nature of the convergence if it takes place. Use the following functions and sets:

a) $f(x,y) = 2xy^2 e^{-x^2 y^2}$, $g(x,y) = (y - 2x^2 y^3)e^{-x^2 y^2}$, $X = [1,+\infty)$, $Y = \mathbb{R}$

b) $f(x,y) = g(x,y) = ye^{-xy^2}$, $X = [1,+\infty)$, $Y = \mathbb{R}$

c) $f(x,y) = 2xy^2 e^{-x^2 y^2}$, $g(x,y) = (2x^2 y^2 - 1)e^{-x^2 y^2}$, $X = [1,+\infty)$, $Y = (0,+\infty)$

d) $f(x,y) = g(x,y) = y^2 e^{-xy^2}$, $X = [1,+\infty)$, $Y = \mathbb{R}$

e) $f(x,y) = \frac{\cos x}{\sqrt[3]{x}} e^{-xy}$, $g(x,y) = \frac{\cos x}{\sqrt[3]{x^2}} e^{-xy}$, $X = [1,+\infty)$, $Y = (0,+\infty)$

f) $f(x,y) = g(x,y) = (y - xy^2)e^{-xy}$, $X = [0,+\infty)$, $Y = [0,+\infty)$

g) $f(x,y) = g(x,y) = \frac{\cos x}{\sqrt[3]{x}} e^{-xy}$, $X = [1,+\infty)$, $Y = (0,+\infty)$

h) $f(x,y) = g(x,y) = \frac{\cos x}{x} e^{-xy}$, $X = [1,+\infty)$, $Y = (0,+\infty)$

i) $f(x,y) = \frac{\cos x}{x} e^{-xy}$, $g(x,y) = \frac{\sin x}{x} e^{-xy}$, $X = [1,+\infty)$, $Y = [0,+\infty)$

j) $f(x,y) = \frac{\cos x}{\sqrt[3]{x}} e^{-xy}$, $g(x,y) = \frac{\sin x}{\sqrt[3]{x}} e^{-xy}$, $X = [1,+\infty)$, $Y = [0,+\infty)$

k) $f(x,y) = g(x,y) = y^2 e^{-xy}$, $X = [0,+\infty)$, $Y = [0,+\infty)$

l) $f(x,y) = \frac{\cos x}{x}$, $g(x,y) = e^{-xy^2}$, $X = [1,+\infty)$, $Y = (0,+\infty)$

m) $f(x,y) = (xy^3 - y^2)e^{-xy}$, $g(x,y) = e^{-xy}$, $X = [0,+\infty)$, $Y = (0,+\infty)$

n) $f(x,y) = \frac{\cos x}{x}$, $g(x,y) = xe^{-xy}$, $X = [1,+\infty)$, $Y = (0,+\infty)$

o) $f(x,y) = \frac{\sin xy}{x^2}$, $g(x,y) = x^2 e^{-xy}$, $X = [1,+\infty)$, $Y = (0,+\infty)$

p) $f(x,y) = y \cos xy$, $g(x,y) = \frac{1}{xy}$, $X = [1,+\infty)$, $Y = [1,+\infty)$

q) $f(x,y) = \frac{\cos xy}{\sqrt[3]{x}}$, $g(x,y) = \frac{\cos xy}{\sqrt[3]{x^2}}$, $X = [1,+\infty)$, $Y = [1,+\infty)$.

Formulate false statements for which the given functions represent counterexamples.

11 Verify what conditions of Dirichlet's theorem are violated for functions $f(x,y)$ and $g(x,y)$ on $X \times Y$. Analyze the nature of the convergence of the integral $\int_X f(x,y)g(x,y)dx$ if it takes place. Do this for the following functions and sets:

a) $f(x,y) = y^2 e^{-xy}$, $g(x,y) = \frac{1}{y^2} e^{-xy}$, $X = [0,+\infty)$, $Y = (0,1]$

b) $f(x,y) = \frac{1}{xy}$, $g(x,y) = \cos x$, $X = [1,+\infty)$, $Y = (1,+\infty)$

c) $f(x,y) = ye^{-xy}$, $g(x,y) = e^{-xy}$, $X = [0,+\infty)$, $Y = (0,+\infty)$

d) $f(x,y) = y^x$, $g(x,y) = \frac{1}{x}$, $X = [1,+\infty)$, $Y = (0,1)$

e) $f(x,y) = e^{-xy}$, $g(x,y) = \sin x$, $X = [0,+\infty)$, $Y = [1,+\infty)$

f) $f(x, y) = y \cos xy$, $g(x, y) = \frac{\cos xy}{x}$, $X = [1, +\infty)$, $Y = (0, +\infty)$

g) $f(x, y) = \cos x$, $g(x, y) = \frac{\cos x}{x} e^{-xy}$, $X = [1, +\infty)$, $Y = (0, +\infty)$

h) $f(x, y) = \sin x$, $g(x, y) = e^{-xy}$, $X = [0, +\infty)$, $Y = (0, 1)$.

Compare with Examples 15–18.

12 Verify what conditions of Abel's theorem are violated for functions $f(x, y)$ and $g(x, y)$ on $X \times Y$. Analyze the nature of the convergence of the integral $\int_X f(x, y)g(x, y)dx$ if it takes place. Do this for the following functions and sets:

a) $f(x, y) = ye^{-xy}$, $g(x, y) = \frac{\sin x}{x^2 y}$, $X = [1, +\infty)$, $Y = (0, +\infty)$

b) $f(x, y) = 2xy^2 e^{-x^2 y}$, $g(x, y) = \frac{e^{-x^2 y}}{y}$, $X = [1, +\infty)$, $Y = (0, +\infty)$

c) $f(x, y) = x^{-y}$, $g(x, y) = y \cos xy$, $X = [1, +\infty)$, $Y = (1, +\infty)$

d) $f(x, y) = \frac{y}{\sqrt{x}} \cos xy$, $g(x, y) = \cos xy$, $X = [1, +\infty)$, $Y = (0, +\infty)$

e) $f(x, y) = \frac{\cos x}{x}$, $g(x, y) = e^{-xy} \cos x$, $X = [1, +\infty)$, $Y = (0, +\infty)$

f) $f(x, y) = y^x$, $g(x, y) = \frac{1}{x}$, $X = [1, +\infty)$, $Y = (0, 1)$

g) $f(x, y) = e^{-xy} \sin x$, $g(x, y) = e^{-xy}$, $X = [0, +\infty)$, $Y = (0, 1)$.

Compare with Examples 19–22.

13 Show that the integrals

a) $\int_0^{+\infty} \sqrt{y} e^{-x^2 y} dx$, $Y = [0, +\infty)$, $y_0 = 0$

b) $\int_0^{+\infty} y^2 e^{-xy^2} dx$, $Y = \mathbb{R}$, $y_0 = 0$

are counterexamples to Example 23.

14 Prove that the integral $\int_1^{+\infty} (y - 2x^2 y^3) e^{-x^2 y^2} dx$, $Y = \mathbb{R}$, $y_0 = 0$ is a counterexample to Example 24.

15 Use the integrals

a) $\int_0^{+\infty} \sqrt{y} e^{-x^2 y} dx$, $Y = [0, +\infty)$, $y_0 = 0$

b) $\int_1^{+\infty} y^2 e^{-xy^2} dx$, $Y = \mathbb{R}$, $y_0 = 0$

c) $\int_0^{+\infty} e^{-xy} \sin xy dx$, $Y = [0, +\infty)$, $y_0 = 0$

to give counterexamples to Example 25.

16 Use the integrals

a) $\int_0^{+\infty} e^{-xy} \cos x dx$, $Y = (0, +\infty)$

b) $\int_1^{+\infty} y^2 e^{-xy^2} dx$, $Y = (0, +\infty)$

c) $\int_0^{+\infty} e^{-xy} \sin xy dx$, $Y = (0, +\infty)$

d) $\int_1^{+\infty} (y - 2x^2 y^3) e^{-x^2 y^2} dx$, $Y = \mathbb{R}$

to illustrate Example 26.

17 Prove that the functions

a) $f(x, y) = \begin{cases} \frac{y \sin xy}{x}, x \neq 0 \\ y^2, x = 0 \end{cases}$, $X = [0, +\infty)$, $Y = \mathbb{R}$

b) $f(x, y) = y^3 e^{-xy^2}$, $X = [1, +\infty)$, $Y = \mathbb{R}$

provide counterexamples to Example 27. Find the conditions in the Theorem on differentiation under the integral sign that are violated by the given functions.

18 Prove that the functions

a) $f(x, y) = (y - 2x^2 y^3)e^{-x^2 y^2}$, $X = [1, +\infty)$, $Y = \mathbb{R}$

b) $f(x, y) = y^2 e^{-xy^2}$, $X = [1, +\infty)$, $Y = \mathbb{R}$

c) $f(x, y) = y^3 e^{-xy^2}$, $X = [1, +\infty)$, $Y = \mathbb{R}$

provide counterexamples to Example 28. Investigate the nature of the convergence of the integrals $\int_1^{+\infty} f(x, y)dx$ and $\int_1^{+\infty} f_y(x, y)dx$ on Y.

19 Show that the functions

a) $f(x, y) = (y - 2x^2 y^3)e^{-x^2 y^2}$, $X = [1, +\infty)$, $Y = (0, +\infty)$

b) $f(x, y) = y^2 e^{-xy^2}$, $X = [1, +\infty)$, $Y = (0, +\infty)$

c) $f(x, y) = y^3 e^{-xy^2}$, $X = [0, +\infty)$, $Y = (0, +\infty)$

d) $f(x, y) = (2xy^2 - 2x^3 y^4)e^{-x^2 y^2}$, $X = [1, +\infty)$, $Y = \mathbb{R}$

e) $f(x, y) = \left(\frac{1}{x} + \frac{1}{x^2 y} \right)e^{-xy}$, $X = [1, +\infty)$, $Y = (0, +\infty)$

satisfy all the conditions of Example 29. Investigate the character of the convergence of the integrals $\int_a^{+\infty} f(x, y)dx$ and $\int_a^{+\infty} f_y(x, y)dx$ on Y.

20 Show that the function $f(x, y) = (2xy - x^2 y^2)e^{-xy}$, $X = [1, +\infty)$, $Y = [0, 2]$ provides a counterexample to Example 30. Investigate the character of the convergence of the integral $\int_1^{+\infty} f(x, y)dx$ on Y.

21 Verify whether the functions

a) $f(x, y) = (y - xy^2)e^{-xy}$, $X = [1, +\infty)$, $Y = [0, 1]$

b) $f(x, y) = xy^2 e^{-xy}$, $X = [1, +\infty)$, $Y = [0, 1]$

satisfy all the conditions of Example 31. Investigate the character of the convergence of the integral $\int_a^{+\infty} f(x, y)dx$.

22 Prove that the functions

a) $f(x, y) = \frac{x^2 - y^2}{(x^2 + y^2)^2}$, $X = [1, +\infty)$, $Y = [0, +\infty)$

b) $f(x, y) = (2xy - x^2 y^2)e^{-xy}$, $X = [1, +\infty)$, $Y = [0, +\infty)$

c) $f(x, y) = \frac{x^2 + 2xy - 2y^2}{(x+y)^4}$, $X = [1, +\infty)$, $Y = [2, +\infty)$

d) $f(x, y) = \frac{x^2 - 2xy}{(x+y)^4}$, $X = [1, +\infty)$, $Y = [1, +\infty)$

provide counterexamples to Example 32. Find the conditions in both Theorems on the interchange of integration that are violated by these functions.

23 Show that the function $f(x, y) = xy^2 e^{-xy}$, $X = [1, +\infty)$, $Y = [0, +\infty)$ provides a counterexample to Example 33.

24 Analyze the following statement: "if $f(x, y)$ is continuous on $[a, +\infty) \times [c, +\infty)$, the functions $F(y) = \int_a^{+\infty} f(x, y) dx$ and $G(x) = \int_c^{+\infty} f(x, y) dy$ are continuous on $[c, +\infty)$ and $[a, +\infty)$, respectively, $\int_c^{+\infty} dy \int_a^{+\infty} f(x, y) dx = \int_a^{+\infty} dx \int_c^{+\infty} f(x, y) dy$ and one of the integrals $F(y)$ or $G(x)$ converges uniformly, then another one also converges uniformly." (Hint: use the function $f(x, y) = (y - xy^2) e^{-xy}$ on $X \times Y = [1, +\infty) \times [0, +\infty)$ to construct a counterexample.)

25 Analyze the following statement: "if $f(x, y)$ is continuous on $[a, +\infty) \times [c, +\infty)$, the integrals $F(y) = \int_a^{+\infty} f(x, y) dx$ and $G(x) = \int_c^{+\infty} f(x, y) dy$ converge uniformly on $[c, +\infty)$ and $[a, +\infty)$, respectively, and $\int_c^{+\infty} dy \int_a^{+\infty} f(x, y) dx = \int_a^{+\infty} dx \int_c^{+\infty} f(x, y) dy$, then $f(x, y)$ keeps the sign on $[a, +\infty) \times [c, +\infty)$." (Hint: use the function $f(x, y) = \frac{x^2 - 4xy + y^2}{(x+y)^4}$ on $X \times Y = [1, +\infty) \times [1, +\infty)$ to construct a counterexample. Check the conditions of Theorem 2 on the interchange of integration.)

26 Show that the function $f(x, y) = xy e^{-xy} \cos xy$, $X = [0, +\infty)$, $Y = [0, +\infty)$ provides a counterexample to Example 34. Check the conditions of both Theorems on the interchange of integration.

Further Reading

B.M. Budak and S.V. Fomin, *Multiple Integrals, Field Theory and Series*, Mir Publisher, Moscow, 1978.

G.M. Fichtengolz, *Differential- und Integralrechnung, Vol. 1–3*, V.E.B. Deutscher Verlag Wiss., Berlin, 1968.

V.A. Ilyin and E.G. Poznyak, *Fundamentals of Mathematical Analysis, Vol. 1,2*, Mir Publisher, Moscow, 1982.

V.A. Zorich, *Mathematical Analysis I, II*, Springer, Berlin, 2004.

Bibliography

1 S. Abbott. *Understanding Analysis*. Springer, New York, 2002.
2 P. Biler and A. Witkowski. *Problems in Mathematical Analysis*. Marcel Dekker, New York, 1990.
3 D. Bressoud. *A Radical Approach to Real Analysis*. MAA, Washington, DC, 2007.
4 T.J.I. Bromwich. *An Introduction to the Theory of Infinite Series*. AMS, Providence, RI, 2005.
5 B.M. Budak and S.V. Fomin. *Multiple Integrals, Field Theory and Series*. Mir Publisher, Moscow, 1978.
6 B. Demidovich. *Problems in Mathematical Analysis*. Mir Publisher, Moscow, 1989.
7 G.M. Fichtengolz. *Differential- und Integralrechnung, Vol.1–3*. V.E.B. Deutscher Verlag Wiss., Berlin, 1968.
8 B.R. Gelbaum and J.M.H. Olmsted. *Counterexamples in Analysis*. Dover Publication, Mineola, NY, 2003.
9 V.A. Ilyin and E.G. Poznyak. *Fundamentals of Mathematical Analysis, Vol.1,2*. Mir Publisher, Moscow, 1982.
10 W.J. Kaczor and M.T. Nowak. *Problems in Mathematical Analysis, Vol.1–3*. AMS, Providence, RI, 2001.
11 K. Knopp. *Theory and Applications of Infinite Series*. Dover Publication, Mineola, NY, 1990.
12 C.H.C. Little, K.L. Teo, and B. Brunt. *Real Analysis via Sequences and Series*. Springer, New York, 2015.
13 B.M. Makarov, M.G. Goluzina, A.A. Lodkin, and A.N. Podkorytov. *Selected Problems in Real Analysis*. AMS, Providence, RI, 1992.
14 G. Polya and G. Szego. *Problems and Theorems in Analysis, Vol.1–2*. Springer, Berlin, 1998.

Counterexamples on Uniform Convergence: Sequences, Series, Functions, and Integrals, First Edition.
Andrei Bourchtein and Ludmila Bourchtein.
© 2017 John Wiley & Sons, Inc. Published 2017 by John Wiley & Sons, Inc.
Companion website: www.wiley.com/go/bourchtein/counterexamples_on_uniform_convergence

15 T.L.T. Radulesku, V.D. Radulesku, and T. Andreescu. *Problems in Real Analysis: Advanced Calculus on the Real Axis.* Springer, New York, 2009.

16 W. Rudin. *Principles of Mathematical Analysis.* McGraw-Hill, New York, 1976.

17 V.A. Zorich. *Mathematical Analysis I, II.* Springer, Berlin, 2004.

Index

Counterexamples on Uniform Convergence: Sequences, Series, Functions, and Integrals, First Edition.
Andrei Bourchtein and Ludmila Bourchtein.
© 2017 John Wiley & Sons, Inc. Published 2017 by John Wiley & Sons, Inc.
Companion website: www.wiley.com/go/bourchtein/counterexamples_on_uniform_convergence